Reviews of Plasma Physics

VOPROSY TEORII PLAZMY

ВОПРОСЫ ТЕОРИИ ПЛАЗМЫ

Translation Editor **J. Hugill**
Culham Laboratory
Abingdon, Oxfordshire, England

Reviews of Plasma Physics

Edited by Acad. M. A. Leontovich

Volume

The Library of Congress cataloged the first volume of this title as follows:

Reviews of plasma physics. v. 1—
  New York, Consultants Bureau, 1965—
    /v. illus. 24 cm.
    Translation of Voprosy teorii plazmy.
    Editor: v. 1— M. A. Leontovich.

  1. Plasma (Ionized gases)—Collected works. I. Leontovich, M. A., ed. II. Consultants Bureau Enterprises, Inc., New York. III. Title: Voprosy teorii plazmy. Eng.
QC718.V63   64-23244

The original text, published by Atomizdat in Moscow in 1982,
has been corrected and updated by the authors.

ISBN 0-306-11001-6

© 1986 Consultants Bureau, New York
A Division of Plenum Publishing Corporation
233 Spring Street, New York, N.Y. 10013

All rights reserved

No part of this book may be reproduced, stored in a retrieval system, or transmitted
in any form or by any means, electronic, mechanical, photocopying, microfilming,
recording, or otherwise, without written permission from the Publisher

Printed in the United States of America

CONTENTS

RUNAWAY ELECTRONS IN A TOKAMAK
V. V. Parail and O. P. Pogutse

1. Introduction	1
2. Generation of Runaway Electrons in a Tokamak	2
3. Trajectories of Runaway Electrons in Tokamaks	11
4. Linear Theory of Runaway Electron Instability	17
5. Quasilinear Stage of the Instability	23
6. Macroscopic Effects Associated with the Development of Instability	39
7. Anomalous Diffusion of Runaway Electrons	47
8. Nonlinear Processes and Ion Heating	55
9. Conclusions	61
References	62

BALOONING EFFECTS AND PLASMA STABILITY IN TOKAMAKS
O. P. Pogutse and E. I. Yurchenko

Introduction	65
Chapter 1. Initial Equations and Methods for Investigation of Ideal Plasma Stability	73
1.1. Plasma Equilibrium and Coordinate System	73
1.2. Method of Normal Mode Equations. Expansion of Equations in a Small Parameter	78
1.3. The Energy Method. Simplification of the Energy Principle	82
Chapter 2. Ballooning Modes of the Flute Instability	88
2.1. Method of Equivalent Harmonics	88
2.2. Asymptotic Variational Method for Solution of Differential Equations	93

CONTENTS

2.3.	Analytical Criterion for the Stability of Ballooning Modes	98
2.4.	Numerical Calculations of Ballooning Modes	110
Chapter 3.	Ballooning Modes of the Kink Instability	120
3.1.	Theory of the Kink Instability of a Toroidal Column	120
3.2.	Numerical Calculations for Low Mode Numbers	129
Chapter 4.	Investigation of the Dissipative Plasma Stability	136
4.1.	Initial Equations and Their Simplification	136
4.2.	Threshold-Free Dissipative Ballooning Modes	142
Conclusions		146
References		148

EQUILIBRIUM OF CURRENT-CARRYING PLASMAS IN TOROIDAL CONFIGURATIONS

L. E. Zakharov and V. D. Shafranov

Introduction		153
Chapter 1.	Equilibrium Equations for Toroidal Plasmas	153
Chapter 2.	General Relations for Equilibrium Plasma Configurations	162
2.1.	Straight Plasma Column with a Circular Cross Section	162
2.2.	Axisymmetric Configurations	164
2.3.	Configurations with Helical Symmetry	169
2.4.	Quasicylindrical Description of Equilibrium Configurations	172
2.5.	Maintaining Field. The Principle of a Virtual Casing	178
2.6.	Integral Relations for a Toroidal Plasma Column	180
Chapter 3.	Exact Solutions of the Equilibrium Equations	189
3.1.	Straight Plasma Column with an Elliptical Cross Section	190

- 3.2. Equilibrium of a Plasma Column with Elliptical Cross Section and Helical Symmetry 196
- 3.3. Plasma Torus with Circular Cross Section 201
- 3.4. Compact Toroidal Plasma Column 206
- 3.5. Magnetostatic Problems Related to Plasma Equilibrium 211
- 3.6. Numerical Methods for Solving Equilibrium Problems 227
- 3.7. Equilibrium with Anisotropic Pressure 245

Chapter 4. Equilibrium of the Plasma Column with Circular Cross Section 248
- 4.1. Approximation of Low Toroidicity for an Axisymmetric Plasma Column 249
- 4.2. Effects of Structural Units on Equilibrium 260
- 4.3. Equilibrium of a Plasma Column with a Circular Cross Section and with an Anisotropic Pressure 269
- 4.4. Stability of the Equilibrium of the Plasma Column 271
- 4.5. Equilibrium of a Plasma Column with a Nonplanar Axis 274
- 4.6. Probe Diagnostics in a Tokamak 281

Chapter 5. The Evolution of a Toroidal Equilibrium ... 285
- 5.1. One-Dimensional Transport Equations 287
- 5.2. Equation for Evolution of Magnetic Fluxes 289

Appendix 293
References 299
Index 303

RUNAWAY ELECTRONS IN A TOKAMAK

V. V. Parail and O. P. Pogutse

1. Introduction

In tokamak devices the heating and confinement of plasma are effected by a current flowing along the plasma column. This current is generated by a toroidal inductive electric field E. The presence of this field is an essential property of devices of this type and inevitably leads to the production of runaway electrons in the plasma of the tokamak. The runaway phenomenon is one of the classical effects of the kinetic theory of ionized gases; it is the direct consequence of the long range for small-angle scattering of Coulomb collisions of charged particles. This specific feature of Coulomb interaction causes a rapid decrease in the cross section for scattering of charged particles, σ, with increasing velocity, $\sigma \sim v^{-4}$. The dynamic friction force for fast electrons cannot compensate the acceleration induced by an electric field, and such particles are continuously accelerated.

Usually, tokamaks operate in regimes of weak electric field when the strong inequality $E \ll E_D$ applies; E_D is the so-called Dreicer field (see below for definition) and $E/E_D \lesssim 0.1$. In this case an exponentially small fraction of the electrons is continuously accelerated. However, the presence of such runaway electrons (especially in discharges with moderate density, where the ratio E/E_D is not too small) leads to numerous experimentally observed effects. First of all, these particles (if they have sufficiently high transverse energy) make a considerable contribution to the "diamagnetic" temperature of the plasma without making any contribution to the temperature measured by the Thomson scattering method. It is for this reason that in early experiments [1], where the Thomson scattering method was not used, the conclusion was reached that the electron temperature increases as the plasma density decreases. Rapid electrons reaching the chamber walls generate intense x-ray radiation, and in runaway regimes in the T-10 tokamak such electrons with relativistic energy cause intense x-ray radiation, leading to the appearance of a large flux of photoneutrons [4]. Since such electrons also have a large directed velocity, they can carry a considerable part of the current; thus, on the T-6 tokamak a study was made of regimes in which practically the entire current was carried by runaway electrons; i.e., the bulk plasma remained cold due to the absence of Joule heating [3]. Apparently, the appearance of the so-called "fan" instability in moderate-density discharges is one of the most striking manifestations of the presence of runaway electrons in tokamaks [4-10]. The excitation of this instability is accompanied, as a rule, by periodic spikes of plasma diamagnetism (with $\tau \sim 1$ msec), x-ray and synchrotron radiation, changes in the loop voltage, and other macroscopic effects, the most dramatic of which was the melting of a hole in the vacuum vessel of the TFR device [4]. The data reviewed here show that runaway electrons in a tokamak can considerably affect the macroscopic characteristics of the discharge. The need for a detailed study of the behavior of runaway electrons in a tokamak is also dictated by the fact that in future devices the number of runaway electrons and their role in the macroscopic behavior of plasma may increase due to an increase of the discharge duration and of the confinement time of fast electrons.

2. Generation of Runaway Electrons in a Tokamak

Before turning to the kinetic description of the escape process, we shall introduce the basic notions treated below.

Let us consider the motion of a test electron in plasma with electron temperature T_e, ion temperature T_i, and density $n_e = \Sigma z_j n_j$ (z_j is an effective ion charge). We assume that electrons and ions have Maxwellian distributions in velocity space. Then the dynamic friction force, i.e., the change in momentum of the test electron in the direction of its motion per unit time, has the following form [11]:

$$\langle F \rangle = \langle m \frac{du}{dt} \rangle = \sum_\alpha \frac{c_\alpha n_\alpha}{T_\alpha} \left(1 + \frac{m}{M_\alpha}\right) G(\xi_\alpha), \tag{2.1}$$

where in (2.1) the summation is extended over all particle species,

$$c_\alpha = 4\pi e^2 e_\alpha^2 \ln \Lambda, \quad \xi_\alpha = u/v_{T_\alpha}, \quad v_{T_\alpha} = \sqrt{2T_\alpha/M_\alpha},$$

u is the velocity of the test electron;

$$G(\xi) = \frac{\Phi(\xi) - \xi \frac{d\Phi}{d\xi}}{2\xi^2}; \quad \Phi(\xi) = \frac{2}{\sqrt{\pi}} \int_0^\xi \exp(-t^2)\, dt;$$

$$G(\xi) \to \frac{1}{2\xi^2} \text{ when } \xi \to \infty,$$

$$G(\xi) \to \frac{2}{3\sqrt{\pi}} \xi \text{ when } \xi \to 0.$$

In the case under consideration, when the plasma has one ion species, from (2.1) one can obtain the following expression [5]:

$$\langle F \rangle = c_e n_e \left[z_i \frac{M_i}{m} \frac{G(\xi_i)}{T_i} + 2 \frac{G(\xi_e)}{T_e} \right]. \tag{2.2}$$

From (2.2) it follows that for an electron with energy $mu^2/2 < T_e$ the dynamic friction increases with velocity, and when $mu^2/2 > T_e$ the friction force decreases. The region with $\partial \langle F \rangle / \partial u > 0$ ensures the maintenance of the stationary classical current, whereas the region with $\partial \langle F \rangle / \partial u < 0$ leads to the generation of runaway electrons. In a plasma with applied electric field E, expressions (2.1) and (2.2), generally speaking, are not valid, due to the deviation of the electron distribution function from Maxwellian. It turns out, however, that for fields which are not too strong ($E/E_D \leq 0.1$), these expressions accurately describe the dynamic friction force for the bulk electrons. The field E_D (the so-called Dreicer field) is determined in the following way:

$$E_D = 4\pi e^3 \frac{n_e}{T_e} \ln \Lambda, \tag{2.3}$$

or

$$E_D = 4 \cdot 10^{-2} \frac{\ln \Lambda}{15} \frac{n_e}{10^{13}} \frac{10^3}{T_e} \left[\frac{V}{cm}, cm^{-3}, eV\right]^*.$$

E_D is the field at which an electron with the energy $mu^2/2 = T_e$ doubles its velocity in the mean free time between two collisions. We also introduce the critical velocity v_{cr} of the test electron, defining it as the velocity at which the acceleration force of the external electric field is equal (with opposite sign) to the dynamic friction force:

$$eE = \langle F(u = v_{cr})\rangle, \quad v_{cr} = v_{T_e}\sqrt{E_D/E}. \qquad (2.4)$$

The critical velocity divides the electron distribution function into two basically different parts. Electrons with velocities $v < v_{cr}$ form a stationary distribution function close to Maxwellian, whereas electrons with $v > v_{cr}$ are not restrained by the dynamic friction force and are continuously accelerated, forming a beam of runaway electrons. In the region $v \sim v_{cr}$ a conversion of the electron distribution function from a symmetrical into a directional one takes place.

As was mentioned earlier, tokamaks operate, as a rule, in regimes of weak electric field when $E/E_D \ll 1$. Under such conditions only superthermal electrons with $v \gg v_{T_e}$ are continuously accelerated, so, for finding the distribution function of such electrons, it is sufficient to use a simplified kinetic equation with a linearized collision integral, which does not take into account the mutual collisions of fast particles. In a fully ionized, infinite plasma with $z = 1$ placed in a homogeneous, constant electric field E, it is convenient to write this expression in spherical coordinates:

$$\frac{\partial f}{\partial t} + \frac{eE}{m}\left(\cos\theta \frac{\partial f}{\partial v} - \frac{\sin\theta}{v}\frac{\partial f}{\partial \theta}\right) = \frac{1}{v^2}\frac{\partial}{\partial v}\left\{v^2 \nu_e(v)\left[\frac{T_e}{m}\frac{\partial f}{\partial v} + vf\right]\right\}$$
$$+ \frac{\nu(v)}{2\sin\theta}\frac{\partial}{\partial \theta}\sin\theta\frac{\partial f}{\partial \theta}. \qquad (2.5)$$

Here θ is the angle between **E** and **v**, $\nu_e(v)$ is the collision frequency of a fast electron with velocity $v \gg v_{T_e}$ with other electrons of plasma: $\nu_e(v) = (4\pi e^4 n_e/m^2 v^3)\ln(mv^2 D/e^2); \; D = (T_e/4\pi e^2 n_e)^{1/2}$

---

*We note that the critical field used by Dreicer [12] is half E_D.

is the Debye radius; $\nu(v) = \nu_i(v) + \nu_e(v)[1 - (T_e/2mv^2)]$; ν_i is given by the same formula as ν_e replacing n_e by n_i. The change of the logarithmic term in ν_e and ν_i will be neglected in what follows, assuming that the logarithm is $mv^2 = mv^2_{Te}\, E_D/E$.

For further calculations it is more convenient to introduce the dimensionless variables

$$\mu = \cos\theta,\ u = v/\sqrt{T_e/m},\ \tau = \nu_e\left(\sqrt{\frac{T_e}{m}}\right)t.$$

Equation (2.5) in these variables assumes the form

$$\frac{\partial f}{\partial \tau} + \frac{E}{E_D}\left(\mu\frac{\partial f}{\partial u} - \frac{1-\mu^2}{u}\frac{\partial f}{\partial \mu}\right) = \frac{1}{u^2}\frac{\partial}{\partial u}\left\{\frac{1}{u}\frac{\partial f}{\partial u} + f\right\} + \frac{1-1/4u^2}{u^3}\frac{\partial}{\partial \mu}(1-\mu^2)\frac{\partial f}{\partial \mu}. \tag{2.6}$$

The first term on the right-hand side of (2.6) describes the change of the energy of a fast electron due to collisions with other electrons (the change in the energy of the electron due to collisions with ions is known to be very small). The second term describes the change of the direction of the momentum in collisions both with electrons and with ions.

The analytical solution of Eq. (2.6) in the region $v \leq v_{cr}$ was first obtained by Gurevich[13]. He suggested finding a solution of (2.6) in the stationary case ($\partial f/\partial \tau = 0$) in the form

$$f = C \exp\{\varphi(u, \mu)\},$$

where C is a certain constant determined by matching conditions. Then the function $\varphi(u, \mu)$ should satisfy the following nonlinear equation:

$$u^2 \frac{E}{E_D}\left[\mu u \frac{\partial\varphi}{\partial u} + (1-\mu^2)\frac{\partial\varphi}{\partial \mu}\right] - \left(\frac{\partial\varphi}{\partial u}\right)^2 - \frac{\partial^2\varphi}{\partial u^2}$$
$$-u\left(1 - \frac{1}{u^2}\right)\frac{\partial\varphi}{\partial u} + 2\left(1 - \frac{1}{4u^2}\right)\mu\frac{\partial\varphi}{\partial \mu}$$
$$-\left(1 - \frac{1}{4u^2}\right)(1-\mu^2)\left(\frac{\partial^2\varphi}{\partial \mu^2} + \left(\frac{\partial\varphi}{\partial \mu}\right)^2\right) = 0. \tag{2.7}$$

Before solving Eq. (2.7) we shall consider qualitatively the behavior of an electron distribution function in a plasma with longitudinal electric field. Under the action of this field electrons attain an additional velocity directed along the field which later is redistributed among other directions due to electron–electron and electron–ion collisions. In a weak electric field (E $\ll$ E$_D$) in the region of thermal velocities, the distribution function is naturally close to Maxwellian. Under these conditions the main contribution to the redistribution of electrons is made by electron–ion collisions leading to isotropization of the electron distribution function in this region of velocities.

When velocities are high, the collision frequency drops, and as a result, due to the effect of the electric field, the distribution function starts to deviate from Maxwellian. Electron–electron collisions leading mainly to a change of the energy turn out to be the most significant in this case. Indeed, since the gradient of the Maxwellian distribution function is proportional to velocity ($\partial f / \partial u \sim u f$), the first term on the right-hand side of (2.6) starts to exceed the second term, which describes isotropization of the electron distribution function. Therefore, at large values of u, when collisions are inefficient and the electric field can considerably increase the component of velocity u_z, this direction in velocity space should be significantly emphasized; i.e., the electron distribution should become directional.

Taking into account the above-noted feature of the distribution function in the region of high velocities, it is natural to seek the solution of Eq. (2.7) in the form of a series in powers close to the value $\mu = 1$ (i.e., near the z axis):

$$\varphi(u, \mu) = \varphi(u, 1) + (\mu - 1) \left.\frac{\partial \varphi}{\partial \mu}\right|_{\mu=1}$$

$$+ \frac{(\mu-1)^2}{2} \left.\frac{\partial^2 \varphi}{\partial \mu^2}\right|_{\mu=1} + \ldots = \varphi_0(u) + (\mu - 1)\varphi_1(u) + (\mu - 1)^2 \varphi_2(u) + \ldots \quad (2.8)$$

Substituting expansion (2.8) into Eq. (2.7) and equating the terms with equal powers (in $\mu - 1$), we come to the following coupled system of equations for the functions $\varphi_0, \varphi_1, \ldots$:

$$u^3 \frac{E}{E_D} \frac{d\varphi_0}{du} - \left(\frac{d\varphi_0}{du}\right)^2 - \frac{d^2\varphi_0}{du^2} - u\left(1 - \frac{1}{u^2}\right)\frac{d\varphi_0}{du} + 2\left(1 - \frac{1}{4u^2}\right)\varphi_1 = 0;$$

$$u^3 \frac{E}{E_D}\left(\frac{d\varphi_0}{du} + \frac{d\varphi_1}{du}\right) - 2u^2 \frac{E}{E_D}\varphi_1 - 2\frac{d\varphi_0}{du}\frac{d\varphi_1}{du}$$

$$- \frac{d^2\varphi_1}{du^2} - u\left(1 - \frac{1}{u^2}\right)\frac{d\varphi_1}{du} + 2\left(1 - \frac{1}{4u^2}\right)(\varphi_1 + \varphi_1^2)$$

$$+ 8\left(1 - \frac{1}{4u^2}\right)\varphi_2 = 0. \qquad (2.9)$$

The solution of system (2.9) in the region $u^2 < E_D/E$ was obtained by Gurevich [13] by means of successive truncation of the chain of equations. Since at $u^2 < E_D/E$ the distribution function does not strongly differ from the symmetrical one, as the first approximation one can assume $\varphi_1 = 0$. Neglecting, for simplicity, small terms (on the order of u^{-2}), we have

$$\varphi_0^{(1)} = -\frac{u^2}{2} + \frac{u^4}{4}\frac{E}{E_D}. \qquad (2.10)$$

In the next approximation, assuming $\varphi_2 = 0$, we find $\varphi_0 = \varphi_0^{(1)} + \varphi_0^{(2)}$, $\varphi_1 = \varphi_1^{(1)}$, where

$$\varphi_0^{(1)} \simeq -\sqrt{\frac{2E_D}{E}}\left[1 - \left(1 - \frac{E}{2E_D}u^2\right)^{1/2}\right];$$

$$\varphi_1^{(1)} = u^2\left[\frac{E}{2E_D}\left(1 - \frac{E}{E_D}u^2\right)\right]^{1/2}, \qquad (2.11)$$

etc. As was noted above, the solution obtained is valid for $1 \ll u^2 \ll E_D/E$. If one evaluates the convergence of a series of successive approximations, it is easy to show that φ_n decreases as $(E/E_D)^{1/2}$, i.e., $\varphi_n/\varphi_{n-1} \sim (E/E_D)^{1/2}$.

Thus, the stationary distribution function in the region of high energies ($v_{Te}^2 < v^2 \lesssim v_{cr}^2$) has the form

$$f(v, \theta) \simeq \frac{n_e}{(\sqrt{\pi}\, v_{Te})^3}\exp\left\{-\frac{mv^2}{2T_e} + \frac{E}{E_D}\frac{mv^2}{T_e} - \left(\frac{2E_D}{E}\right)^{1/2}\right.$$

$$\times\left[1 - \left(1 - \frac{E}{E_D}\frac{mv^2}{T_e}\right)^{1/2}\right] - \left(\frac{E}{2E_D}\right)^{1/2}\frac{mv^2}{T_e}$$

$$\left.\times\left(1 - \frac{E}{E_D}\frac{mv^2}{T_e}\right)^{1/2}(1 - \cos\theta)\right\}. \qquad (2.12)$$

It follows from (2.12) that in the region of thermal velocities the electron distribution function is close to Maxwellian; in the region of high velocities f_e differs considerably from the latter: thus, for example, with increasing velocity f_e is considerably smaller than Maxwellian. Besides, in the region of high velocities the distribution function (2.12) becomes directional. This is most strikingly manifested when $v^2 = (2T_e/m)(E_D/3E)$. The average angle between **v** and **E**,

$$\bar{\theta} = \int \theta f(v, \theta)\, d\theta / \int f(v, \theta)\, d\theta,$$

in this case is minimal:

$$\bar{\theta}_{min} = \left(\frac{27\pi^2}{8}\right)^{1/4} \left(\frac{E}{E_D}\right)^{1/4}. \tag{2.13}$$

It is interesting to note that at higher velocities the directivity of the distribution function decreases.

As was noted above, the solution obtained by Gurevich [13] is valid only in the region $v < v_{cr}$. This is connected with the choice of the approximation $\varphi_1 \ll \varphi_0$, assuming a sufficiently great symmetry of the distribution function with respect to θ. Lebedev [14] obtained the solution of system (2.9) for the region $v > v_{cr}$ ($u^2 > E_D/E$). In this case it was assumed that φ_0 and φ_1 are of the same order of magnitude and $\varphi_2 \ll \varphi_0, \varphi_1$. As a result Lebedev obtained the following expression for f_e valid at $v \gtrsim v_{cr}$:

$$\left.\begin{aligned}
f &= f_0 \exp\{\varphi_1(u^2) + \varphi_2(u^2)(\mu - 1)\}; \\
\varphi_1(u^2) &= \ln E_{i1}\left(\frac{1}{u^2}\right); \\
\varphi_2(u^2) &\simeq u^2 \frac{2E}{E_D} E_{i1}^{-1}\left(\frac{1}{u^2}\right) \exp\left(-\frac{1}{u^2}\right).
\end{aligned}\right\} \tag{2.14}$$

To obtain the undetermined constant one needs to match the solution of (2.14) with the solution of Gurevich in the region $u^2 < E_D/E$. After this, one can finally write the solution of (2.14) in the region $v \gtrsim v_{cr}$:

$$f = 2^{5/6}(2\pi)^{-3/2} v_{Te}^{-3} \sqrt[4]{\frac{E_D}{E}} \exp\left[-\frac{E_D}{4E} -\right.$$

$$-\sqrt{\frac{2E_D}{E}} - \frac{1}{2} + \varphi_1(u^2) + \varphi_2(u^2)(\mu - 1)\bigg]. \qquad (2.15)$$

Knowing the values of the distribution function in the runaway region, we can find the value S of the flow of runaway electrons, i.e., the number of electrons per unit volume entering the regime of continuous acceleration per unit time.

For this purpose we first of all integrate (2.15) over the transverse velocities, i.e., evaluate the one-dimensional distribution function in the runaway region:

$$f_z \simeq \frac{E_D}{E} 2^{4/3} \frac{n_e}{\sqrt{\pi}\, v_{Te}} \sqrt[4]{\frac{E_D}{2E}} \exp\left[-\frac{E_D}{4E} - \sqrt{\frac{2E_D}{E}} - \frac{1}{2}\right]. \qquad (2.16)$$

Then we substitute (2.16) into (2.5) and integrate it over v_z from $-\infty$ to $v_0 \gg v_{cr}$. Taking into account that Coulomb collisions conserve the number of particles (when $v_z > v_0$ the collision frequency is negligibly small), we can obtain the value of the flow S:

$$S \simeq 2^{1/3} \frac{1}{\sqrt{\pi}} n_e v_{ee} \sqrt[4]{\frac{2E_D}{E}} \exp\left[-\frac{E_D}{4E} - \sqrt{\frac{2E_D}{E}} - \frac{1}{2}\right]. \qquad (2.17)$$

It should be noted that the value of the runaway electron flow was obtained for the first time in [7]; it differs from Eq. (2.17) only in the multiplier of the exponential. Since that time numerous attempts have been made to revise the value S to make it closer to the flow obtained by numerical methods [9].

The point is that in Lebedev [13] [Eq. (2.17)] the procedure of matching solutions cannot be considered correct, since the regions of their applicability do not overlap. The most consecutive solution for f_e and S was first obtained in [16] by dividing the velocity space into five regions, each of which contained an approximate expression for the distribution function using its small parameters. Generalization of the solution [16] for the case of plasma having an effective charge different from unity is given in [17]. Finally, [18] investigates the effect of relativistic corrections for the value of the flow S.

Nowadays it is common practice to use the following expression for the runaway flow:

$$\Gamma = \frac{S}{v_e n_e} = K(z_i) \left(\frac{E_D}{E}\right)^{\frac{3}{16}(z_i+1)} \exp\left[-\frac{E_D}{4E} - \sqrt{\frac{(z_i+1)E_D}{E}}\right], \quad (2.18)$$

where $z_i = \frac{1}{n_e}\sum_j z_{ij}^2 n_{ij}$. The value of $K(z_i)$ is on the order of unity.

As mentioned above, all the expressions for the runaway flow (without taking into consideration the effective charge) obtained in later works differ from the expression derived by Gurevich[13] only in the factor multiplying the exponential. It should be borne in mind, however, that the value E_D entering into the exponent of (2.18) is known only with logarithmic accuracy, since the Coulomb logarithm for collisions of electrons is described with this accuracy only. Therefore, to a certain extent the refinement of the preexponential factor in the formula for the runaway flow is exceeding the accuracy.

<u>Problem 1.</u> Find, on the basis of a one-dimensional model, the distribution function for runaway electrons [19].

A one-dimensional kinetic equation can be obtained from a three-dimensional one with a collision integral in the Landau form by integrating it over transverse velocities. For electrons with velocities higher than thermal, it assumes the following form:

$$\frac{\partial f_z}{\partial t} + \frac{eE}{m}\frac{\partial f_z}{\partial v_z} = \frac{\partial}{\partial v_z} v_0 \frac{v_{Te}^3}{|v_z|^3}\left(v_z f_z + \frac{T_e}{m}\frac{\partial f_z}{\partial v_z}\right) + A\delta(v_z). \quad \text{(I)}$$

The source of particles is included into the right-hand side of (I). The magnitude of this source (value A) is unknown beforehand and will be found below from the condition for existence of the steady-state solution (it is clear that in the case under consideration the steady-state solution can be obtained only if there is a constant source of particles). The boundary conditions for Eq. (I) are $f_z(v_z = -\infty) = 0$, $f_z(v_z = +\infty) = $ const. The last constant is naturally associated with the value A and eventually with the value E_D/E. We integrate Eq. (I) over v_z from $-\infty$ to $v_z > 0$:

$$\frac{eE}{m} f_z = v_0 \frac{v_{Te}^3}{|v_z|^3} \left(v_z f_z + \frac{T_e}{m} \frac{\partial f_z}{\partial v_z} \right) + A. \tag{II}$$

When $v_z \to +\infty$, the first term on the right-hand side of (II) can be neglected; i.e., we obtain the relation between A and $f_z \times (v_z \to +\infty)$:

$$\frac{eE}{m} f_z (v_z = +\infty) = A. \tag{III}$$

After integrating Eq. (II) over v_z once more and using the condition that the distribution function at $v_z \to +\infty$ is finite, we obtain

$$f_z(v_z > 0) = \frac{A}{v_0 u} \left[1 + \sqrt{\frac{2v_{Te}}{u}} \exp(\rho^2) \int_\rho^\infty \exp(-z^2) \, dz \right];$$

$$\rho = \sqrt{\frac{u}{2v_{Te}} \left(\frac{v_z^2}{v_{Te}^2} - \frac{v_{Te}}{u} \right)} = \sqrt{\frac{v_{Te}}{2u} \left(\frac{v_z^2}{v_{cr}^2} - 1 \right)}; \tag{IV}$$

$$u = \frac{eE}{mv_0}, \quad v_{cr}^2 = \frac{v_{Te}^3}{u}.$$

In the region $|v_z| < v_{cr}$ the distribution function becomes Maxwellian. From the normalizing condition (as a rough approximation we assume that the obtained solution is also valid at $|v_z| \leq v_{Te}$) $f_z(0) = n_e/\sqrt{\pi} v_{Te}$ we obtain

$$A = \frac{v_0 u n_e}{\pi v_{Te}} \sqrt{\frac{u}{2v_{Te}}} \exp\left(-\frac{v_{Te}}{2u} \right). \tag{V}$$

As follows from (III), the value of A gives the value of the flow of electrons into the runaway region, and finally from (IV) and (V) there follows the expression for the electron distribution function in the region $v_z \geq v_{cr}$:

$$f_z(v_z \geq v_{cr}) \simeq \frac{A}{v_0 u} \left(1 + \frac{v_{cr}^2}{v_z^2} \right) = \frac{n_e}{\pi v_{Te}} \sqrt{\frac{u}{2v_{Te}}} \exp\left(-\frac{E_D}{4E} \right) \left(1 + \frac{v_{cr}^2}{v_z^2} \right). \tag{VI}$$

3. Trajectories of Runaway Electrons in Tokamaks

In this section we discuss the influence of the geometry of the tokamak on the dynamics of runaway electrons. First of all we con-

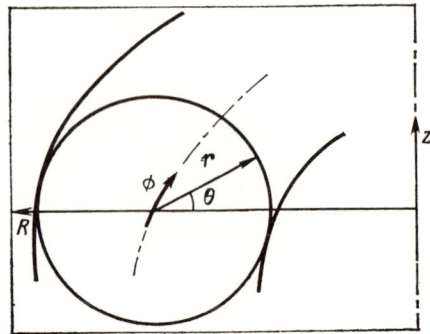

Fig. 1. Toroidal system of coordinates.

sider the dynamics of free electrons in a tokamak. These electrons have an energy in the range of >1 MeV; they practically do not collide and are accelerated by the electric field up to relativistic energies.* This group of electrons is responsible for the experimentally observed hard x-ray radiation in tokamaks. In certain cases such electrons lead to the generation of a large flow of neutrons induced by photonuclear reactions [2].

We consider, first of all, the process of acceleration of electrons in a toroidal electric field. The motion of electrons along the toroidal axis ϕ (see Fig. 1) is described by the following equation:

$$\frac{d}{dt}(\gamma m_0 v_\phi) = -e\dot{E}_\phi$$

or

$$\frac{d}{dt}\gamma\beta = \frac{e}{m_0 c}\frac{dA_\phi}{dt}\ ;\ \mathrm{curl}\ \mathbf{A} = \mathbf{B}, \tag{3.1}$$

where $\beta = v_\phi/c = (\gamma^2 - 1)^{1/2}/\gamma$; A_ϕ is the toroidal component of the vector potential. The solution of (3.1) is derived directly from

$$W = m_0 c^2(\gamma - 1);\ \beta^2 = \left(\frac{eE_\phi t}{m_0 c}\right)^2 \bigg/ \left(1 + \frac{eE_\phi t}{m_0 c}\right)^2; \tag{3.2}$$

---

*The evaluations given below do not naturally describe the case when relativistic electrons leave the plasma due to the destruction of magnetic surfaces by MHD activity.

$$\gamma(t) = \left\{1 + \left[\frac{e}{m_0 c} A_\phi(t)\right]^2\right\}^{1/2}. \tag{3.3}$$

The value of the vector potential in (3.3) is proportional to the change in the magnetic flux enclosed by the electron trajectory. It is clear that the change in the energy of a fast electron leads to a change in its trajectory since eventually it crosses the limiter. This trajectory is extreme for the electron; it is associated with the maximum electron energy which can be held in the discharge. To evaluate this energy we consider the motion of the electron in a tokamak in the drift approximation (the motion of the guiding center). Such motion in a tokamak, as is known [15], is described by the following equation:

$$\mathbf{v}_{dr} = \mathbf{e}_\theta v_\theta + \mathbf{e}_\phi v_\phi + v_c \mathbf{e}_z, \tag{3.4}$$

where $v_\theta = v_\| B_\theta / B_0$ is the velocity in the θ-direction due to the helicity of the magnetic field; $v_c = (v_\| p_\| + \tfrac{1}{2} v_\perp p_\perp)/mR\omega_{ce}$ is the vertical drift of electrons due to the centrifugal force and the gradient of the toroidal magnetic field (the system of coordinates is shown in Fig. 1). The analysis of Eq. (3.4) shows that in the case of a homogeneous current distribution, $J(r) = $ const, the projection of the electron orbit on the plane R, z is a circle with its center shifted a distance Δ along R relative to the magnetic surface, which is also assumed to be a circle. To evaluate the value of the shift we can make use of the fact that the change in the electron energy under the influence of the longitudinal electric field during one turn round the torus is small. In this case we can assume that the longitudinal adiabatic invariant is approximately constant:

$$J = \gamma m_0 R v_\phi - \frac{e}{c} \int_0^r B_\theta R dr. \tag{3.5}$$

In (3.5) $R = R_0 + r\cos\theta$. When the absolute value of Δ is small compared to the drift orbit radius r, one can assume that $r = r_0 + \Delta\cos\theta$, where r_0 is the radius of the magnetic surface. In this approximation $B_\theta(r_0) = B_\theta(r) - \Delta\cos\theta(\partial B_\theta/\partial r_0)|_r$ and from (3.5) it follows that in the simplest case of a homogeneous current distribution the value of Δ is described by the ratio

$$\Delta = \frac{a^2}{2R_0} \frac{I_A}{I}, \tag{3.6}$$

where a is the radius of the limiter (the current radius); $I_A = \beta\gamma m_0 c^3/e = 17\sqrt{\gamma^2-1}$ [kA] is the Alfvén current. From (3.6), considering that $\Delta_{max} = a$, we obtain the estimation of the maximum energy of the electron retained in a tokamak with the given current:

$$(\beta\gamma)_{max} = \frac{eI}{m_0 c^3} 2 \frac{R_0}{a};$$

$$\gamma^2 - 1 < \frac{4}{17} I^2 \left(\frac{R_0}{a}\right)^2 \text{ [kA]}. \qquad (3.7)$$

From (3.2) and (3.7) it is also possible to obtain an estimation of the maximum confinement time of relativistic electrons in a tokamak in the stationary stage of the discharge, when E_ϕ = const:

$$\tau_{max} \approx \frac{2}{17} \frac{Imc}{eE} \frac{R}{a} = \frac{R^2}{a} \frac{I}{V} 10^{-5} \text{ [kA, cm, V, sec]}. \qquad (3.8)$$

From (3.8) it follows, for example, that in the T-10 tokamak with parameters R_0 = 150 cm, a = 30 cm, I = 400 kA, and V = 2 volts the confinement time of relativistic electrons is τ_{max} = 1.5 sec, which may be compared with the time of current duration.

It will be recalled that Eqs. (3.7) and (3.8) have been obtained for the electron which is initially (at t = 0) in the center of the column r(0) = 0. For electrons with r(0) = r_0 the value of a in Eqs. (3.7) and (3.8) should be replaced by $a - r_0$.

It should also be noted that a peaked current distribution confines particles better than the homogeneous current considered above.

And, finally, we consider the influence of an inhomogeneity of the magnetic field in a tokamak on the movement of runaway electrons. In a perfect tokamak (i.e., in a tokamak in which the magnetic field is constant along the ϕ axis) due to the presence of the current, the magnitude of the magnetic field changes along a line of force. In such a system there is a group of particles trapped in the region where the magnetic field is minimum, i.e., on the outer side of the torus. Because of the conservation of the magnetic moment of the electron ($\mu = p_\perp^2/B$ = const) and of its energy $\varepsilon = mc^2$, it follows that those particles will be trapped for which the following inequality is satisfied: $p_\parallel^2/p_\perp^2 < (B_{max} - B_{min})/B_{max} \leq r/R = \varepsilon$. Such particles move in a tokamak along banana trajectories [21] which deviate from the magnetic surface a distance $\Delta_t \leq \Delta/\sqrt{\varepsilon}$.

It is easier to find the deviation of electrons which are just trapped for which it may be assumed that $p_{\parallel}(\theta) \simeq p_{\parallel}(0)(1 + \cos\theta)/2$, $p_{\parallel}(0) \leq p\sqrt{\varepsilon}$. In this case the ratio $\Delta_t = \Delta/\sqrt{\varepsilon}$ follows immediately from (3.5). Naturally they do not make a contribution to the runaway flow. However, the number of such fast particles is relatively small, first of all, because these particles have a large transverse energy, and as shown above, the longitudinal energy of the runaway electrons, as a rule, is larger than the transverse energy. It is true that the number of such particles may sharply increase during the development of the kinetic instability, leading to quasielastic scattering of particles in velocity space (see below).

In a real tokamak, due to the discreteness of the coils of the longitudinal magnetic field, there also exists the so-called locally trapped particles, i.e., particles trapped in the local minima of the longitudinal magnetic field (the inequality $p_{\parallel}^2/p_{\perp}^2 \leq \delta = \Delta B_\phi/B_\phi$ is fulfilled for them, where δ is the depth of modulation of the toroidal magnetic field). Such particles, unlike the trapped ones, are not confined in tokamaks, since there is no rotational transform for them, and the vertical drift with velocity $v_c = (v_\parallel p_\parallel + \frac{1}{2} v_\perp p_\perp) \times (mR\omega_{ce})^{-1}$ should lead to their escape from the tokamak in a time $\tau_d \simeq a/v_c = aR\omega_{ce}/(v_\parallel p_\parallel + \frac{1}{2}v_\perp p_\perp)$. It should be borne in mind, of course, that in practice only sufficiently fast electrons will leave the system for which the time τ_d is less than the characteristic time of detrapping from the loss cone due to electron−electron and electron−ion collisions τ_c. The last quantity is evaluated as $\tau_c \simeq \delta[\nu_{ee}(v) + \nu_{ei}(v)]^{-1}$. From the condition $\tau_d < \tau_c$ it is possible to evaluate the characteristic energy required by locally trapped electrons to leave the system due to vertical drift:

$$\left(\frac{v}{v_{Te}}\right)^2 \geqslant \left[\frac{aR\omega_{ce}(\nu_{ee}(T_e)+\nu_{ei}(T_e))}{v_{Te}^2 \delta}\right]^{2/5}. \quad (3.9)$$

Here $\nu_{ee}(T_e)$ and $\nu_{ei}(T_e)$ are the electron−electron and electron−ion collision frequencies for electrons with thermal energy. Relation (3.9) is written for nonrelativistic electrons. The evaluations show that for the T-10 tokamak, for example, at $n_e = 3 \cdot 10^{13}$ cm^{-3}, $B_\phi = 30$ kG, $a = 30$ cm, $R_0 = 150$ cm, $T_e = 1$ keV, and $z_{eff} = 1$, particles leave the system for energies $\varepsilon \geq 15$ keV. An evaluation analogous to (3.9) shows that such a mechanism of departure is more significant for ions, since they have a lower collision frequency $\nu_{ii}(T_i) \sim 10^{-2} \cdot [\nu_{ee}(T_e) + \nu_{ei}(T_e)]$.

We note that a rigorous consideration of the influence of the inhomogeneity of the magnetic field on the runaway flow in tokamaks given in [22] revealed that the exponential factor G [S $\sim \exp(-G)$] changes slightly compared to the case of a homogeneous field and is expressed by the following relation:

$$G = \frac{1}{4} \frac{E_D}{\langle E \rangle} + \sqrt{\langle B \rangle \left\langle \frac{1}{B} \right\rangle \frac{(1+z_i) E_D}{\langle E \rangle}}, \qquad (3.10)$$

where $\langle ... \rangle$ indicates an average over the magnetic surface.

Problem 2. Evaluate the influence of binary collisions on the number of runaway electrons in a tokamak [13].

Usually, when calculating the rate at which electrons enter the regime of continuous acceleration, only those collisions leading to a small change in the energy and momentum of a test particle are taken into account (see Section 2). However, if the lifetime of fast electrons is sufficiently great, the rate of entry to the acceleration regime can increase considerably due to "close" collisions of fast and slow electrons during which the latter receive an energy $\Delta W > mv_{cr}^2 / 2$. This process leads to an exponential growth of the number of runaway electrons and can considerably change the value of the runaway flow.

Let us consider the case which is most favorable for the multiplication process, when the lifetime of the runaway electron in a tokamak τ_E equals the time of free acceleration up to the energy W_{max} at which the shift of the drift orbit is equal to the radius of the plasma column:

$$\tau_E^{max} = \frac{m_0 c}{eE} \gamma_{max}, \qquad (I)$$

where $\gamma_{max} = 0.47 \cdot I \cdot R/4a$ [kA].

The cross section of the Coulomb interaction of two electrons with energy transfer more than ΔW equals

$$\sigma = 2\pi r_e^2 \frac{\gamma^2}{\gamma^2 - 1} \frac{m_0 c^2}{\Delta W}, \tag{II}$$

where $r_e = e^2/m_0 c^2$ is the classical electron radius.

Taking into account that the collision frequency of fast particles $\nu = \sigma n_e c$, it is easy to derive the "multiplication coefficient of runaway electrons for the time τ_E^{max}:

$$K = \nu \tau_E^{max}. \tag{III}$$

From relation (III) it follows that in large-scale tokamaks of the T-10 type, with good confinement of runaway particles, the "multiplication" coefficient K can reach values $K \geq 10$.

4. Linear Theory of Runaway Electron Instability

As was noted above, the most obvious experimental manifestation of the presence of runaway electrons in tokamaks is the so-called "fan" instability observed in many tokamaks [4, 5, 8-10, 24]. Before turning to its theoretical description we shall briefly enumerate the peculiarities of its development as observed experimentally. The most detailed study of this instability was made in the TM-3 tokamak.

As a rule, the instability developed in discharges with moderate density ($n_e \leq 10^{13}$ cm^{-3}) and comparatively small current ($q \geq 6$). The instability manifested itself as short ($\tau_1 \sim 10$-100 μsec) periodic bursts of plasma diamagnetism, the interval between two neighboring bursts greatly exceeding the duration of each of them and amounting to the order of $\tau_2 \sim 1$ msec. Simultaneously with the change in diamagnetism, there was a sharp increase in intensity of x-ray and synchrotron radiation and a change in the loop voltage (whose sign ΔV was, generally speaking, different for different devices). Each burst of instability was accompanied by oscillations with a wide frequency spectrum $\omega_{pi} \leq \omega \leq \omega_{pe}$. Simultaneously with these oscillations, bolometers registered the arrival on the chamber wall of energetic locally trapped electrons and ions. Each burst of instability was accompanied by anomalous ion heating, the increase of the ion temperature ΔT_i being $\Delta T_i \sim \frac{1}{3} \Delta T_e$, where ΔT_e is the increase of the transverse (diamagnetic) temperature of the electrons. It should be stressed that from the theoretical

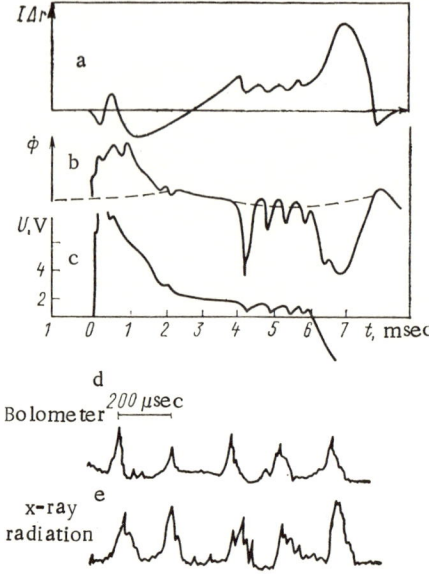

Fig. 2. Experimental manifestation of the "fan" instability: a) plasma column displacement; b) plasma diamagnetism; c) loop voltage; d) bolometer indication; e) hard x-ray radiation.

point of view the relaxation character of the instability seems most strange. It should be remembered that the characteristic interval between successive bursts of instability has the scale $\tau_2 \sim 10^{-3}$ sec, whereas the characteristic period of excited oscillations is $f \leq 10^{-10}$ sec. We do not comprehend the "explosive" character of the instability when the burst time is considerably less than the characteristic time for the system to cross the stability boundary.

The dependence of the values given above on time in the TM-3 tokamak is shown in Fig. 2.

Now we turn to a theoretical investigation of the stability of plasma with a weak longitudinal electric field. It was shown above that the presence of such a field in a plasma leads to the emergence of a quasi-one-dimensional tail of runaway electrons on the electron distribution function. The presence of such a nonequilibrium can, generally speaking, lead to the development of various instabilities. Below, the most dangerous electrostatic instability will be considered.

As usual, for finding the condition for the onset of instability, it is necessary to study the dispersion law of the oscillations which are of interest to us. In the case under consideration when the plasma is placed in a strong magnetic field B_0 (directed along the z axis), the dispersion law for potential oscillations is derived by equating the longitudinal component of the dielectric tensor ε_l to zero:

$$\varepsilon_l = 1 + \frac{4\pi e^2}{Mk^2} \int \frac{\mathbf{k} \frac{df_i}{\partial \mathbf{v}} d\mathbf{v}}{\omega - \mathbf{k}\mathbf{v} + i\delta}$$

$$+ \frac{4\pi e^2}{mk^2} \sum_{n=-\infty}^{\infty} \int \frac{\left(k_z \frac{\partial f_e}{\partial v_z} + \frac{n\omega_{ce}}{v_\perp} \frac{\partial f_e}{\partial v_\perp}\right)}{\omega - n\omega_{ce} - k_z v_z + i\delta} J_n^2 \left(\frac{k_\perp v_\perp}{\omega_{ce}}\right) d\mathbf{v}. \quad (4.1)$$

Here J_n is Bessel's function, δ is a positive constant tending to zero, and f_i and f_e are the ion and electron distribution functions, respectively.

In the region where the imaginary part ε_l'' is much less than the real part ε_l', the eigenfrequency ω_k is defined by the equation $\varepsilon_l' = 0$, and the growth rate of a small perturbation $\gamma_k = -\varepsilon_l'' \times (\partial \varepsilon_l'/\partial \omega)^{-1}$ is found from the condition $\varepsilon_l = 0$. From expression (4.1) it is easy to find ε_l'':

$$\varepsilon_l'' = -\pi \frac{4\pi e^2}{Mk^2} \int \mathbf{k} \frac{\partial f_i}{\partial \mathbf{v}} \delta(\omega - \mathbf{k}\mathbf{v}) d\mathbf{v}$$

$$- \pi \frac{4\pi e^2}{mk^2} \sum_{n=-\infty}^{n=+\infty} \int J_n^2 \left(\frac{k_\perp v_\perp}{\omega_{ce}}\right) \delta(\omega - n\omega_{ce} - k_z v_z)$$

$$\times \left(k_z \frac{\partial f_e}{\partial v_z} + \frac{n\omega_{ce}}{v_\perp} \frac{\partial f_e}{\partial v_\perp}\right) d\mathbf{v}. \quad (4.2)$$

Below we shall consider the case of a strong magnetic field, for which the inequality $\omega_{ce} > \omega_{pe}$ holds everywhere within the plasma (such an inequality holds in most regimes of tokamak operation). It follows from the linear theory [25] that in this case, in a homogeneous high-temperature plasma, there are three branches of eigenoscillations:

1. ion-acoustic modes with the dispersion law

$$\omega_s \approx kc_s, \quad c_s = \sqrt{T_e/M}; \quad (4.3)$$

$$\omega_s \lesssim \omega_{pi};$$

2. magnetized Langmuir oscillations (Trivelpiece−Gould mode):

$$\omega_l \simeq \omega_{pe} k_z/k,$$
$$\omega_{pi} < \omega_l \leq \omega_{pe}; \quad (4.4)$$
$$k_z v_{Te} < \omega_l;$$

3. upper hybrid oscillations (Bernstein mode):

$$\omega_h = \omega_{ce}^2 + \omega_{pe}^2 k_\perp^2/k^2, \quad (4.5)$$
$$k_z v_{Te} \leq \omega_{ce}$$

(generally speaking, there are also higher Bernstein modes with frequencies close to $n\omega_{ce}$, but we shall not consider them for the reasons pointed out below).

From the expression for γ_k it follows that it is necessary to satisfy the inequality $\omega \varepsilon_l'' < 0$ to excite oscillations. It is seen from (4.2) that excitation or damping of waves is caused only by those particles for which the resonance condition $\omega - n\omega_{ce} - k_z v_z = 0$ and $\omega - \mathbf{k v}_i = 0$ is satisfied.

It should be noted, first of all, that since the distribution function for ions is Maxwellian, the ion Cerenkov resonance leads to damping of the oscillations. The same applies to the electron Cerenkov resonance for oscillations whose longitudinal phase velocity is $\omega/k_z > v_{Te}$ (in this region it is $\partial f_e/\partial v_z < 0$).

Now we turn to Doppler resonances when $n \neq 0$. It should be emphasized, first of all, that since, in the case we are considering, the inequalities $\partial f_e/\partial v_\perp < 0$ and $|\partial f_e/\partial v_\perp^2| \geq |\partial f_e/\partial v_z^2|$ are satisfied, the Doppler resonance cannot lead to the excitation of oscillations with $\omega \simeq n\omega_{ce}$. Indeed, in this case we have $k_z v_z = \omega - n\omega_{ce} \ll \omega$ for resonant particles, and it follows from (4.2) that $\omega \varepsilon_l'' \sim -\int \frac{(n\omega_{ce})^2}{v_\perp} \frac{\partial f_e}{\partial v_\perp} d\mathbf{v} > 0$ (the excitation of such oscillations is possible only for distribution functions with a loss cone, or when the longitudinal temperature of resonant particles is much less than the transverse temperature). For oscillations with $\omega \ll \omega_{ce}$, it follows from (4.2) that $k_z v_z \simeq n\omega_{ce}$ and the Doppler resonance with n < 0 will lead to their excitation if the inequality $|\partial f_e/\partial v_z| < |\partial f_e/\partial v_\perp|$ is satisfied, which occurs for a beam of runaway electrons [26, 27] when the Doppler resonance is satisfied for particles with $v_z > v_{cr} \gg v_{Te}$.

A more detailed analysis based on a quasilinear theory (see Section 5) shows that for the resonance with n < 0 (the so-called anomalous Doppler effect) the interaction of waves with resonant electrons leads to a decrease of their longitudinal energy and to an increase of their transverse energy, as a result of which the system tends to a stable equilibrium with an isotropic beam distribution function. The situation is reversed for the resonance with n > 0 (normal Doppler effect) and instability would lead to a still greater deviation of the system from equilibrium.

Below we shall consider the excitation of unstable Trivelpiece-Gould oscillations without analyzing in detail the stability of ion-sound oscillations. If the latter are excited, it is due first to the Cerenkov resonance on electrons in the region $\omega/k_z \le u_0 = eE/m\nu_{ei}$; the Doppler contribution to their growth rate from the beam particles is exponentially small, and it is unlikely to play a significant role (unless the quasilinear relaxation in the region of the Cerenkov resonance decreases the growth rate to a level where the Doppler term is important).

We now find the condition for instability of Langmuir oscillations. To do this we have to substitute in the expression for ε_l'' the expression derived earlier [see (2.16)] for the electron distribution function and equate ε_l'' to zero. However, since we have to consider the process of the growth of the tail of runaway electrons in time, we should cut off the runaway electron distribution function at $v_z = v_b(t) = eEt/m$ (t is the time passed from the application of the field E). Taking this into account, one can write

$$f_l(v_z \ge v_{cr}) \approx \Gamma \frac{n_e}{\pi^{3/2} v_{Te}^3} \frac{E}{E_D} \left(1 + \frac{v_{cr}^2}{v_z^2}\right) \theta$$

$$\times \left(\frac{v_b(t)}{v_z} - 1\right) \exp\left(-\frac{v_\perp^2}{v_{Te}^2}\right), \qquad (4.6)$$

where $\theta(x) = \begin{cases} 1 \text{ if } x > 0 \\ 0 \text{ if } x \le 0 \end{cases}$ is the step function. Substitution of (4.6) into the expression for the growth rate gives

$$\gamma_k = \frac{\omega_{pe}^2}{k^2} \frac{\omega_k}{v_{Te}} \frac{E}{E_D} \Gamma \left[-\frac{2v_{cr}^2 |k_z|^3}{\omega_k} + \frac{1}{4} \frac{k_\perp^2}{k_z \omega_{ce}} \right. \\ \left. \times \left(1 + \frac{k_z^2 v_{cr}^2}{\omega_{ce}^2}\right) \theta\left(\frac{k_z v_b(t)}{\omega_{ce}} - 1\right) \right]. \quad (4.7)$$

The first term on the right-hand side of (4.7) describes the damping of oscillations in the region of the Cerenkov resonance (this resonance is also in the runaway region $\omega_k \ge k_z v_{cr}$); the second term on the right-hand side of (4.7), written on the assumption $\omega_k \ll \omega_{ce}$, describes the excitation of oscillations on the Doppler resonance.

At marginal stability the following equalities should be satisfied: $\gamma_k = 0$, $\partial\gamma_k/\partial k_z = 0$, $\partial\gamma_k/\partial k_\perp = 0$ (the latter two conditions refer to oscillations with maximum growth rate).

It follows from these conditions that instability develops when the velocity of the beam v_b exceeds the following value:

$$v_b \ge 3 \left(\frac{\omega_{ce}}{\omega_{pe}}\right)^{3/2} v_{cr} = v_b^*; \quad (4.8)$$

in this case oscillations with $\omega_{ce}/k_z = v_b^*$ and

$$k_\perp^2/k_z^2 \simeq 3 \quad (4.9)$$

are excited. It should be noted that relations (4.8) and (4.9) have been obtained on the assumption of a fairly strong beam, when electron−ion collisions do not make a substantial contribution to the damping of oscillations. Oscillations defined by relations (4.8) and (4.9) have a Cerenkov resonance in the region

$$v_1 = \frac{\omega_k}{k_z} \simeq \frac{\omega_{pe}}{k} = \sqrt{3}\, v_{cr} \left(\frac{\omega_{ce}}{\omega_{pe}}\right)^{1/2}. \quad (4.10)$$

If the condition for instability development (4.8) is fulfilled, the oscillation amplitude grows exponentially with time; the evolution of the instability, with allowance for the influence of the oscillation on the distribution function of resonant electrons, should be described by the quasilinear approximation.

Problem. Find the minimum value of E/E_D (minimum beam density) at which the instability is possible.

In a collisionless high-temperature plasma the minimum beam density at which the instability occurs is defined from (4.7), where the value ω_{ce}/k_z would be replaced by the light speed and $k_\perp \simeq k$. Hence, provided that $v_{cr}^2 = v_{Te}^2 E_D/E$ we obtain (T_e is expressed in keV):

$$\left(\frac{E_D}{E}\right)_{max} \simeq 10^{-1} \frac{c^2}{v_{Te}^2} \frac{\omega_{pe}^3}{\omega_{ce}^3} \simeq \frac{25}{T_e} \left(\frac{\omega_{pe}}{\omega_{ce}}\right)^3. \tag{I}$$

Electron–ion collisions may be another factor limiting the instability development. In this case the stability boundary is defined by the equation

$$\gamma_k \simeq \frac{\omega_{pe}^2}{k^2} \frac{\omega}{v_{Te}} \frac{E}{E_D} \Gamma \frac{k_\perp^2}{4 k_z \omega_{ce}} - \nu_{ei}.$$

Taking into account (2.18), the condition for the onset of instability becomes

$$\frac{\omega_{pe}^3}{4\omega_{ce}^2} \frac{c}{v_{Te}} K(z_i) \left(\frac{E_D}{E}\right)^{\frac{3}{16}(z_i+1)-1} \exp\left(-\frac{E_D}{4E} - \sqrt{\frac{(z_i+1)E_D}{E}}\right) \geq \frac{\omega_{pe} \ln \Lambda}{n_e D^3} \tag{II}$$

or

$$\frac{E_D}{4E} + \sqrt{\frac{(z_i+1)E_D}{E}} \leq \ln \Lambda. \tag{III}$$

The threshold is the minimum value of E/E_D from (I) and (III).

5. Quasilinear Stage of the Instability

To describe the influence of unstable oscillations on the distribution function of runaway electrons it is necessary to solve a system of quasilinear equations. For a collisionless plasma this has the following form [28]:

$$\frac{dW_k}{dt} = 2\gamma_k W_k, \tag{5.1}$$

$$\frac{\partial f}{\partial t} + \frac{eE}{m}\frac{\partial f}{\partial v_z} = \int d\mathbf{k}\left(-\frac{\omega_{ce}}{v_\perp}\frac{\partial}{\partial v_\perp} + k_z\frac{\partial}{\partial v_z}\right)D_i$$

$$\times \left(-\frac{\omega_{ce}}{v_\perp}\frac{\partial f}{\partial v_\perp} + k_z\frac{\partial f}{\partial v_z}\right) + \int d\mathbf{k}\frac{\partial}{\partial v_z}D_0\frac{\partial f}{\partial v_z}; \tag{5.2}$$

$$D_1 = 4\left(\frac{\pi e}{m}\right)^2 \frac{W_k k_\perp^2 v_\perp^2}{\omega_{ce}^2}\delta(\omega_k + \omega_{ce} - k_z v_z);$$

$$D_0 = 16\left(\frac{\pi e}{m}\right)^2 \frac{k_z^2}{k^2} W_k \delta(\omega_k - k_z v_z);$$

$W_k = k^2|\varphi|_k^2/4\pi$ is the spectral energy density of unstable oscillations;

$$\gamma_k = \frac{\pi}{2}\omega_k \frac{4\pi e^2}{mk^2}\int d\mathbf{v}\,\delta(\omega_k - k_z v_z)\frac{\partial f}{\partial v_z}k_z + \frac{\pi}{8}\omega_k\frac{4\pi e^2}{mk^2}$$

$$\times \int d\mathbf{v}\delta(\omega_k + \omega_{ce} - k_z v_z)\frac{k_\perp^2 v_\perp^2}{\omega_{ce}^2}\left(k_z\frac{\partial f}{\partial v_z} - \frac{\omega_{ce}}{v_\perp}\frac{\partial f}{\partial v_\perp}\right) - \frac{\nu_{ei}}{2}. \tag{5.3}$$

In obtaining (5.1) and (5.2) it was assumed that the transverse wavelength is greater than the electron Larmor radius ($k_\perp v_\perp \times \omega_{ce}^{-1} \ll 1$) and only resonances with n = 0, 1 are taken into account. The first term on the right-hand side of (5.2) describes resonant interaction of runaway electrons with oscillations due to the anomalous Doppler effect and the second due to the Cerenkov resonance.

From Eq. (5.2) it is possible to understand the physics of the excitation of instability due to the anomalous Doppler effect. From the form of the quasilinear operator in (5.2) it follows that interaction in the region of an anomalous Doppler resonance leads to the diffusion of particles in velocity space along lines $(v_z - \omega/k_z)^2 + v_\perp^2$ = const; i.e., they isotropize in the coordinate system moving with the phase velocity of a wave ω/k_z. If the phase velocity is small (which will be further assumed and is valid for $\omega_{pe} \ll \omega_{ce}$), the isotropization of electrons will occur at almost constant energy; i.e., only a small portion of the energy of the runaway electrons will be transferred to the waves. Since the initial distribution function is such that the characteristic transverse energy of the beam particles is much less than the longitudinal energy, the development of instability should be accompanied by an increase in the transverse energy of the beam particles and a decrease in the longitudinal energy. A quantum-mechanical consideration of the elementary process of plasmon excitation due to the anomalous Doppler effect leads to the same conclusion. Indeed, from the quantum-mechanical standpoint the resonance condition can be considered as a law of conservation of energy of plasmon radiation $\hbar\omega_k = n\hbar\omega_{ce} + \hbar k_z v_z$, where $\hbar\omega_k$ is the plasmon energy, n is the change in the magnetic quantum number of electrons, and $\hbar k_z v_z \simeq -(1/2m)\{(mv_z - \hbar k_z)^2 - mv_z^2\}$ is the decrease in the longitudinal energy of the electron due to the radiation of the plasmon with a longitudinal momentum $\hbar k_z$. In the anomalous Doppler effect (n < 0) the radiation of the plasmon ($\omega_k > 0$) is accompanied by a decrease in the longitudinal energy of the electron ($k_z > 0$) and an increase in its transverse energy (n < 0).

We now turn to a quantitative description of the quasilinear stage of the "fan" instability. It should be noted, first of all, that since the regions of the Cerenkov and Doppler resonances are separated in phase space, they may be considered independently. Qualitatively the linear and quasilinear stages of the instability development can be conceived in the following way [29, 30]. When a longitudinal electric field is switched on, a "tail" of runaway electrons starts to grow on the electron distribution function; i.e., the value of $v_b(t)$ increases [see (4.6)]. When v_b exceeds the critical value v_b^* [see (4.8)], oscillations having the maximum growth rate are excited. Quasilinear interaction of such oscillations with resonant electrons leads, first of all, to the isotropization of the distribution function in the region $v_z \simeq \omega_{ce}/k_z \sim v_b^*$ of the Doppler resonance and, second, to the formation of a narrow one-dimensional plateau in the region $v_z = \omega_k/k_z \simeq \omega_{pe}/k = v_1$ of the Cerenkov resonance. It is easy to see that after the plateau is formed, all the oscillations with $\omega/k_z \simeq \omega_{pe}/k = v_1 = $ const becomes unstable, since in the expression for the growth rate a term, stabilizing the instability, disappears. As the condition k = const actually does not set any limitations on the possible value of k (and this value precisely is important for the Doppler resonance), the subsequent stage of the instability development leads to the generation of a wide spectrum of oscillations in k_z and ω space leading to the isotropization of practically the entire "beam" region of the electron distribution function (Fig. 3). This fact can account for the

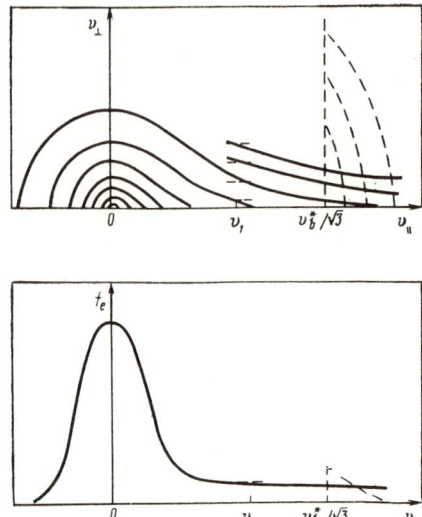

Fig. 3. Change of the electron distribution function at the first stage of the "fan" instability.

sharp burst of instability observed experimentally, which ends when the transverse energy of the beam equals its longitudinal energy. After that the growth ceases and oscillations damp, for example, due to electron−ion collisions.

Subsequent development of the process in time can be conceived in the following way. Since the region of velocities with $v_z = v_1$ (the plateau region) is sufficiently narrow, it rapidly relaxes to an equilibrium state after the oscillations disappear, due to collisional diffusion, and the decrement of oscillation due to the Cerenkov effect increases to its initial value (the driving term being absent, since in the region of the Doppler resonance the distribution function is isotropic). Hence, the system will "wait" until the longitudinal electric field accelerates the tail particles sufficiently and gives rise to the temperature anisotropy necessary for the excitation of the subsequent burst of instability. Such a qualitative picture is in agreement with the experimentally observed relaxation character of the instability.

A quantitative description of the dynamics of the instability development is possible with the help of a simplified system derived from a quasilinear equation by taking moments of the electron distribution function. The "quasi-isotropic" part of the electron distribution function may be described only by two moments — the transverse $T_\perp$ and longitudinal $T_\parallel$ "temperatures" $\left(T_\alpha = \int_{v_{cr}}^{\infty} \frac{mv_\alpha^2}{2} f\partial v;\right.$ the "temperature" is actually the mean energy of the beam particles). Multiplying Eq. (5.2) successively by $mv_\perp^2/2$ and $mv_z^2/2$ and integrating from v_{cr} to ∞, we obtain the following equations for the moments [31]:

$$\frac{dT_\perp}{dt} = \nu_T \frac{T_\parallel - T_\perp}{T_\parallel} \frac{W}{n_e}; \qquad (5.4)$$

$$\frac{dT_\parallel}{dt} = -\nu_T \frac{T_\parallel - T_\perp}{T_\parallel} \frac{W}{n_e} + Q. \qquad (5.5)$$

Here $\nu_T = \omega_{pe}^2 \frac{n_b}{n_e} \left\langle \frac{1}{v_z} \right\rangle \left\langle\!\!\left\langle \frac{1}{k_z} \right\rangle\!\!\right\rangle$; $W = \int W_k dk$ is the total energy of excited waves per unit volume, $\langle\!\langle ... \rangle\!\rangle$ represents averaging over the spectrum, $\langle ... \rangle$ represents averaging over the distribution function, and $n_b = \int_{v_{cr}}^{\infty} f\, dv$ is the beam density. The order of magnitudes are given by $\nu_T \sim (\omega_{pe}^2/\omega_{ce})(n_b/n_e)$; $Q \sim eE\langle v_z \rangle$. An equation for the excited waves should be added to the system (5.4)-(5.5). It is obtained by integrating (5.1) over k. As a result we have

$$\frac{dW}{dt} = \gamma W \frac{T_\parallel - T_\perp}{T_\parallel} - \gamma_1 \frac{W}{1 + W/W_1} - \nu_{ei} W. \qquad (5.6)$$

The first term in Eq. (5.6) describes the excitation of oscillations by the anomalous Doppler effect, the second — by Landau damping taking into account the formation of a quasilinear plateau in the process of instability development (W is the level of thermal noise). Within an order of magnitude $\gamma \sim \frac{\omega_{pe}^3}{\omega_{ce}^2} \frac{\langle\!\langle k_z \rangle\!\rangle}{k} \frac{n_b}{n_e} \sim \nu_T \frac{\langle\!\langle \omega \rangle\!\rangle}{\omega_{ce}}$, this value containing an additional multiplier ω/ω_{ce} as compared to ν_T. This corresponds to the fact that scattering of electrons on waves occurs almost elastically and only in the next order with respect to ω/ω_{ce} does there appear a term describing a change in the number of quanta $\gamma_1 \approx \gamma \left(\frac{T_\| - T_\perp}{T_\|}\right)_{th} \gg \nu_{ei}$, $\left(\frac{T_\| - T_\perp}{T_\|}\right)_{th}$ being the "temperature" anisotropy at threshold.

The system (5.4)-(5.6) describes the following processes taking place in a plasma. At the initial stage, when the level of noise is very small (W ~ W_T), the transverse "temperature" $T_\perp$ does not change, whereas the longitudinal "temperature" $T_\|$ increases (this increase corresponding to the acceleration of electrons by the electric field E). When the inequality $(T_\| - T_\perp)/T_\| > (\gamma_1 + \nu_{ei})/\gamma$ is fulfilled [see (5.6)], oscillations begin to build up; with the increase of W the term describing the isotropization of the electron distribution function in the region $v_z \sim v_b$ increases in Eqs. (5.4) and (5.5). The transverse "temperature" starts increasing, the longitudinal "temperature" dropping. Simultaneously the second term on the right-hand side of (5.6) begins to decrease, which corresponds to the formation of the plateau in the Cerenkov part of the distribution function. The oscillation energy sharply increases and isotropization of electrons in velocity space takes place, as a result of which the first term in (5.6) starts decreasing; when it becomes smaller than $\nu_{ei}W$, oscillations begin to damp due to collisions. Following the decrease in the oscillation level, a rapid relaxation of the plateau occurs, and for the subsequent burst of instability it is necessary for a finite temperature anisotropy to reappear. The process then repeats.

Equations (5.4)-(5.6) should naturally be written in dimensionless variables:

$$(T_\| - T_\perp)/T_\| = \Delta \tilde{T}; \quad \tilde{W} = W/nT_\|; \quad \tau = t\nu_T;$$

$$\tilde{\gamma}_1 = \gamma_1/\nu_T; \quad \tilde{\nu}_{ei} = \nu_{ei}/\nu_T;$$

$$\tilde{Q} = Q/(\nu_T nT_\|); \quad \tilde{W}_T = W_T/nT_\|.$$

In addition, we shall assume that the change in the beam temperature is smaller than the temperature proper $\Delta \tilde{T} \ll 1$. This is valid in all bursts of instability except the first.

In this case equations for $\Delta \tilde{T}$ and $\tilde{W}$ assume the following form:

$$\frac{d}{d\tau}\Delta \tilde{T} = -2\Delta \tilde{T}\tilde{W} + \tilde{Q}; \tag{5.7}$$

$$\frac{d\widetilde{W}}{d\tau} = \widetilde{\gamma}\,\widetilde{W}\,\Delta\widetilde{T} - \widetilde{\gamma}_1\,\frac{\widetilde{W}}{1+\widetilde{W}/\widetilde{W}_T} - \widetilde{\nu}_{ei}\,\widetilde{W}. \quad (5.8)$$

The system (5.7)-(5.8) was solved numerically for values of parameters close to those obtained in the experiment. However, before proceeding to the description of the results of such a computation, we show that the system (5.7)-(5.8) has unstable solutions. To prove it, we show that small perturbations about the equilibrium solution of the system (5.7)-(5.8) are unstable. Equilibrium values $\Delta\widetilde{T}_0$ and $\widetilde{W}_0$ equal

$$\Delta\widetilde{T}_0 = \widetilde{Q}/2\widetilde{W}_0; \quad (5.9)$$

$$\widetilde{W}_0 \simeq \frac{\widetilde{\gamma}\widetilde{Q}/2 - \widetilde{\gamma}_1 \widetilde{W}_T}{\widetilde{\nu}_{ei}}. \quad (5.10)$$

Linearizing Eqs. (5.7) and (5.8) with respect to perturbations and taking into account equilibrium values [(5.9) and (5.10)], it is easy to obtain a dispersion equation describing the behavior in time of small deviations from equilibrium; it follows from this equation that the equilibrium is unstable when the following inequality is satisfied:

$$\widetilde{\gamma}\widetilde{Q} < 2\widetilde{\gamma}_1\,\widetilde{W}_0\,\frac{(\widetilde{W}_0/\widetilde{W}_T)^2}{(1-\widetilde{W}_0/\widetilde{W}_T)^2}, \quad (5.11)$$

or in dimensional variables considering the inequality $W_0/W_T \gg 1$:

$$\gamma Q < 2\gamma_1 \nu_T W_T. \quad (5.12)$$

If characteristic values of the parameters γ, γ_1, ν_T, and Q are substituted into (5.12), it turns out that autooscillation instability should decrease with increasing magnetic field and longitudinal electric field (the first dependence is, apparently, justified by the experiment). When the inverse inequality (5.12) is satisfied, the system (5.7)-(5.8) has no autooscillation solutions. In this case, in plasma with a sufficiently high beam density, instability due to anomalous Doppler effect would be excited continuously.

The system (5.7)-(5.8) was solved numerically for values of parameters close to those obtained in experiments on the TM-3 device [5]. The results of the solution are given in Fig. 4. At the chosen parameter values, the characteristic time between two bursts

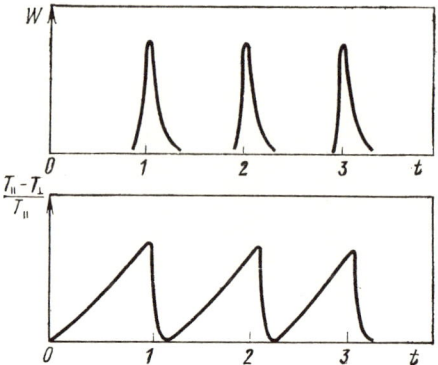

Fig. 4. Results of numerical modelling of the "fan" instability.

of instability agrees within an order of magnitude with the value obtained experimentally; the whole picture of the instability behavior in time coincides qualitatively as well.

We now turn to a more accurate kinetic description of the instability based on the analysis of Eq. (5.2) [19, 31, 32]. As was shown above, the instability arises when $v_b \geq 3v_{cr}(\omega_{ce}/\omega_{pe})^{3/2}$, the oscillations with $k^2/k_z^2 \simeq 3$, for which the Doppler resonance is at $\omega_{ce}/k_z = v_b$ and the Cerenkov resonance at $\omega/k_z \simeq \omega_{pe}/k \simeq (v_b\sqrt{3}/3)(\omega_{pe}/\omega_{ce}) = v_1$, being excited first in this case. When the instability arises in the region $v_z = v_1$, a one-dimensional plateau is rapidly formed, which leads to the generation of all oscillations with $k_z = k$; i.e., all the beam particles in the velocity interval $v_b\sqrt{3}/3 \leq v_z \leq v_b$ resonate due to the anomalous Doppler effect. The final form of the distribution function in this region can be obtained from Eq. (5.2), noting that the number of particles along the line $\eta = v_\perp^2 + v_z^2 = \text{const}$ remains constant as $E \to 0$ (see [23, 24]):

$$f_e = \frac{1}{\sqrt{\eta - v_2}} \int_{v_2}^{\sqrt{\eta}} f_0(v_z, v_\perp(\eta, v_z))\, dv_z$$

$$= \Gamma \frac{E}{E_D} \frac{n_e}{\sqrt{\pi}\, v_{Te} 2\pi} \frac{1 - \exp\left(-\dfrac{\eta - v_b^2/3}{v_{Te}^2}\right)}{\sqrt{\eta}\,(\sqrt{\eta} - v_b/\sqrt{3})} \theta\left(\frac{v_b}{\sqrt{\eta}} - 1\right). \quad (5.13)$$

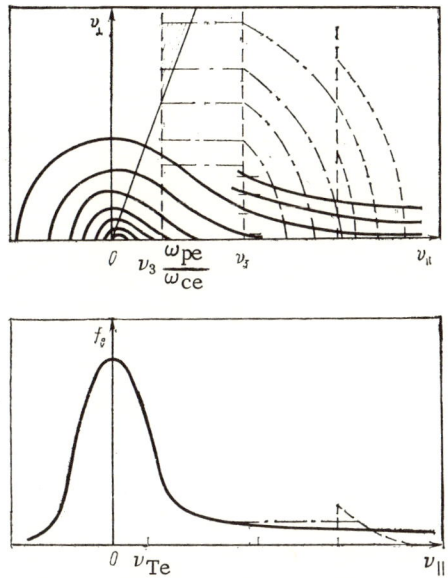

Fig. 5. Change of the electron distribution function at the second stage of instability.

Here $v_2 = v_b/\sqrt{3}$ is the minimum velocity prior to which the relaxation process proceeds.

It is easy to see that the integral of this function over transverse velocities has a maximum with respect to v_z for $v_z = v_2$:

$$f_e(v_z) = 2\pi \int_0^{v_b^2 - v_z^2} v_\perp \, dv_\perp f_e$$

$$\simeq \Gamma \frac{E}{E_D} \sqrt{\frac{n_e}{\pi}} v_{Te} \begin{cases} \ln \dfrac{v_b(1-1/\sqrt{3})}{\sqrt{v_{Te}^2 + v_z^2} - v_b/\sqrt{3}} & \text{if } v_z \geqslant v_b/\sqrt{3} \\ 1 + \dfrac{v_{cr}^2}{v_z^2} & \text{if } v_z < v_b/\sqrt{3}. \end{cases} \qquad (5.14)$$

Such a distribution function is unstable to the Trivelpiece–Gould mode in the region $v_z \leq v_2$ due to the Cerenkov effect. As is usual [6], the quasilinear relaxation of this instability is accompanied by the formation of a one-dimensional plateau on the electron distribution function; the boundary relaxation points are determined from the law of conservation of particle number along the trajectory and from continuity of the distribution function. In addition, the final distribution function should not destabilize oscillations due to the anomalous Doppler effect in the region $v_z \geq v_3$ (v_3 will be determined below) or due to the Cerenkov mechanism for $v_3 \omega_{pe}/\omega_{ce} \leq v_z < v_3$. This function has the following form (see Fig. 5):

$$f = \Gamma \frac{E}{E_D} \frac{n_e}{(\sqrt{\pi}\, v_{Te})^3} \times$$

$$\times \begin{cases} \left(1 + \frac{v_{cr}^2}{v_z^2}\right) \exp\left(\frac{v_\perp^2}{v_{Te}^2}\right) & \text{if } v_z \leq v_3 \frac{\omega_{pe}}{\omega_{ce}}; \\[2ex]
\dfrac{v_{Te}^2 \left(1 + \dfrac{v_{cr}^2}{v_\perp^2 + v_3^2}\right)\left(1 - \exp\left\{-\dfrac{v_\perp^2 + v_3^2\left(1 - \dfrac{\omega_{pe}^2}{\omega_{ce}^2}\right)}{v_{Te}^2}\right\}\right)}{\sqrt{v_\perp^2 + v_3^2}\left(\sqrt{v_\perp^2 + v_3^2} - v_3^2 \dfrac{\omega_{pe}^2}{\omega_{ce}^2}\right)} \\[2ex]
\times \theta\left(\dfrac{v_b^2}{v_\perp^2} - 1\right) & \text{if } v_3 \dfrac{\omega_{pe}}{\omega_{ce}} < v_z \leq v_3; \\[2ex]
v_{Te}^2 \dfrac{\left(1 + \dfrac{v_{cr}^2}{\eta}\right)\left(1 - \exp\dfrac{\eta - v_3^2 \dfrac{\omega_{pe}^2}{\omega_{ce}^2}}{v_{Te}^2}\right)}{\sqrt{\eta}\left(\sqrt{\eta} - v_3^2 \dfrac{\omega_{pe}^2}{\omega_{ce}^2}\right)} \\[2ex]
\text{if } v_b \geq v_z \geq v_3. \end{cases} \quad (5.15)$$

Formula (5.15) is obtained analogously to (5.13) by using the law of conservation of particle number along the trajectory η = const for $v_3 \leq v_z \leq \eta \leq v_b$ and $v_\perp$ = const for $v_3 \omega_{pe}/\omega_{ce} \leq v_z \leq 3$.

The velocity v_3 in expression (5.15) is determined from the continuity of the integral $\int_0^\infty f_\infty v_\perp dv_\perp$ at the point $v_z = v_3 \omega_{pe}/\omega_{ce}$:

$$\ln \frac{v_b^2 + v_3^2}{v_3^2} \simeq 2 \frac{v_{cr}^2}{v_3^2} \frac{\omega_{ce}^2}{\omega_{pe}^2}. \qquad (5.16)$$

From Eq. (5.16) it follows, in particular, that a quasilinear relaxation occurs in the absence of collisions (i.e., for a sufficiently large number of runaway electrons) right up to velocities $v_z \leq v_{cr}$.

It was pointed out above that the isotropic part of the distribution function of runaway electrons formed after the development of instability due to the anomalous Doppler effect is unstable as before. The Trivelpiece−Gould mode is excited in the region of higher phase velocities due to the Cerenkov mechanism.

In this section we consider the quantitative theory of this process. We should emphasize, first of all, that the dynamics of the process of quasilinear relaxation at this stage differs substantially in two limiting cases. The first case occurs when the radial distribution of the beam of runaway electrons is fairly wide and the convection of oscillations away from the region of beam localization is insignificant. In the opposite limiting case of narrow beam distribution, oscillations irreversibly escape from the generation region at the group velocity and lose their energy elsewhere (for example, in the cold plasma at the column periphery). The criterion for transition from one regime to another will be given below.

Let us first consider the situation where the convection of oscillations from the region of beam localization is not significant. In the velocity range of the Cerenkov resonance $v_3 \omega_{pe}/\omega_{ce} \leq v_z \leq v_b \sqrt{3}/3$ Eqs. (5.1) and (5.2) can be rewritten in the following manner:

$$\frac{\partial f_z}{\partial t} = \frac{\partial}{\partial v_z} D \frac{\partial f}{\partial v_z}, \qquad (5.17)$$

$$\frac{\partial W}{\partial t} = \pi \omega_{pe} |\mu| v_z^2 \frac{1}{n_e} \frac{\partial f_e}{\partial v_z} W. \qquad (5.18)$$

Here

$$D = \pi \frac{e^2}{m^2} \frac{\omega_{pe}^2}{v_z^2} \int_{-1}^{1} d\mu |\mu| W\left(\frac{\omega_{pe}}{v_z}, \mu\right);$$

$$\mu = \frac{k_z}{k} = \cos\theta.$$

The system (5.17)−(5.18) differs from the usual one-dimensional system of quasilinear equations in the dependence of the spectral energy density on the angle in wave vector space. Averaging Eq. (5.18) over θ ($\langle W \rangle = \int_{-1}^{1} d\mu\, W$) we can by the standard method obtain the relation between f and $\langle W \rangle$ from Eqs. (5.17) and (5.18):

$$f_z(t) - f_z(t=0) = \frac{e^2}{m^2} \omega_{pe} n_e \frac{\partial}{\partial v_z} \frac{\langle W \rangle - \langle W_0 \rangle}{v_z^5}, \qquad (5.19)$$

where $f_z(t=0)$ is the isotropic initial (with respect to Cerenkov generation) distribution function of resonance electrons, and $\langle W_0 \rangle$ is the initial noise level. From relation (5.19) there follows at once the expression for $\langle W \rangle$ as a function of v_z (in other words, of $k = \omega_{pe}/v_z$):

$$\langle W \rangle = \left(\frac{m}{e}\right)^2 \frac{v_z^5}{2\pi \omega_{pe} n_e} \int_{v_3 \omega_{pe}/\omega_{ce}}^{v_z} \left(f_z(t) - f_z(t=0)\right) dv_z'. \qquad (5.20)$$

Since the final distribution function has a plateau, it can be concluded from Eq. (5.20) that $\langle W \rangle \sim v_z^6$. To determine the behavior of $\langle W \rangle$ at the left instability boundary for $v_z \simeq v_3 \omega_{pe}/\omega_{ce}$, we require a more accurate determination of the integration limits and of the difference $f_z - f_z(t=0)$, than is usually done in the one-dimensional quasilinear theory [6]. The dependence $\langle W \rangle \sim v_z^6$ shows that the main oscillation energy is concentrated in the region of minimum $k_{min} \sim \omega_{pe}/v_{z\,max}$.

Knowing the average value of the oscillation spectrum $\langle W \rangle$, we can also determine the dependence of the spectrum W on μ. For this purpose, we integrate Eq. (5.18) over time from 0 to ∞:

$$W = W_0 \exp(B\mu). \qquad (5.21)$$

Here $B = \int_0^{\infty} dt \pi \omega_{pe} v_z^2 \frac{1}{n_e} \frac{\partial f_z(t)}{\partial v_z}$ is a function of v_z, obtained as a result of integrating $\partial f_z/\partial t$ over time (actually B is the wave amplification factor). We recall that, within the framework of applicability of the quasilinear theory, $B \gg 1$. Averaging Eq. (5.21) over μ, we obtain $\langle W \rangle$ which is already known from previous calculations [see Eq. (5.20)]:

$$\langle W \rangle = 2 \frac{W_0}{B} (\exp B - 1). \qquad (5.22)$$

Taking into account that $B \gg 1$, from Eq. (5.22) we can easily express B, thus also W, in terms of $\langle W \rangle$:

$$W = W_0 \exp\left\{\mu \ln \frac{\langle W \rangle}{W_0}\right\} = W_0 \left(\frac{\langle W \rangle}{W_0}\right)^\mu. \qquad (5.23)$$

It follows from Eq. (5.23) that since $\ln(\langle W \rangle / W_0) \gg 1$ the spectrum of the oscillations excited in μ is almost one-dimensional (with half-width $\Delta \mu \sim [\ln(\langle W \rangle / W_0)]^{-1} \ll 1$). In fact, the spectrum in μ can be much wider. The point is that, in a plasma limited in r, the quantity $k_\perp \geq 2\pi/a$ (where a is the plasma dimension across the magnetic field), so that $\Delta \mu_{min} \sim \Delta(k_z/k_\perp) \sim 1/k_z^2 a^2 \sim [v_z \times (\omega_{pe} a)^{-1}]^2$; this quantity can, generally speaking, be greater than $\Delta \mu = [\ln(\langle W \rangle / W_0)]^{-1}$. Taking this situation into account, the spectrum of oscillations in μ can be evaluated in the following manner: $W(\mu = 1) = 0$; then with decrease of μ inside the region $\mu \gtrsim 1 - \Delta \mu_{min}$ the oscillation intensity increases and for $\mu < 1 - \Delta \mu_{min}$ diminishes rapidly (exponentially) with decreasing μ.

The characteristic time of quasilinear relaxation is completely determined by $\langle W \rangle$ and, by analogy with the usual one-dimensional theory [6], can be evaluated as

$$\tau \sim 10^{-1} \ln \Lambda \frac{1}{\omega_{pe}} \frac{n_e}{n_b}. \qquad (5.24)$$

Thus, we have shown that, in the case of the absolute instability due to the Cerenkov effect, oscillations with $\mu = 1$, i.e., $\omega \simeq \omega_{pe}$, are mainly excited.

Let us now consider the opposite limiting case where the runaway electron beam is localized radially, and the oscillations escape irreversibly from the resonance region. (Such a situation can occur, for example, in the case of a hollow beam, the oscillations moving to the plasma center in this case being effectively dissipated by Landau damping.) To take into account the convection of oscillations we have to add the term $u_{gr} \partial W / \partial r$ in Eq. (5.18):

$$\frac{\partial W}{\partial t} + u_{gr} \frac{\partial W}{\partial r} = 2\gamma W, \qquad (5.25)$$

where $u_{gr} = \partial \omega_k / \partial k_r \simeq \omega_k k_r / k^2 \simeq \omega_{pe} \mu k_r / k^2$. The convection of oscillations becomes important if $u_{gr} \partial W / \partial r \gg \partial W / \partial t$. Using the estimate (5.24) we can rewrite this condition in the form

$$\frac{\omega_{pe}}{k} \frac{\tau}{a_b} \sim \frac{v_b \tau}{a_b} \gg 1. \tag{5.26}$$

Here a_b is the effective beam radius. When condition (5.26) is satisfied, the time derivative in Eq. (5.25) can be neglected:

$$\frac{\omega_{pe}}{k} \mu \frac{\partial W}{\partial r} = \pi \omega_{pe} \mu v_z^2 \frac{1}{n_e} \frac{\partial f_z}{\partial v_z} W. \tag{5.27}$$

Equation (5.27) [or rather the expression for the group velocity $u_{gr} \simeq (\omega_{pe}/k)\mu$] is valid everywhere except in the narrow region of frequencies $\omega \to \omega_{pe}$ and $\omega \to \omega_{pi}$. Since in an inhomogeneous plasma these singular points cannot make a substantial contribution to the relaxation process, we shall henceforth neglect them. We introduce the average oscillation energy over the beam localization region; from Eq. (5.27) we obtain

$$W\left(\frac{\omega_{pe}}{v_z}, \mu\right) = W_0\left(\frac{\omega_{pe}}{v_z}, \mu\right) \exp\left\{\Gamma_0 v_z \frac{\partial f_z}{\partial v_z}\right\}, \tag{5.28}$$

where W_0 is the thermal (initial) level of oscillations. Using Eq. (5.28) we can write Eq. (5.27) in the following manner:

$$\frac{\partial f_z}{\partial t} = \frac{\partial}{\partial v_z} B \exp\left(\Gamma_0 v_z \frac{\partial f_z}{\partial v_z}\right) \frac{\partial f_z}{\partial v_z}, \tag{5.29}$$

where $B = 2\pi(e/m)^2(\omega_{pe}^2/v_{Te}^3)W_1$.

The solution of the system (5.28)-(5.29) is given below. Now we shall consider what oscillation spectrum is excited when the instability is convective. From Eq. (5.28) it follows that the oscillation amplification factor Γ does not depend on frequency. This results from the fact that $\Gamma = \gamma/u_{gr}$ and the group velocity, like the growth-rate, is proportional to μ, so that low-frequency oscillations stay longer in the beam localization region (unlike the case of an absolute instability, where τ does not depend on frequency). This is, apparently, the most significant difference between the convective and absolute instabilities; in the first case a broad oscillation spectrum from $\omega \sim \omega_{pi}$ to $\omega \sim \omega_{pe}$ is excited, in the second oscillations are localized close to the frequency $\omega \sim \omega_{pe}$. Here

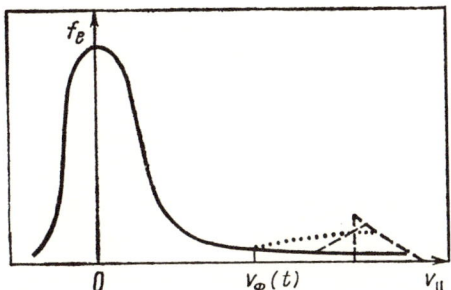

Fig. 6. Beam relaxation in the convective phase of instability.

we should emphasize that excitation of oscillations with $\omega \sim \omega_{pi}$ should lead to a linear (in W/nT) mechanism of ion heating.

<u>Problem.</u> Investigate the dynamics of quasilinear relaxation of the beam instability for the convective limiting case.

The system of quasilinear equations for this case has been described above: (5.28)-(5.29).

It is clear that there is sense in considering the instability when the following inequality is satisfied:

$$1/\alpha = \Gamma_0 v_z \, \partial f_z \partial v_z \gg 1. \tag{I}$$

When the strong inequality (I) is satisfied (under the actual experimental conditions $1/\alpha \sim 10^1$), Eq. (5.29) can be solved by the method of successive approximations using a small parameter $\alpha \ll 1$. For simplification, we shall also assume below that the velocity range within which $W \neq 0$ is sufficiently narrow for Eq. (5.29) to be written in the following manner:

$$\frac{\partial f_z}{\partial t} = \frac{\partial}{\partial v_z} A \exp\left\{\Gamma_1 \frac{\partial f_z}{\partial v_z}\right\} \frac{\partial f_z}{\partial v_z}, \tag{II}$$

where $A = B/v_b^3$, $\Gamma_1 = \Gamma_0 v_b$.

In the zeroth approximation in α it follows from Eq. (II) that $\partial^2 f_z / \partial v_z^2 = 0$. In other words the function $f_z = c_1 v_1 + c_2$, where c_1 and c_2 are functions of time. Higher-order expansions in α should be used for their determination. We find one relationship between them from the condition of conservation of the number of particles in the quasilinear relaxation process. As a result, the distribution function in the zeroth approximation can be written in the form

$$f'_z = \frac{2(v_z - v_F)}{(v_b - v_F)^2} = \frac{f_z}{n_b v_b}. \tag{III}$$

Here f'_z is normalized to 1: $\int f'_z \, dv_z = 1$; v_F is a time-dependent (to first approximation) integration constant characterizing the position of the front; the beam distribution function f'_z becomes zero when $v_F \geq v_z$ (see Fig. 6). Solution (III) is valid for $v_F \geq v_3 \omega_{pe}/\omega_{ce}$. The inverse limiting case will be considered below. To determine $v_F(t)$ we use the equation for the moments of f'_z. We substitute Eq. (III) into Eq. (II) and integrate Eq. (II) over v_z with weight v_z from v_F to v_b:

$$\frac{dv_F}{dt} = -\frac{6A}{v_b - v_F} \exp\left[\frac{2\Gamma_1 n_b}{(v_b - v_F)^2}\right]. \tag{IV}$$

Equation (IV) has the solution

$$(v_b - v_F)^4 \exp\left[-\frac{2\Gamma_1 n_b}{(v_b - v_F)^2}\right] = 24 A \Gamma t \tag{V}$$

or using Eqs. (5.28) and (V):

$$f_z = \frac{v_z - v_F(t)}{\Gamma_1} \ln \frac{v_b^4}{24 A \Gamma_1 t}; \tag{VI}$$

$$W = \frac{W_1}{\mu} \begin{cases} \dfrac{v_b^4}{24 A \Gamma_1 t} & \text{if } v_z > v_F; \\ 0 & \text{if } v_z \leq v_F \end{cases} \tag{VII}$$

$$v_F(t) = v_b - \sqrt{\frac{2\Gamma_1}{\ln v_b^2/24 A \Gamma_1 t}}.$$

From Eqs. (VI) and (VII) it follows in particular that the quantity $\partial f_z/\partial t$ decreases logarithmically with time, and, consequently, the oscillation energy density diminishes with t.

A solution, similar to Eqs. (VI) and (VII) can also be obtained for $v_F < v_3 \omega_{pe}/\omega_{ce}$. In this case, f_z and W will differ from zero in the whole range $v_3 \omega_{pe}/\omega_{ce} \leq v_z \leq v_b$; their dependence on time is described by formulas similar to Eqs. (VI) and (VII).

Let us also evaluate the degree of accuracy of the solution obtained. For this purpose we differentiate Eq. (II) with respect to v_z and with the help of Eqs. (VI) and (VII) we find the ratio of the neglected terms to those considered:

$$\frac{\partial^2 f_z}{\partial v_z^2}\left(\frac{1}{v_b}\frac{\partial f_z}{\partial v_z}\right)^{-1} \sim \left(\ln\frac{v_b^4}{24 A\Gamma_1 t}\right)^{-1} \sim \alpha \ll 1.$$

From this inequality it follows that our solution is valid for a sufficiently large amplification factor.

Here we should emphasize the basic qualitative difference between the quasilinear relaxation of an inhomogeneous beam and the case of homogeneous plasma discussed earlier. In the absolute instability case the oscillations are concentrated about $\mu = 1$; i.e., the spectrum is practically one-dimensional while in the case of the convective instability, the oscillations are distributed almost uniformly over the entire frequency range $1 \geq \mu \geq (m/M)^{1/2}$ ($\omega_{pi} \leq \omega \leq \omega_{pe}$).

6. Macroscopic Effects Associated with the Development of Instability

It has already been mentioned that the instability bursts are accompanied by experimentally recorded macroscopic changes in plasma parameters [4-10, 24]. In the course of the instability, the plasma diamagnetism increases, high-energy electrons trapped in local magnetic mirrors appear, the loop voltage diminishes, the synchrotron radiation increases, and, at the same time, epithermal ions are heated. All these effects can be explained consistently within the framework of the above theoretical model.

Changes in many of the above parameters occurring in the process of the instability development can be described by using the values of the electron distribution function before and after the instability obtained in Section 5. Thus changes in plasma dimagnetism are expressed by the following formula:

$$\Delta E_\perp = 2\pi \int_{v_3\frac{\omega_{pe}}{\omega_{ce}}}^{v_b} dv_z \int \frac{mv_\perp^2}{2} v_\perp dv_\perp (f_\infty - f_0) \simeq n_b \frac{mv_b^2}{9} \simeq \quad (6.1)$$

$$\simeq \frac{2}{3} E_\parallel (t = 0).$$

Here

$$E_\parallel (t = 0) = \int \frac{mv_\parallel^2}{2} f_0 \, d\mathbf{v}.$$

Thus the increase in the transverse energy significantly exceeds its initial value and can be comparable to the entire beam energy. This additional energy is experimentally recorded as an increase in plasma diamagnetism.

Knowing the electron distribution function, we can also estimate the number of electrons trapped in local magnetic mirrors and leaving the system because of toroidal drift. The trapping of the electrons occurs when the following condition is fulfilled:

$$v_z/v_\perp \leqslant \sqrt{\delta B/B}. \tag{6.2}$$

Here $\delta B/B$ is the magnitude of the inhomogeneity of the toroidal magnetic field due to the finite distance between the coils. Integrating the distribution function (5.15) over the phase space region defined by inequality (6.2) and considering that the mirrors occupy a fraction $\delta V/V$ of the total plasma volume V, we can determine the number of locally trapped electrons generated in the instability (see Fig. 6):

$$\frac{\Delta n_b}{n_b} \simeq \frac{\delta V}{V} \frac{2\pi}{n_b} \int_{v_3 \omega_{pe}/\omega_{ce}}^{v_b} dv_z \int_{v_z \sqrt{\frac{B}{\delta B}}}^{v_b} v_\perp \, dv_\perp f_\infty$$

$$\leqslant \left(\sqrt{\frac{\delta B}{B}} - \frac{v_3}{v_b} \frac{\omega_{pe}}{\omega_{ce}} \right) \frac{\delta V}{V}, \tag{6.3}$$

where $v_3/v_b \sim 1/3$ at the first stage of the instability development and $v_3/v_b \sim \frac{1}{3}(\omega_{pe}/\omega_{ce})^{3/2}$ at the second stage. It should be noted that in systems with $\delta B/B < 1/3$ (this condition is indeed fulfilled in all existing tokamaks) locally trapped electrons appear only at the second stage of instability development due to the Cerenkov mechanism of beam-particle retardation.

Lastly, knowing f_∞ we can evaluate the energy transmitted by the beam to the Trivelpiece-Gould mode:

$$W = \int (f_0 - f_\infty) \frac{mv^2}{2} \, d\mathbf{v}. \tag{6.4}$$

We shall give here a simple estimate for W. First, the beam loses part of its energy during angular scattering of particles by the anomalous Doppler effect $\Delta W_D \simeq (\omega/\omega_{ce}) mn_b v_b^2 \leq (\omega_{pe}/\omega_{ce}) mn_b v_b^2$. Second, the beam loses energy due to the excitation of oscillations at the second stage of instability due to the Cerenkov mechanism. This energy can be evaluated in the following way. Energy loss occurs in electrons with longitudinal velocity in the interval $v_3 \omega_{pe} \times \omega_{ce}^{-1} \leq v_z \leq v_3$. The total number of such electrons is $\Delta n_b/n_b \simeq v_3/v_b$; their energy loss is $\Delta W/mn_b v_b^2 \sim (v_3/v_b)^2$. It can be shown that under experimental conditions no less than 10% of the beam energy goes into oscillations. Since the beam energy is comparable with the plasma energy, it follows from this estimate that the ratio $W/n_e T_e \geq 10^{-1}$. It is clear that under such conditions it is important to take into account various nonlinear processes leading, first of all, to anomalous ion heating. This aspect will be treated in more detail in Section 8. Here we describe the linear mechanism of ion heating which can be important when the instability has a convective character. (It is just under these conditions that nonlinear mechanisms are insignificant since oscillations do not accumulate in the plasma.) As was shown above, a wide spectrum of oscillations from $\omega \geq \omega_{pi}$ to $\omega \leq \omega_{pe}$ is generated in the plasma when the instability has a convective character. The oscillations can interact with the ions if the inequality $\omega/k \leq 4 v_{Ti}$ is satisfied. Since the condition $\omega/k_z \simeq \omega_{pe}/k > v_{Te}$ must be fulfilled for the Trivelpiece–Gould mode to exist, only waves of frequency close to ω_{pi} can interact with ions. Hence it follows at once that the oscillations generated due to the anomalous Doppler effect cannot interact with the ions, since otherwise the inequality $\omega_{pi}/k_z > v_{Te}$ will be satisfied for them and thus $\omega_{ce}/k_z > v_{Te}(\omega_{ce}/\omega_{pe})(M/m)^{1/2}$, which is generally greater than the velocity of light.

It has been shown above that if the Cerenkov stage of the instability has a convective character, the amplification factor does not depend on the frequency, and the wave energy will be distributed over the spectrum in the following manner:

$$W_\omega = \frac{\langle W_T \rangle}{\omega} \exp \tilde{\Gamma}. \qquad (6.5)$$

Here $\tilde{\Gamma} = \int (2\gamma/u_{gr}) dr$; $\langle W_T \rangle$ is the thermal noise energy averaged over frequency. In deriving Eq. (6.5) account was taken of the fact that the spectral density of the thermal noise energy is inversely proportional to the oscillation frequency. To obtain the fraction of the energy of oscillations with a frequency close to ω_{pi} we can use the transport equation

$$\frac{\partial W_\omega}{\partial t} + \frac{\omega}{k}\frac{\partial W_\omega}{\partial r} = 2\gamma_\omega W_\omega. \qquad (6.6)$$

Upon integrating Eq. (6.6) over the beam localization region, we obtain the energy flow carried by oscillations with a given frequency:

$$\frac{d}{dt}\int W_\omega dV = \frac{\omega}{k}\frac{\langle W_T\rangle}{\omega}\exp\tilde{\Gamma} = \frac{\omega}{k_z}\frac{\langle W_T\rangle}{\omega_{pe}}\exp\tilde{\Gamma}. \qquad (6.7)$$

By integrating Eq. (6.7) over the spectrum we can obtain the relation of the energy flow transmitted by oscillations to ions to the total energy flow into oscillations:

$$\frac{d}{dt}\int_{\omega_{pi}}^{2\omega_{pi}} d\omega \int W_\omega dV \left(\frac{d}{dt}\int_{2\omega_{pi}}^{\omega_{pe}} d\omega \int W_\omega dV\right)^{-1} \sim \omega_{pi}/\omega_{pe}. \qquad (6.8)$$

Consequently, we can say that in the regime of convective instability, only a small fraction of the energy of the excited oscillations should go into heating epithermal ions.

Now we consider the question of the change in loop voltage accompanying each burst of instability. To explain these spikes, let us consider in greater detail the equation of an equivalent secondary circuit containing a plasma loop with a current

$$\frac{1}{c^2}\frac{d}{dt}\frac{LI^2}{2} = UI - RI^2. \qquad (6.9)$$

Here L is the plasma loop self-inductance factor, R its active resistance, U the loop voltage measured at the gap in the shell, and I the total plasma current. It follows from Eq. (6.9) that a change in voltage at a given current can be associated either with inductive effects or with effects of a change in the active resistance. Let us consider the first of these processes. For this purpose we neglect the second term on the right-hand side of Eq. (6.4):

$$\Delta U \simeq \frac{I}{c^2} \frac{\Delta L}{2\tau}, \qquad (6.10)$$

where ΔL is the change in inductance during instability development and τ is the characteristic time of this change. It turns out that the change in plasma inductance is basically connected with a change in the minor radius of the column resulting from a sharp increase in the transverse beam energy. The radial expansion of the column leads to a decrease in self-inductance by an amount

$$\Delta L \simeq -4\pi R \Delta a/a. \qquad (6.11)$$

The change in the minor radius of the column Δa can be associated with a change in the transverse energy of the beam $\Delta E_\perp$ in the following way. From the condition of freezing-in of the longitudinal field and pressure balance we find

$$\pi a^2 \Delta E_\perp = 2\pi a \Delta a B_0^2 \qquad (6.12)$$

or $\Delta a/a \simeq \Delta E_\perp / 2B_0^2$.

Substituting Eq. (6.12) into Eqs. (6.11) and (6.10), we obtain

$$\Delta U \simeq -\frac{I}{c^2} \frac{4\pi R}{4\tau} \frac{\Delta E_\perp}{B_0^2}. \qquad (6.13)$$

Actually, the integral $\int_0^\tau \Delta U dt$ enters into Eq. (6.13). The true (i.e., experimentally observed) value of ΔU can be much lower because of the screening effect of the liner. To take the latter into account, we should replace τ by the liner integration time constant. It is to be noted that expansion of the column minor radius leads to a negative voltage spike.

Now we shall evaluate the change in voltage associated with a change in the active resistance of the column due to the loss of longitudinal momentum by runaway electrons during the instability development. It is natural that the sign of ΔU in this case will be positive:

$$\Delta U = -R \Delta I_b \qquad (6.14)$$

or $\Delta U/U = -\Delta I_b/(I - I_b)$.

The change of I_b can be evaluated as $\Delta I_b = e n_b \Delta v_b \pi a_b^2 \leq I_b$ so that

$$\Delta U/U \leqslant I_b/(I - I_b). \qquad (6.15)$$

As a rule, in experiments on TM-3 $I_b \ll I - I_b$; therefore, the effect of the change in active resistance was insignificant. In experiments on T-6 [3] where the beam carried a major fraction of the current, the quantity described by Eq. (6.14) could exceed the inductive effect of a decrease in voltage, i.e., the total ΔU could be positive. It is to be noted, however, that in any case the active resistance of the plasma cannot be greater than the classical value, since the instability considered concerns only epithermal electrons.

Lastly, knowing the evolution of the electron distribution function during instability development, we can evaluate the change in time of the plasma synchrotron radiation. We shall consider this process using a simple model described in [35]. At the initial stage of instability development (when the beam anisotropy is sufficiently great) it may be assumed that quasilinear effects lead only to an increase in the transverse energy of resonance electrons. In this case the quasilinear equation can be rewritten in the following manner:

$$\frac{\partial f_e}{\partial t} \simeq \frac{1}{v_\perp} \frac{\partial}{\partial v_\perp} \frac{D}{v_\perp} \frac{\partial f_e}{\partial v_\perp}, \qquad (6.16)$$

where

$$D \simeq \frac{8\pi^2 e^2 \omega_{ce}^2}{m^2} \int d\mathbf{k} \frac{W_\mathbf{k}}{k^2} \delta(\omega_{ce} - k_z v_z) J_1^2\left(\frac{k_\perp v_\perp}{\omega_{ce}}\right); \quad W_\mathbf{k} = |E|_\mathbf{k}^2/8\pi.$$

We shall assume that the inequality $k_\perp v_b/\omega_{ce} \ll 1$ is satisfied for the beam particles throughout the whole stage of instability. This permits us to simplify the equation for the diffusion coefficient:

$$D \simeq \frac{2\pi^2 e^2 v_\perp^2}{m^2} \int d\mathbf{k}\, W_\mathbf{k}(t)\, \delta(\omega_{ce} - k_z v_z) \frac{k_\perp^2}{k^2}.$$

Furthermore, it is appropriate to introduce a new time-like variable τ:

$$\tau = \frac{2\pi^2 e^2}{m^2} \int_0^t dt' \int d\mathbf{k}\, \frac{k_\perp^2}{k^2} W_\mathbf{k}(t')\, \delta(\omega_{ce} - k_z v_z). \qquad (6.17)$$

As a result, Eq. (6.16) takes the form of a diffusion equation:

$$\frac{\partial f_e}{\partial \tau} = \frac{1}{v_\perp} \frac{\partial}{\partial v_\perp} v_\perp \frac{\partial f_e}{\partial v_\perp}.$$

The electron distribution function before the moment of instability excitation

$$f_e = \frac{f_z(v_z = v_{cr})}{2\pi v_{Te}^2} \left(1 + \frac{v_{cr}^2}{v_z^2}\right) \exp\left(-\frac{v_\perp^2}{v_{Te}^2}\right) \quad (6.18)$$

is the initial condition for this equation. Here $f_z(v_z = v_{cr})$ is the one-dimensional distribution function in the region $v_z \simeq v_{cr}$. In this case, the solution of Eq. (6.16) can be written in the following form:

$$f(v_z, v_\perp, \tau) = \frac{f_z(v_z = v_{cr})}{2\pi} \left(1 + \frac{v_{cr}^2}{v_z^2}\right) \frac{1}{4\tau + v_{Te}^2} \exp\left(-\frac{v_\perp^2}{4\tau + v_{Te}^2}\right). \quad (6.19)$$

To express Eq. (6.19) in real time t, we have to solve the equation for the energy level of excited oscillations. For simplicity, we shall consider that the instability has a convective character. In this case, the oscillation level is given by

$$W_k \simeq \frac{T_e}{2} \exp(2\gamma_k L/u_{gr}). \quad (6.20)$$

where L is the characteristic size of the beam localization region and u_{gr} is the wave group velocity. Considering that the instability excites oscillations with characteristic value $(kr_D)^{-1} \sim 3$ and $\mu \sim 0.2$ we obtain from (6.17)

$$t \simeq \frac{\tau}{v_z^2} \frac{n_e r_D^3}{\pi^2 \omega_{pe}} \left(\frac{v_z}{v_{Te}}\right)^3 10 \exp\left(-\frac{2L\gamma_k k v_{Te}}{\mu(1-\mu^2) v_{Te}}\right)$$

$$\sim 10^3 \frac{\tau}{v_z^2} \frac{n_e r_D^3}{\omega_{pe}} \exp\left(-4 \cdot 10^{-3} \frac{L}{r_D}\right). \quad (6.21)$$

A simple estimate shows that for $L/r_D \sim 10^3$, $t \sim 1$–10 μsec for $\tau_0 \simeq v_b^2$. It is precisely in this period of time ($\tau \sim \tau_0$) that the transverse energy increases to the level of the beam longitudinal energy, and saturation of the instability takes place.

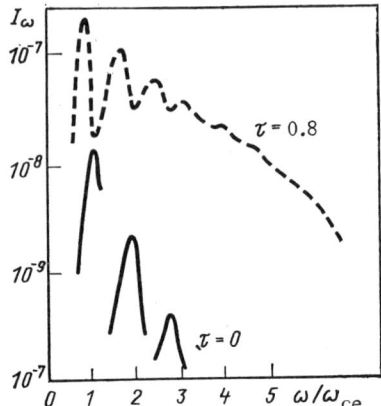

Fig. 7. Synchronous beam radiation during the "fan" instability.

Knowing the time dependence of the distribution function of the accelerated electron beam we can evaluate the time dependence of the synchrotron radiation. For this purpose, we have to insert the value obtained for $f(\tau)$ into the Scott-Trubnikov formula [36]. The total power emitted in the m-th harmonic of an extraordinary wave propagating strictly across the magnetic field, has the following form:

$$\frac{dp}{d\Omega} \simeq \delta \frac{e^2 m^2 \omega_{ce}^2}{2\pi c} \frac{n_b}{n_e} \left(\frac{v_{Te}}{c}\right)^4 m^2 \frac{(4\tau + v_{Te}^2)^2}{v_{Te}^4}$$
$$\times \exp\left(-\frac{m^2 v_{Te}^2}{c^2}\left(\frac{4\tau}{v_{Te}^2} + 1\right)\right) \left\{\left(1 + \frac{2c^2/v_{Te}^2}{m(4\tau/v_{Te}^2 + 1)} + \frac{2c^4/v_{Te}^4}{m^2(4\tau/v_{Te}^2 + 1)^2}\right) I_m - I_{m-1}\right\}, \qquad (6.22)$$

where $I_m = I_m\left(\frac{m^2(4\tau/v_{Te}^2+1)}{c^2/v_{Te}^2}\right)$ and δ is the fraction of runaway electrons. It is to be noted that formula (6.22) is valid for all harmonics with m > 1. From Eq. (6.22) it follows that a characteristic increase in the intensity of the synchrotron radiation occurs during an interval on the order of 1 μsec, which was observed under experimental conditions. The time dependence of the characteristic spectrum of the synchrotron radiation given by Eq. (6.22) for typical experimental conditions is shown in Fig. 7.

7. Anomalous Diffusion of Runaway Electrons

As was shown above (see Sections 2 and 3) electrons having velocity $v_z > v_{cr}$ are freely accelerated by the electric field, and in the absence of anomalous effects, they manage to reach relativistic energy before leaving the system due to displacement of the drift surface from the magnetic surface. The value of this energy depends on the size of the device; its measurement (for example, with the help of hard x-ray radiation) can be used to determine how classical the behavior of runaway electrons in a tokamak is.

Experiments performed so far [37-39] showed that in discharges with moderate density (i.e., in discharges with the maximum value of E/E_D) the behavior of runaway electrons is not classical. First, the characteristic energy of fast electrons is less than the value predicted by the classical theory, so that their energy containment time exceeds the energy containment time of background electrons by no more than one order of magnitude. Moreover, it was shown [40] that when E/E_D becomes greater than a certain critical value, the intensity of hard x-ray radiation from the discharge begins to decrease.

All these data allow the conclusion that in discharges with a large value of E/E_D there is an anomalous diffusion mechanism for runaway electrons which prevents them from accumulating in the system and from being accelerated up to relativistic energies. One of the possible mechanisms for such an anomaly can be the instability due to the anomalous Doppler effect considered above. It is known [41] that any kinetic instability in an inhomogeneous plasma should lead to the anomalous diffusion of resonance particles across the magnetic field. In this case, oscillations serve as a kind of a third body with which electrons collide; these collisions lead to enhanced diffusion of the runaway electrons across the magnetic field. For a joint description of the process of instability development and diffusion of resonance electrons one can use a system of quasilinear equations which have the following form in inhomogeneous plasma:

$$\frac{\partial f}{\partial t} + \frac{eE}{m}\frac{\partial f}{\partial v_z} = \int d\mathbf{k}\left(-\frac{v_z}{v_\perp}\frac{\partial}{\partial v_\perp} + \frac{\partial}{\partial v_z} + \frac{k_\theta}{k_z\omega_{ce}}\frac{\partial}{\partial r}\right)$$

$$\times 4\left(\frac{\pi e}{m}\right)^2 \frac{k_z^2}{k^2}\frac{k_\perp^2 v_\perp^2}{\omega_{ce}^2}\frac{|E|_\mathbf{k}^2}{4\pi}\delta(\omega_{ce}-k_z v_z)$$

$$\times\left(-\frac{v_z}{v_\perp}\frac{\partial f}{\partial v_\perp} + \frac{\partial f}{\partial v_z} + \frac{k_\theta}{k_z\omega_{ce}}\frac{\partial f}{\partial r}\right)$$

$$+ \int d\mathbf{k}\left(\frac{\partial}{\partial v_z} + \frac{k_\theta}{k_z\omega_{ce}}\frac{\partial}{\partial r}\right) 16\left(\frac{\pi e}{m}\right)^2 \frac{k_z^2}{k^2}\frac{|E|_\mathbf{k}^2}{4\pi}\delta(\omega-k_z v_z)$$

$$\times\left(\frac{\partial f}{\partial v_z} + \frac{k_\theta}{k_z\omega_{ce}}\frac{\partial f}{\partial r}\right); \qquad (7.1)$$

$$\frac{\partial}{\partial t}\frac{|E|_\mathbf{k}^2}{4\pi} + \frac{\partial\omega}{\partial\mathbf{k}}\frac{\partial}{\partial r}\frac{|E|_\mathbf{k}^2}{4\pi}$$

$$= \left\{\frac{\pi}{4}\omega_\mathbf{k}\frac{4\pi e^2}{mk^2}\int d\mathbf{v}\,\delta(\omega_{ce}-k_z v_z)\frac{k_\perp^2 v_\perp^2}{\omega_{ce}^2}\right.$$

$$\times\left(-\frac{k_z v_z}{v_\perp}\frac{\partial f}{\partial v_\perp} + k_z\frac{\partial f}{\partial v_z} + \frac{k_\theta}{\omega_{ce}}\frac{\partial f}{\partial r}\right)$$

$$+ \pi\omega\frac{4\pi e^2}{mk^2}\int d\mathbf{v}\,\delta(\omega-k_z v_z)\left(k_z\frac{\partial f}{\partial v_z} + \frac{k_\theta}{\omega_{Be}}\frac{\partial f}{\partial r}\right) - \nu_{ei}\left.\right\}\frac{|E|_\mathbf{k}^2}{4\pi}. \quad (7.2)$$

Equation (7.1) is written for collisionless plasma since we will only be interested in particles with velocities $v_z > v_{cr}$; k_r is the r-component of the wave vector and k_θ the θ-component.

The first term on the right-hand side of Eq. (7.1) [and the first term on the right-hand side of Eq. (7.2) corresponding to it] describes the interaction of resonant particles with oscillations due to the anomalous Doppler effect both in velocity and coordinate space. The second term on the right-hand side of Eqs. (7.1) and (7.2) describes the interaction of electrons with oscillations due to the Cerenkov effect.

Since in this section we will be interested in comparatively slow diffusion of resonance electrons in coordinate space rather than the dynamics of instability development [the velocity of the first process is approximately $(\rho_{Le}/a)^2$ times less than that of the second], we can seek a solution of the system (7.1)-(7.2) averaged over a time greater than the period between the two subsequent bursts of instability (if the latter has a relaxation character). This permits us, first, to neglect the region of the Cerenkov resonance, since at $W/W_T \gg 1$ Landau damping in the region $v_z \simeq \omega_{pe}/k$ is absent, and, second, in this case we can seek a quasistationary solution of the system (7.1)-(7.2) in the region $v_z > v_{cr}$.

In the following we will confine ourselves to the case where the instability has an absolute character (the conditions for which are given above). This permits us to exclude from Eq. (6.1) cross derivatives of the type $\partial^2 f/\partial r \partial v$ because of the spectral symmetry of the excited oscillations about k_θ. We shall also assume that $\langle\langle (k - \langle\langle k \rangle\rangle)/k \rangle\rangle \ll 1$ ($\langle\langle ... \rangle\rangle$ means frequency average) for all excited oscillations. This inequality is practically always satisfied and corresponds to narrowing the zone of the Cerenkov resonance. In this case, we can assume that $k_\perp \simeq k = k_0 \simeq \omega_{pe}/v_{cr}$, i.e.,

$$\frac{|E|_k^2}{4\pi} = W_0(\cos\theta) \frac{\delta(k-k_0)}{2\pi k^2} \delta(\omega - \omega_{pe}\cos\theta), \tag{7.3}$$

where $\cos\theta = k_z/k$, i.e., $W_0 = W_0(v_z)$.

Using the relationships above, the system (7.1)-(7.2) can be rewritten in the following way:

$$\frac{\partial f}{\partial t} + \frac{eE}{m}\frac{\partial f}{\partial v_z} = \left(-\frac{v_z}{v_\perp}\frac{\partial}{\partial v_\perp} + \frac{\partial}{\partial v_z}\right)\pi \frac{e^2}{m^2}$$
$$\times \frac{v_\perp^2}{k_0 v_z^3} W_0 \left(-\frac{v_z}{v_\perp}\frac{\partial f}{\partial v_\perp} + \frac{\partial f}{\partial v_z}\right) + \frac{c^2 k_0}{B^2 \omega_{ce}^2}$$
$$\times W_0 \frac{v_\perp^2}{v_z^2} \frac{1}{r}\frac{\partial}{\partial r} r \frac{\partial f}{\partial r}, \tag{7.4}$$

$$\frac{dW_0}{dt} = W_0\left[\pi\omega_{pe}\frac{\omega_{pe}^2}{\omega_{ce}n_e}\frac{1}{k_0 v_z}\left(\frac{\partial}{\partial v_z}\langle v_\perp^2 f\rangle - 2v_z \langle f\rangle\right) - v_{ei}\right], \tag{7.5}$$

where $\langle f \rangle = \int f \, d\mathbf{v}_\perp$.

Below we shall consider the limiting case of strong instability when a sufficiently large number of runaway electrons is present in plasma. In this case, it follows from (7.5) that at the stationary stage of instability development in the range of velocities $v_{cr} \leq v_z \leq v_b$ (mv_b^2 is the characteristic energy of the beam particles) the electron distribution function in velocity space is almost isotropic.

The existence of the small parameter

$$\frac{T_{\|} - T_{\perp}}{T_{\|} + T_{\perp}} = \frac{\int (v_z^2 - v_{\perp}^2) f_e \, d\mathbf{v}}{\int (v_z^2 + v_{\perp}^2) f_e \, d\mathbf{v}} \ll 1 \qquad (7.6)$$

permits us to trace qualitatively the diffusion of runaway electrons using moments of the electron distribution function similar to those introduced in Section 5. For this purpose we shall multiply Eq. (7.4) successively by $v_z^2 + v_{\perp}^2$ and $v_z^2 - v_{\perp}^2$ and integrate it over the region of instability localization. As a result, we obtain a system of equations similar to (5.4)-(5.5):

$$\frac{d}{dt}(T_{\|} + T_{\perp}) = Q - \nu_T \frac{k_0 v_b^4}{\omega_{ce}^4 a^2} \frac{T_{\|} + T_{\perp}}{T_{\|}} \frac{\overline{W}_0}{n_e}; \qquad (7.7)$$

$$\frac{d}{dt}(T_{\|} - T_{\perp}) = Q - \nu_T \frac{T_{\|} - T_{\perp}}{T_{\|}} \frac{\overline{W}_0}{n_e} \left(1 + \frac{k_0^2 v_b^4}{\omega_{ce}^4 a^2}\right), \qquad (7.8)$$

where $T_{\|} \approx mv_b^2$.

In obtaining (7.7)-(7.8), we replaced terms with derivatives with respect to r by $\partial/\partial r \to 1/a$; $\overline{W}_0 = \int W_0(v_z) dv_z / v_b$. To obtain a closed system of equations, we shall integrate Eq. (7.5) over v_z (in the limits of $v_{cr} \le v_z \le v_T$, where $W_0 \gg W_T$):

$$\frac{d\overline{W}_0}{dt} \simeq \overline{W}_0 \left[\gamma \frac{T_{\|} - T_{\perp}}{T_{\|}} - \nu_{ei}\right]. \qquad (7.9)$$

In the system (7.7)-(7.9) the quantities $\overline{W}_0$, $T_{\|}$, and $T_{\perp}$ are unknown. The equality

$$n_b = f_z(v_z = v_{cr}) v_b$$

is used to determine the quantity n_b.

The stationary solution of the system (7.7)-(7.9) permits us to trace qualitatively the dependence of the beam energy, its anisotropy and oscillation energy density on the main plasma parameters

$$\left(\frac{v_b}{v_{Te}}\right)^2 \simeq \frac{1}{\sqrt{n_e r_D^3}} \left(\frac{\omega_{ce}}{\omega_{pe}}\right)^{3/2} \frac{a}{\rho_{Le}} \left(\frac{E_D}{E}\right)^{3/4} \frac{1}{\sqrt{\Gamma}}; \qquad (7.10)$$

$$\overline{W}_0 \approx \frac{eE}{m} \frac{B_0^2}{c^2} \frac{\omega_{ce}^2 a^2}{k_0 v_b^2} \sim \sqrt{\Gamma}; \qquad (7.11)$$

$$\frac{T_\parallel - T_\perp}{T_\parallel} \approx \frac{\nu_{ei}}{\omega_{pe}} \frac{\omega_{ce}}{\omega_{pe}} \frac{1}{\Gamma} \sqrt{\frac{E}{E_D}}. \qquad (7.12)$$

The value Γ characterizes the number of electrons entering the runaway regime and is determined by formula (2.18).

From (7.10)-(7.12) it follows that with an increase in the ratio E/E_D (i.e., with an increase in the number of particles entering the regime of continuous acceleration), their characteristic energy decreases. This is a direct consequence of the fact that the efficiency of the anomalous diffusion process leading to the loss of fast electrons from the system is proportional to the oscillation energy density $\overline{W}_0$. The last value is, naturally, proportional to the number of runaway electrons, i.e., to the value Γ. The characteristic energy of the beam particles is inversely proportional to the number of runaway electrons.

The relations obtained permit us to explain the anomalies in the behavior of runaway electrons in discharges with moderate density which were discovered experimentally and described above. Indeed, when E/E_D is large, an instability develops in the system leading to a rapid drift of runaway electrons from the system, whose magnitude is proportional to E/E_D.

Now we shall briefly discuss the influence of the runaway electron beam on the energy balance of plasma in tokamaks. It is known that regimes with moderate density are the most anomalous with respect to the electron component. The point is that the magnitude of the known classical mechanisms of electron energy losses (heat transfer to ions by collisions, radiation due to impurities and neoclassical thermal conductivity), unlike the Joule energy input, decreases with decreasing plasma density (or, which is the same, with increasing E/E_D). However, the experiments prove that the temperature of the main electron component does not increase with decreasing density. It is just for this reason that most empirical formulas for the electron thermal conductivity include an anomaly factor inversely proportional to the plasma density. First, we have to remind ourselves how the anomalous electron thermal conductivity is determined experimentally. We write an equation for the electron energy balance of the whole plasma column:

$$\frac{d}{dt}\overline{n_e T_e} = \frac{IE}{\pi a^2} - \frac{\overline{n_e T_e}}{\tau_E}. \qquad (7.13)$$

Here $\overline{n_e T_e} = \frac{2}{a^2}\int_0^a n_e T_e r\,dr$, τ_E is the energy confinement time for electrons, I the total current. In the stationary stage of the discharge

$$\tau_E = \frac{\pi a^2 \overline{n_e T_e}}{IE}. \qquad (7.14)$$

The anomaly of the electron thermal conductivity is given by

$$A = \frac{a^2}{\overline{\chi}_e^{neo}\tau_E} = \frac{IE}{\pi \overline{n_e T_e}\, \overline{\chi}_e^{neo}}, \qquad (7.15)$$

where $\overline{\chi}_e^{neo}$ is the neoclassical coefficient of electron thermal diffusion (see, e.g., [21]).

The presence of the runaway electron beam in the plasma can have an effect on the energy balance of the main electron component. First, such electrons can carry a significant fraction of the current without making a contribution to the Joule energy input. Hence, in Eqs. (7.13)-(7.15) I should be replaced by $I - I_b$, where I_b is the current carried by the beam. Using relations (7.10)-(7.12) we can show that

$$\frac{I - I_b}{I_b} \sim \sqrt{\frac{\omega_{pe}}{\nu_{ei}}} \left(\frac{\omega_{pe}}{\omega_{ce}}\right)^{3/2} \frac{\rho_{Le}}{a}\left(\frac{E}{E_D}\right)^{3/4}\frac{1}{\sqrt{\Gamma}}. \qquad (7.16)$$

From Eq. (7.16) it follows that when E/E_D is sufficiently large ($E/E_D \geq 0.1$), runaway electrons can carry a significant fraction of the current, thereby reducing (at a given value of the safety factor q) the Joule heating. It has already been noted above that in experiments on T-6 [3] we observed regimes in which the beam carried almost the whole current so that the background plasma remained cold.

And, lastly, the presence of the beam can lead to the development of a specific mechanism for energy loss from the main electron component. Let us imagine that the runaway electron energy irreversibly leaves the discharge. This may result from the loss from the plasma of Trivelpiece–Gould oscillations excited by the beam [42] or of the anomalous diffusion of runaway electrons considered above. In this case, the thermal electrons will also lose energy due to the presence of a continuous flow of particles and energy from the thermal distribution to the region of continuous acceleration. Quantitatively, the value of the cooling effect of the main electron component can be determined in the following way. Let us write a one-dimensional kinetic equation for electrons assuming that above a certain velocity $v_0 \gg v_{T_e}$ there exists in the system a drain of magnitude ν_1:

$$\frac{\partial f}{\partial t} = -\frac{eE}{m}\frac{\partial f}{\partial v_z} + \frac{\partial}{\partial v_z} \nu_{ei}(T_e) \frac{V_{T_e}^3}{|v_z|^3}$$

$$\times \left(v_z f + \frac{T_e}{m}\frac{\partial f}{\partial v_z}\right) - \nu_1 f \theta(v_z - v_0) \qquad (7.17)$$

Here θ is the step function. Let us multiply Eq. (7.17) by $mv_z^2/2$ and integrate it over velocity space. Since the integral of electron–electron collisions does not change the electron energy, we derive

$$\frac{3}{2}\frac{d}{dt} n_e T_e = eE \int_{-\infty}^{\infty} v_z f \, dv_z - \nu_1 m \int_0^{\infty} \frac{v_z^2}{2} f \, dv_z \qquad (7.18)$$

To find the electron distribution function in the region $v_z > v_0$, we can make use of the fact that in the region $v_z > v_0 \geq v_{cr}$ the main contribution to the collision integral is given by the dynamic friction force (it may be shown that neglecting velocity space diffusion in the region $v_z \geq v_0$ is valid if the inequality $eEv_0/m\nu_{ei}v_{T_e}^2 > 1$ is satisfied). In this case the stationary solution of Eq. (6.17) in the region $v_z > v_0$ has the following form:

$$f = f(v_0) \exp\left\{-\frac{v_z - v_0}{u_1} + \frac{v_{Te}^3}{u}\left(\frac{1}{v_z^2} - \frac{1}{v_0^2}\right)\right\}, \qquad (7.19)$$

where $u_1 = eE/m\nu_1$, $u = eE/m\nu_{ei}$.

Substituting (7.19) into (7.18) we obtain

$$\frac{3}{2}\frac{d}{dt} n_e T_e = eE \int_{-\infty}^{v_0} v_z f dv_z - \frac{eE}{m}\frac{mv_0^2}{2}\frac{f(v_0)}{1+2\frac{v_{Te}^2 \nu_{ei}}{v_0^3 \nu_1}}$$

$$+ eE \int_{v_0}^{\infty} 2\frac{v_{Te}^3}{v_z^3}\frac{\nu_{ei}}{\nu_1}\left(1+2\frac{v_{Te}^3}{v^3}\frac{\nu_{ei}}{\nu_1}\right)^{-1} v_z f dv_z. \quad (7.20)$$

Let us consider the limiting case of a powerful energy flow; in this case it follows from Eq. (7.20) that

$$\frac{3}{2}\frac{d}{dt} n_e T_e = eE \int_{-\infty}^{v_0} vf dv - \frac{eE}{m}\frac{mv_0^2}{2} f(v_0). \quad (7.21)$$

The first term on the right-hand side of (7.21) describes the Joule heating of plasma, the second term electron cooling due to a continuous energy flow to the runaway region (to the region of $v_z > v_0$). Equation (7.21) gives the maximum cooling effect when all the electrons with velocity $v_z > v_0$ leave the system without having enough time to transfer part of their energy to the main component by collisions. Consideration of the finite collision frequency in the runaway region leads to a decrease in the cooling effect — the second term on the right-hand side of Eq. (7.20) describes the decrease of electron cooling, due to collisions, and the last term on the right-hand side of Eq. (7.20) describes the contribution made by the electron tail to the Joule heating. If in (7.21) we take $f(v_0) = f_z(\infty) = (n_e/\sqrt{\pi} v_{Te})\Gamma$ as an estimate, it turns out [43] that the cooling effect under consideration, in discharges with moderate density, greatly exceeds all the classical mechanisms of electron energy loss and can be a decisive factor in the electron energy balance in such regimes of tokamak operation. And, lastly, we have to point out another convective mechanism for the loss of runaway electrons from the discharge due to their trapping in local minima of the toroidal field and subsequent vertical drift. It was shown above (see Section 3) that locally trapped electrons leave the system if their energy exceeds the critical value $E_{cr} \simeq E_\perp \geq m(\nu_{ei} v_{Te}^3 aR \times \omega_{ce} B/\delta B)^{2/5}$. It is clear that a sufficiently large number of fast electrons trapped in local minima of the magnetic field can appear only when there is instability due to the anomalous Doppler effect.

In this case a convective flow of energy from the column is possible due to the toroidal drift.

8. Nonlinear Processes and Ion Heating

It has been shown above that instability development leads to the excitation of Trivelpiece−Gould oscillations; the density of their energy might be high enough $W_l \leq mn_b v_b^2 \omega_{pe}/\omega_{ce}$. Usually, in discharges with instability the beam energy is comparable to the energy of the main electron component $mn_b v_b^2 \sim n_e T_e$, hence $W_e \times (n_e T_e)^{-1} \leq 1$. It is clear that under these conditions nonlinear processes can exert a substantial effect on the dynamics of instability development.

In this section we shall treat in detail only one (evidently, the most striking) manifestation of nonlinear processes, explaining the anomalous heating of ions observed experimentally. We recall that, in experiments on TM-3, it was shown that in each instability burst the ion temperature increased by the value $\Delta T_i \leq 0.3 \cdot \Delta T_{e\perp}$ ($T_{e\perp}$ is the diamagnetic temperature of electrons).

As is known, nonlinear processes with the lowest threshold (with respect to $W/n_e T_e$) are three-plasmon interactions involving decay or induced wave scattering on resonant particles.

In the range of frequencies under consideration $\omega_{pi} \leq \omega \leq \omega_{pe}$ the two most effective processes are [41, 45, 46]: decay of the Trivelpiece−Gould mode into two Trivelpiece−Gould modes at lower frequency and decay of the Trivelpiece−Gould mode into a Trivelpiece−Gould mode and an ion-acoustic wave.

Using the laws of conservation of energy and momentum, we can describe these processes in the following manner:

$$\omega_l = \omega_l' + \omega_l''; \quad k_l = k_l' + k_l''; \tag{8.1}$$

$$\omega_l = \omega_l' + \omega_s''; \quad k_l = k_l' + k_s''. \tag{8.2}$$

If $\omega \leq \omega_{pe}$, then $\omega_s'' \ll \omega_l, \omega_l'$. In isothermal plasma the damping rate of ion-sound oscillations is greatly increased, and the efficiency of the second process, described by Eq. (8.2), is comparable with the efficiency of scattering of the Trivelpiece−Gould mode on ions. In this case, the frequency ω_s'' ceases to be an eigenfrequency, and $\omega_s'' \simeq |\mathbf{k}_l - \mathbf{k}_l'|\mathbf{v}_l$, where $\mathbf{v}_l$ is the velocity of resonant ions. Moreover, in inhomogeneous plasma the decay process rapidly changes to an induced scattering process because the need to satisfy the laws of conservation (8.2) requires that the ion-sound oscillation frequency $\omega_s'' = \omega_l - \omega_l'$ ceases to be an eigenfrequency.

Plasma inhomogeneity also leads to a sharp decrease in the efficiency of the process described by Eq. (8.1) since resonance conditions are satisfied only in a small part of the plasma volume (see, e.g., [45]). For this reason, we shall only consider one nonlinear process in detail — an induced scattering of Langmuir waves on ions. Besides, it will be shown below that under real experimental conditions consideration of the first mechanism only permits us to explain effective heating of the ions.

Thus, we shall consider the nonlinear mechanism of induced scattering of the Trivelpiece–Gould mode on ions. It describes the successive decrease in the frequency of the mode quanta for $\omega_s'' \sim |\mathbf{k}_l - \mathbf{k}_l'|\mathbf{v}_{Ti}$, the energy $\hbar\omega_s''$ being expended on ion heating. In its general form, the equation describing the development of the spectral density of wave quanta n_k is written in the following manner:

$$\frac{\partial n_k}{\partial t} + \gamma_k n_k = \int T_{kk'} n_{k'} n_k dk', \qquad (8.3)$$

where $\gamma_k = \nu_{ei} + \sqrt{\frac{\pi}{2}} \frac{\omega_k}{(kr_D)^3} \exp\left(-\frac{1}{(kr_D)^3}\right)$ is the linear damping on electrons due to collisions and Landau damping. We neglect ion Landau damping because ω_l' is far from ω_{pi}. $T_{kk'}$ is the matrix element of nonlinear interaction. For further consideration it is appropriate to introduce a dimensionless variable μ: $\mu = \omega_l/\omega_{pe} = \cos\theta = k_{zl}/k_l$. Each act of induced scattering leads to a change in ω_l (i.e., in the Trivelpiece–Gould mode frequency) by a small value on the order of $\Delta\mu \sim |\mathbf{k}_l - \mathbf{k}_l'|v_{Ti}/\omega_l \ll \mu$. This means that only Trivelpiece–Gould waves with neighboring frequencies interact in a nonlinear manner, which allows us to write Eq. (8.3) in a diffusion approximation:

$$\frac{\partial}{\partial t} n_l(\mu, k) + \gamma_k n_l(\mu, k) = \int 2\pi k'^2 \, dk' \, T(k, k', \mu)$$
$$\times (\Delta\mu)^2 \frac{\partial}{\partial \mu} n_l(\mu, k'). \qquad (8.4)$$

Equation (8.4) is valid only in regions sufficiently far away from the source of waves excited by the beam. Below, for simplicity, we shall suppose that the beam excites a spectrum of oscillations which is monochromatic in μ and k:

$$N_{kl} = N_l \, \delta(\mu - \mu_0) \frac{\delta(k-k_0)}{2\pi k^2}. \tag{8.5}$$

This assumption does not affect the results by an order of magnitude: the calculations considered in this section cannot pretend to a higher accuracy. With the assumption referred to above, the equation for N_l can be written in the following manner:

$$\frac{\partial N_l}{\partial t} = \gamma_l N_l - \int 2\pi k'^2 \, dk' \, \Delta\mu n_l(\mu_0, k') \, N_l \, T(k_0, k', \mu_0) - \nu_{ei} N_l. \tag{8.6}$$

The application of (8.5) also enables us to include into (8.4), in explicit form, the wave source at $\mu = \mu_0$:

$$\frac{\partial}{\partial t} n_l(k, \mu) + \gamma_k n_l(k, \mu) = \int 2\pi k'^2 \, dk' \, T(k, k', \mu)$$

$$\times (\Delta\mu)^2 n_l(k, \mu) \frac{\partial}{\partial \mu} n_l(k', \mu) + T(k, k_0, \mu_0)$$

$$\times \Delta\mu n_l(k, \mu) N_l \, \delta(\mu - \mu_0 + \Delta\mu). \tag{8.7}$$

Equations (8.6) and (8.7) represent a closed system describing the behavior of Trivelpiece–Gould waves. Here it is necessary to make a remark. Equation (8.6) does not include quasilinear effects leading to instability saturation. We can consider them qualitatively assuming that $\gamma_l = \gamma_l^0 \theta(\tau - t)$, where γ_l^0 is a linear increment determined by Eq. (4.7); τ, the characteristic time of the burst, is determined from the condition

$$\int_0^\tau \gamma_l^0 N_l \omega_0 \, dt = W_l^{\max} \simeq m n_b v_b^2 \frac{\omega_0}{\omega_{pe}} \sim n_e T_e,$$

$$\tau \simeq n_e T_e / \gamma_l^0 N_l \omega_0. \tag{8.8}$$

N_l should be derived from a self-consistent solution of the system (8.6)-(8.7). Before proceeding to its solution, let us note the following. We will not need the concrete form of the matrix element $T(k, k', \mu)$; we will only need its functional dependence on parameters k, k', μ. It is known [47] that in plasma with $\omega_{pe} > \omega_{ce}$ the matrix element is of the order of magnitude

$$T_{kk'} \sim \omega_k \omega_{k'}/n_e T_e. \tag{8.9}$$

Generally speaking, the form of the matrix element changes in a magnetized plasma because of the motion of electrons across the magnetic field [48]: $T_{kk'} \sim \frac{\omega_k \omega_{k'}}{n_e T_e}\left(1 + \frac{\omega_{pe}^2}{\omega_{ce}^2}\frac{\omega_{pe}^2}{\omega_k^2}\right)$. However, apart from this the change is insignificant. Substituting (8.9) into (8.6) and (8.7), we obtain

$$\frac{\partial n_l}{\partial t} + \gamma_{\tilde{k}} n_l = \int 2\pi k'^2 \, dk' \, \frac{(k'^2 + k^2) v_{Ti}^2}{n_e T_e} n_l \frac{\partial}{\partial \mu} n_l'$$
$$+ \frac{\omega_0}{n_e T_e} k v_{Ti} n_l(\mu_0) N_l \, \delta(\mu - \mu_0 + \Delta \mu); \tag{8.10}$$

$$\frac{\partial N_l}{\partial t} = \gamma_l n_l - N_l \int 2\pi k'^2 \, dk' \, n_l' \frac{\omega_0 k' v_{Ti}}{n_e T_e} - \nu_{ei} N_l. \tag{8.11}$$

The system (8.10), (8.11) determines stationary states with a high degree of arbitrariness. This is removed by the condition for stability of the stationary state to the excitation of oscillations in those regions of **k**-space where $n_l = 0$ [47]:

$$\left.\begin{array}{ll}\gamma_k = \gamma_{Nl} & \text{if } n_l \neq 0; \\ \gamma_k > \gamma_{Nl} & \text{if } n_l = 0,\end{array}\right\} \tag{8.12}$$

where

$$\gamma_{Nl} = \int 2\pi k'^2 \, dk' \, \frac{(k'^2 + k^2) v_{Ti}^2}{n_e T_e} \frac{\partial}{\partial \mu} n_l'.$$

Equations (8.12) have a straightforward geometrical interpretation. They show that at fixed μ, the curve $\gamma_k(k)$ is located above the curve $\gamma_{Nl}(k)$ and touches the latter at those points where oscillations are excited. Since the curve $\gamma_{Nl}(k)$ is a parabola, it can touch the curve $\gamma_k(k)$ at one point $k = k_m(\mu)$ determined by the conditions

$$\gamma_{Nl}(k_m) = \gamma_k(k_m) \quad \text{and} \quad \frac{d}{dk}\gamma_{Nl}(k_m) = \frac{d}{dk}\gamma_k(k_m). \tag{8.13}$$

Using explicit expressions for $\gamma_k(k)$ and $\gamma_{Nl}(k)$ we can obtain the following condition for $k_m(\mu)$:

$$d\gamma_k/dk_m = \gamma_k(k_m)/k_m. \tag{8.14}$$

Hence it follows that k_m does not depend on the form of the spectrum and is determined only by linear damping. Moreover, from Eq. (8.14) it follows that when $k = k_m$ the contribution made by Landau damping is small; i.e., the oscillation damping is determined mainly by electron–ion collisions:

$$\gamma_k(k_m) \simeq \nu_{ei}(1 + k_m^2 r_D^2). \tag{8.15}$$

The value of k itself depends logarithmically on plasma parameters and turns out to be on the order of $k_m \sim (1/5 - 1/7) r_D^{-1}$.

The above considerations permit us to draw the conclusion that the oscillation spectrum, at least in the region which is not very close to the source, has the form of a jet extended into the region of small μ:

$$n_l(\mu, k) = n(\mu)\,\delta(k - k_m)/2\pi k^2. \tag{8.16}$$

Finally, using Eq. (8.16), we rewrite the system (8.10)-(8.11) in the following way:

$$\partial n/\partial t + \nu_{ei} n = \gamma_1 n\, \partial n/\partial \mu + \beta n N_l \delta(\mu - \mu_0); \tag{8.17}$$

$$\partial N_l/\partial t = -\nu_{ei} N_l - \beta n(\mu_0) N_l + \gamma_l N_l, \tag{8.18}$$

where $\gamma_1 = k_m^2 v_{Ti}^2/n_e T_e$, $\beta = \gamma_1 \omega_0/k_m v_{Ti} \gg \gamma_1$.

The stationary solution of Eq. (8.18) determines the value $n(\mu_0)$ which is the boundary condition for Eq. (8.17):

$$n(\mu_0) = (\gamma_l - \nu_{ei})/\beta. \tag{8.19}$$

Now we shall integrate Eq. (8.17) over a narrow layer from $\mu_1 = \mu_0 - \Delta\mu$ to $\mu_2 = \mu_0 + \Delta\mu$. Considering that $n(\mu + \Delta\mu) \equiv 0$, we find

$$N_l = \frac{\gamma_1}{\beta} n(\mu_0) = \frac{\gamma_1(\gamma_l - \nu_{ei})}{\beta^2}. \tag{8.20}$$

Let us now assume that collisions are infrequent and consider the solution of Eq. (8.17) in the region $\mu < \mu_0$:

$$\frac{\partial n}{\partial t} = \gamma_1 n \frac{\partial n}{\partial \mu}; \quad n(\mu = \mu_0) = \frac{\gamma_l}{\beta} = n_0. \tag{8.21}$$

Equation (8.21) is exactly similar to the nonlinear equation of motion of an incompressible liquid and describes the process of the wave front twisting. The characteristic time of the wave front twisting $\tau_1 \sim \Delta\mu / \gamma_1 n(\mu_0)$ is much less than the time for the wave to travel a distance $\mu \sim 1$ (which corresponds to a change in ω from ω_0 to ω_{pi}). Therefore, we can find at once the solution of Eq. (8.21) in the form

$$n(\mu) = n_0(\mu - \mu_0(t))\,\theta(\mu_0 - \mu). \qquad (8.22)$$

To find the time dependence of the position of the front $\mu_F(t)$ we integrate (8.17) over μ from $\mu_1 = 0$ to $\mu = \mu_0 + \delta$:

$$n_0 d\mu_F/dt = -\beta N_i n_0/2 \qquad (8.23)$$

or

$$\mu_F = \mu_0 - \gamma_1 \frac{\gamma_l}{2\beta} t = \mu_0 - \frac{\gamma_l\, t}{2} \frac{k_m v_{Ti}}{\omega_0}. \qquad (8.24)$$

Let us now assume that the characteristic time of instability is less than the time for the wave to travel a distance μ_0. The solution of Eq. (8.21) for the spectral quantum density $n(\mu)$ in this case has the form

$$n(\mu) = n_0 \theta(\mu - \mu_F(t))\,\theta(\mu_0 - \mu_F(t-\tau) - \mu). \qquad (8.25)$$

The velocity of the wave front displacement to the region of smaller μ remains the same in this case.

In the opposite limiting case, when the time of the quasilinear instability relaxation $\tau > (1/\gamma_l)(\omega_0/k_m v_{Ti})$, we can construct a stationary solution of Eq. (8.17):

$$n(\mu) = n_0 - \frac{\nu_{ei}}{\gamma_l}(\mu_0 - \mu). \qquad (8.26)$$

Lastly, from Eq. (8.18) we can evaluate the fraction of energy transferred by each quantum of Trivelpiece–Gould waves to electrons due to electron–ion collisions. Assuming that collisions only carry away a small fraction of energy (which is confirmed by the result), we can write

$$\left(\frac{d}{dt} W\right)_{\nu_{ei}} = -\nu_{ei} W(t), \qquad (8.27)$$

where $W(t) = \omega_0 \mu_F(t) n_0$. From Eq. (8.27) it follows that

$$\frac{(\Delta W)_{\nu_{ei}}}{W(\mu_0)} = \frac{\nu_{ei}}{2\gamma_l \dfrac{k_m v_{Ti}}{\omega_0}}. \qquad (8.28)$$

Formula (8.28) is an expression for the fraction of the wave energy lost due to ν_{ei} and expended in heating nonresonant electrons. Under actual experimental conditions $(\Delta W)_{\nu_{ei}}/W(\mu_0)$ is very small (no more than 1%). The remaining energy of the waves is expended in ion heating. Since $W_l^{max} \sim mn_b v_b^2 \omega_0/\omega_{ce}$, we can state that the fraction $\sim \omega_{pe}/\omega_{ce}$ of the beam energy is expended in ion heating, which was indeed observed in the experiment.

In conclusion, we note that consideration of other nonlinear mechanisms, e.g., of the type described by Eq. (8.1), can only increase the efficiency of ion heating, since it is entirely similar to the above process of induced scattering on ions. The only difference between them is that, for equal intensity, this process leads to more rapid energy transfer to the frequency region $\omega \sim \omega_{pi}$ (the energy transfer step size is greater); hence there are smaller losses of the wave energy on its way to the ions.

9. Conclusions

Above we considered the behavior of runaway electrons in present-day tokamaks. It is natural to ask whether such runaway electrons would lead to noticeable macroscopic effects in future tokamak reactors. We cannot give a unique answer to this question as yet. On the one hand, the increase in the size of the tokamak at a constant safety factor q should lead to a decrease in the ratio E/E_D since $E/E_D \sim B_0 T_e^{5/2}/Rqn_e$. The same result is achieved with an increase in plasma density at a constant magnetic field. On the other hand, the electron temperature in tokamak reactors should be considerably higher than the values of $T_e \sim 1$ keV obtained so far, and an increase in temperature sharply increases the ratio E/E_D. Thus the question of the formation of runaway electrons in tokamak reactors remains open and is therefore not treated in detail in the present survey. For this reason the survey did not treat the problem of the formation of runaway electrons at the initial nonstationary stage of the discharge. And, lastly, we have to mention a suggestion which has been recently made, to use runaway electrons for producing in a tokamak a quasistationary current — an idea which is still far from being developed.

REFERENCES

1. G. A. Bobrovskii, E. I. Kuznetsov, and K. A. Razumova, Zh. Eksp. Teor. Fiz., $\underline{59}$, 1103 (1970).
2. A. B. Berlizov et al., At. Energ., $\underline{43}$, 90 (1977).
3. V. S. Vlasenkov et al., Nucl. Fusion, $\underline{13}$, 509 (1973).
4. TFR Group, Nucl. Fusion, $\underline{16}$, 473 (1976).
5. V. V. Alikaev, K. A. Razumova, and Yu. A. Sokolov, Fiz. Plazmy, $\underline{1}$, 303 (1975).
6. A. A. Vedenov, in: Reviews of Plasma Physics, Vol. 3, Consultants Bureau (1967).
7. A. A. Omens et al., Phys. Rev. Lett., $\underline{36}$, 255 (1976).
8. TFR Group, in: 5th Intern. Conf. on Plasma Phys. and Controlled Nucl. Fusion Res., Tokyo, Vol. 1 (1974), p. 135.
9. TFR Group, in: 7th European Conf. on Controlled Fusion and Plasma Phys., Lausanne, Vol. 1 (1975), p. 132.
10. P. Brossier et al., in: 6th Intern. Conf. on Plasma Phys. and Controlled Nucl. Fusion Res., Berchtesgaden, Vol. 1 (1976), p. 403.
11. B. A. Trubnikov, in: Reviews of Plasma Physics, Vol. 1, Consultants Bureau (1965).
12. H. Dreiser, Phys. Rev., $\underline{115}$, 238 (1959); $\underline{117}$, 239 (1960).
13. A. V. Gurevich, Zh. Eksp. Teor. Fiz., $\underline{39}$, 1296 (1960).
14. A. N. Lebedev, Zh. Eksp. Teor. Fiz., $\underline{48}$, 1393 (1965).
15. R. M. Kulsrud et al., Phys. Rev. Lett. $\underline{31}$, 690 (1973).
16. M. Kruskal and J. B. Bernstein, Phys. Fluids, $\underline{7}$, 407 (1964).
17. R. H. Cohen, Phys. Fluids, $\underline{19}$, 239 (1976).
18. J. W. Connor and R. I. Hastie, Nucl. Fusion, $\underline{15}$, 415 (1975).
19. V. V. Parail and O. P. Pogutse, Nucl. Fusion, $\underline{18}$, 303 (1978).
20. A. I. Morozov and A. S. Solov'ev, in: Reviews of Plasma Physics, Vol. 2, Consultants Bureau (1966).
21. A. A. Galeev and R. Z. Sagdeev, in: Reviews of Plasma Physics, Vol. 7, Consultants Bureau (1979).
22. A. V. Gurevich and Ya. S. Dimant, Nucl. Fusion, $\underline{18}$, 629 (1978).
23. Yu. A. Sokolov, Pis'ma Zh. Eksp. Teor. Fiz., $\underline{29}$, 244 (1979).
24. TFR Group, in: 3rd Intern. Congress on Waves and Instabilities in Plasmas, Palaiseau, France, 1977.
25. A. B. Mikhailovskii, Theory of Plasma Instabilities, Vol. 1, Consultants Bureau (1974).
26. B. B. Kadomtsev and O. P. Pogutse, Zh. Eksp. Teor. Fiz., $\underline{53}$, 2025 (1967).

27. V. D. Shapiro and V. I. Shevchenko, Zh. Eksp. Teor. Fiz. 54, 1187 (1968).
28. V. L. Yakimenko, Zh. Eksp. Teor. Fiz., 44, 1534 (1963).
29. V. V. Parail and O. P. Pogutse, Fiz. Plazmy, 2, 125 (1976).
30. Duk In Choi and W. Horton, Fusion Res. Center Rep. No. 120, December, 1976.
31. K. Papadopulous, B. Hui, and W. Winsor, Nucl. Fusion, 17, 1087 (1977).
32. C. S. Liu et al., Phys. Rev. Lett., 39, 701 (1977).
33. D. J. Rowlands, V. L. Sizonenko, and K. N. Stepanov, Zh. Eksp. Teor. Fiz., 50, 994 (1966).
34. D. J. Rowlands, V. D. Shapiro, and V. I. Shevchenko, Zh. Eksp. Teor. Fiz., 50, 979 (1966).
35. C. S. Liu and Y. Mok, Phys. Rev. Lett., 38, 162 (1977).
36. G. Bekefi, Radiation Processes in Plasmas, Wiley (1966).
37. J. D. Strachan, Nucl. Fusion, 16, 743 (1976).
38. S. Sesnic and G. Fussmann, Max Planck Institute for Plasma Physics Report, March 29, 1976.
39. G. Fussmann, E. Glock, and H. P. Zehrfeld, 7th Intern. Conf. on Plasma Phys. and Controlled Nucl. Fusion Res., Innsbruck, IAEA-CN-37/T-4, 1978.
40. D. G. Bulyginskii et al., Fiz. Plazmy, 6, 860 (1980).
41. B. B. Kadomtsev, in: Reviews of Plasma Physics, Vol. 4, Consultants Bureau (1966).
42. K. Molving, M. S. Tekula, and A. Bers, Phys. Rev. Lett. 38, 1404 (1977).
43. V. V. Parail and O. P. Pogutse, Nucl. Fusion, 18, 1357 (1978).
44. A. V. Gurevich et al., Pis'ma Zh. Eksp. Teor. Fiz., 26, 733 (1977).
45. A. A. Galeev and R. Z. Sagdeev, in: Reviews of Plasma Physics, Vol. 7, Consultants Bureau (1979).
46. V. N. Tsytovich, Theory of Turbulent Plasma, Consultants Bureau (1977).
47. B. N. Breizman, V. E. Zakharov, and S. L. Musher, Zh. Eksp. Teor. Fiz., 64, 1297 (1973).
48. A. M. Rubenchik, Zh. Eksp. Teor. Fiz., 68, 1005 (1975).

BALLOONING EFFECTS AND PLASMA STABILITY IN TOKAMAKS

O. P. Pogutse and E. I. Yurchenko

INTRODUCTION

A thermonuclear reactor based on a tokamak can be of practical interest only at sufficiently large plasma pressure, when the ratio of the gas kinetic pressure to the pressure of the magnetic field $\beta = 8\pi\bar{p}/B^2$ is on the order of 5-6% [1]. At this pressure the thermal energy of the plasma is of the same order as the magnetic energy of the longitudinal current. Until recently the main source of plasma instability in a tokamak has been associated with the energy of the longitudinal current which feeds the most powerful instability, i.e., a kink. For correct choice of the current profile and in the presence of a conducting wall or feedback stabilization, the kink instability can be suppressed. As the pressure increases, instabilities for which the thermal energy of plasma is the energy reservoir gain in importance. These are the so-called ballooning modes of the flute and kink instabilities, which manifest themselves in the development of perturbations on the outer (low field) side of the torus.

Ballooning modes, associated with the curvature of field lines and the pressure gradient, can limit the maximum plasma pressure in tokamaks. The physical nature of these limitations will be discussed in the present survey, and this will permit us to see what possibilities there are for avoiding these limitations and what maximum pressures can be achieved.

Historically, the first paper dealing with the problem under consideration was that of Suydam [2] (1958), who obtained the stability criterion for a cylindrical plasma column carrying current, for flute perturbations parallel to field lines:

$$S^2/4 + 8\pi p' r/B_0^2 > 0, \qquad (1)$$

where $S = q'r/q$, $q = rB_0/RB_J$ is the "safety factor" with respect to helical instability, B_0 is the longitudinal field, B_J is the field of the current, and R is the major radius of the equivalent torus.

This criterion has an obvious physical meaning: the first term describes the stabilizing effect of the shear (changes in the angle of field lines with the radius), whereas the second term shows that when the field lines are curved, the plasma pressure gradient is the cause of instability.

In 1960, on the basis of the energy principle, Mercier obtained the stability criterion for a toroidal plasma column of arbitrary cross section against local flute perturbations [3]. This criterion was not obvious from a physical point of view; therefore, much time and the efforts of a great number of theoreticians were needed to elucidate its role in stability studies [12-21].

The analysis of plasma stability in toroidal geometry requires the use of a quite complicated mathematical apparatus, which considerably hinders the comprehension and interpretation of the results obtained. For this reason, in parallel with full-scale investigations for elucidating concrete physical mechanisms for instability, model calculations were made as well [4, 5].

In 1966 Furth, Killeen, Rosenbluth, and Coppi examined plasma stability in slab geometry, modelling the effect of curvature of the field lines by a variable gravitational force, and discovered ballooning instability [4]. It turned out that in this model flute perturbations are increased in the region of unfavorable curvature and suppressed in the favorable region rather than being constant along field lines, which is the case with cylindrical geometry. The critical value of pressure, above which the ballooning instability starts, depends on the ratio of the length of a field line with favorable curvature and the length with unfavorable curvature.

In tokamaks the curvature of field lines is variable, being unfavorable on the outer side and favorable on the inner (high field) side of the torus.

The ratio of lengths of the corresponding sections of field lines is fixed, but the evaluation of the critical pressure, according to a model formula, is of no interest for tokamaks. This results from the shortcomings of the model, which does not take into consideration the effect of the finite plasma pressure upon the depth of the magnetic well. In tokamaks, a magnetic well appears due to the relative shift of magnetic surfaces outside the toroidal axis of symmetry and it plays a stabilizing role. The tokamak magnetic well depth was calculated by Solov'ev and Shafranov in 1966 [16].

To investigate ballooning instability in tokamaks, Kadomtsev and Pogutse in 1966 developed a method based on an expansion of the MHD equations into a power series in the small parameter B_J/B_0. This method considerably simplifies calculations in toroidal geometry because expansion of the normal mode equations in the field ratio has deep physical meaning: magnetosonic oscillations, leading to compression and rarefaction of field lines and suppressed by a strong longitudinal field, are neglected, ion-acoustic modes, which do not disturb the magnetic field, are separated, and only the most dangerous Alfvén oscillations associated with the curvature of the magnetic field lines are taken into consideration.

The development of this method enabled Kadomtsev and Pogutse to obtain a stability criterion which is a generalization of the Suydam criterion for the case of a torus with circular magnetic surfaces [7]:

$$S^2/4 + 8\pi p'r/B_0^2 + \alpha U - \alpha^2/2 > 0, \qquad (2)$$

where U is the magnetic well depth and $\alpha = -8\pi p'Rq^2/B^2$.

In this criterion the first two terms are cylindrical and the third and fourth are toroidal, characterizing the effect of the magnetic well and the destabilizing ballooning effect respectively.

In 1967 Shafranov and Yurchenko, using the method of Kadomtsev and Pogutse, took into consideration the dependence of the rotational transform on poloidal angle and demonstrated that the stability criterion can be written in a compact form [8]:

$$\frac{S^2}{4} + \frac{8\pi p'r}{B_0^2}(1-q^2) > 0. \qquad (3)$$

Ware and Haas obtained the same criterion independently by using the energy principle [9].

Comparison of (2) and (3) shows that in tokamaks with circular surfaces the ballooning effect and the deepening of the magnetic well due to finite plasma pressure are mutually compensating. The torodial term which remains in (3) and is proportional to q^2 characterizes the stabilizing effect of the geometrical part of the magnetic well, which is present in a tokamak at zero pressure. With $q(r) > 1$ the ballooning instability does not limit the permissible plasma pressure. The complete disappearance of the ballooning effect caused certain doubts, but since the calculation was only valid provided $\varepsilon \beta_J \ll 1$ ($\varepsilon = r/R$ is the toroidal curvature, $\beta_J = 8\pi \bar{p}/B_J^2$), which was used for obtaining criterion (3), it was supposed that the paradox might be resolved if the highest terms of the expansion in the parameter $\varepsilon \beta_J$ were taken into account [10, 11].

Applying Mercier's general geometric criterion [3] to a tokamak with circular surfaces, Shafranov and Yurchenko showed that in this case also the criterion takes the form of Eq. (3) [18]. Since the use of the method of normal modes and the Mercier criterion led to a similar result in the configuration studied, it became possible to use the general geometric criterion in the investigation of plasma stability in more complicated magnetic systems (tokamaks with noncircular magnetic surfaces and stellarators) for which the method of normal modes entails considerable mathematical difficulties [22-28].

From its deviation the Mercier criterion is a necessary criterion for plasma stability to small-scale perturbations in toroidal magnetic configurations, but since for many years nobody found a more stringent criterion, several authors stated that it was sufficient.

The fact that these statements are erroneous became evident in 1977 when the Princeton group of theoreticians published the results of numerical calculations which demonstrated that when the Mercier criterion is satisfied, plasma with a fixed boundary in a tokamak is unstable at a pressure less than the critical value [29]. The critical plasma pressure in a tokamak with circular magnetic surfaces turned out to be on the order of 1-3%, depending on current profiles and pressure. Thus, there seemed to be a theoretical limitation to the possibility of achieving ignition in a tokamak. The result of the Princeton group was immediately confirmed by Bateman and Peng [30] and removed any doubts. Analyses carried out by several authors showed that the development of ballooning modes of the flute instability manifests itself in the concentration of perturbations on the outer side of the torus [31, 32]. To elucidate the physical nature of this instability and its difference from the ballooning instability studied earlier, it was necessary to obtain an analytical criterion for the stability of such modes.

Apart from purely technical difficulties associated with an analytical solution, there was a basic problem inherent in the character of the normal mode equations describing ballooning modes. Since the flute perturbations under discussion are strongly extended along the field lines, which in the presence of shear do not close (except on rational magnetic surfaces), the method used led, as a rule, to ordinary differential equations in real space having both periodic and nonperiodic coefficients. This created a problem for the interpretation and solution of such equations since every physically realized perturbation should be periodic.

The problem was solved by Connor, Hastie, and Taylor [19] and independently by Glasser [82] and Lee and Van Dam [59]. They developed a method for the investigation of normal mode equations, being a set of partial differential equations, which leads to the solution of ordinary differential equations. These equations have both periodic and nonperiodic coefficients, but they must be solved in a Fourier space rather than in real space, the physical perturbations being naturally periodic in this case. For the analytical solution of such equations, Pogutse and Yurchenko [33] developed a variational asymptotic method which permitted the stability criterion for ballooning modes of the flute instability to be obtained for a tokamak with circular surfaces (1978):

$$\frac{1}{2} S^2 + \frac{8\pi p' r}{B_0^2} (1-q^2) - \frac{3}{2} \alpha \exp\left(-\frac{1}{|S|}\right) - \frac{1}{2} S \alpha^2 > 0. \qquad (4)$$

With a peaked current profile, criterion (4) is more stringent than criterion (3) and limits the threshold of plasma pressure since it takes into account two new physical effects associated with shear. An exponential term which is nonanalytical with respect to shear describes the effect of branch intersection of oscillations localized on neighboring rational surfaces. The last term, proportional to $S\alpha^2$, characterizes the ballooning effect associated with shear. This new ballooning effect plays an essential role since the usual ballooning effect, proportional to α^2 in the case under consideration, has been compensated by the magnetic well as in criterion (3).

The importance of the problem of ballooning modes stimulated by achievements in additional methods of heating, which allowed plasma pressure in experimental devices to be raised considerably, attracted the attention of theoreticians. In a short time a great number of articles dealing with the calculation of stability in real configurations have appeared [34-40]. An interesting result was obtained independently by Mercier [20], Fielding and Haas [35], and Zakharov [38]. In 1978 they discovered numerically a second zone of stability at high plasma pressure.

Above we discussed small-scale perturbations which are more easily analyzed by analytical methods.

Large-scale plasma oscillations, unlike small-scale ones, essentially depend on the concrete distribution of parameters. So far no instability criteria have been written for them to fit arbitrary distributions; therefore, numerical methods are most often used for investigation of large-scale modes.

The first results indicating the destabilizing influence of toroidal effects on kink oscillations were obtained by Bateman in 1971 and Freidberg, Haas, and Grossman in 1974-75 in numerical calculations of the stability of a toroidal plasma column with a skin current [41-43]. The authors of these works came to the conclusion that kink perturbations in this model have a ballooning character; i.e., they develop on the outer side of the torus. Calculations have also been made of the dependence of critical β on the toroidicity and the safety factor.

In 1977 in Princeton and Oak Ridge, numerical calculations of toroidal plasma stability with free boundary and arbitrary current distribution across the column cross section were performed. It was demonstrated that when β increases, stability becomes worse, and when β exceeds a critical value zones of stability disappear altogether [29].

This result was immediately explained using an analytical model by Pogutse and Yurchenko. It turned out that the degenerate modes of the cylindrical plasma column obtained by Shafranov in 1970 [45] are resolved by the toroidal curvature. This is analogous to the splitting of the zonal structure of a solid body electronic spectrum. The splitting is determined by the ballooning effect $\pm\varepsilon\beta_J$ [44]; i.e., one of the new levels is always less stable than the initial degenerate one.

Detailed calculations of the effect of the cross-sectional shape of the plasma column and plasma boundary on the value of the pressure threshold were performed in Princeton in 1978 [46].

Above we have discussed the stability of a toroidal plasma with infinite conductivity. Real plasma always has finite conductivity; i.e., there is no complete freezing-in of plasma into the magnetic field in this case. This leads to dissipative instabilities which develop considerably more easily than ideal instabilities.

The history of the study of plasma stability to dissipative flute perturbations dates from the work of Furth, Killeen, and Rosenbluth in 1963 [47]. Modelling the curvature in slab geometry by the introduction of an equivalent gravitational force, they showed that taking into account finite conductivity led to instability in plasma which was stable under conditions of ideal hydrodynamics.

In 1967 Green and Johnson, investigating the stability of a toroidal plasma column at low pressure, obtained a criterion with dissipation [48] which was analogous to Mercier's general criterion for ideal plasma. The main effect of taking into account finite conductivity was the disappearance of the stabilizing effect of shear.

Consistent consideration of finite plasma pressure in arbitrary geometry, i.e., consideration of the ballooning effect and deepening of the well due to the pressure, in the presence of dissipation resulting from finite conductivity, was carried out by Glasser, Green, and Johnson in 1975 [49]. Mikhailovsky independently derived the stability criterion for a tokamak with circular surfaces [50]:

$$8\pi p'r/B_0^2 + \alpha U - \alpha^2/2 > 0. \tag{5}$$

Later it was shown that this criterion followed from [49] as well. Mikhailovskii, from obvious physical considerations, evaluated the depth of the magnetic well at high pressure and pointed out its saturation [51].

When the plasma pressure becomes larger than the critical value, the ballooning effect overcomes the deepening of the magnetic well and the plasma becomes unstable, which follows from criteria (2) and (5). The dissipative instability (5) appears earlier than the ideal one (2) and, due to the absence of shear stabilization, criterion (5) sets less restrictions on the value of the pressure threshold than criterion (2). This result is quite natural because dissipative instabilities should develop under conditions when ideal instabilities are forbidden.

Progress in the study of ideal ballooning modes of the flute instability in 1977-1978 placed emphasis again on the need to consider dissipation. Moreover, comparison of the criterion for ideal ballooning modes (4) with the dissipative criterion (5) resulted in a paradox: the pressure gradient threshold above which the instability develops proved to be lower for ideal ballooning modes than that which was obtained from the dissipative criterion (5).

The first attempt to solve the paradox was made by Bateman and Nelson in 1978 [52]. They showed that in a dissipative toroidal plasma, at an arbitrarily small pressure gradient, there is an instability with a growth rate on the order of the inverse skin time $\dot{\gamma} \sim 1/\tau_s$ ($\tau_s = 4\pi\sigma a^2/c^2$).

However, the paradox remained since the dissipative instability which develops on violating criterion (5) had a much greater increment $\gamma \sim 1/\tau_s (\tau_s/\tau_\theta)^{2/3}$ ($\tau_\theta = \sqrt{4\pi\rho_0} a/B_j$ is the poloidal Alfén time defined in the field of the current. In high-temperature plasma the ratio of the skin time to the Alfén time is large.)

In 1979 Pogutse and Yurchenko, using the method of expansion of normal mode equations in a small parameter, found threshold-free dissipative ballooning modes with a large growth rate $\gamma \sim 1/\tau_s(\tau_s/\tau_\theta)^{2/3}$ and thus resolved the paradox [53].

The present survey has the following structure. Chapter 1 considers the basic equations and methods for study of ideal plasma stability. Current development of the analytical theory and numerical calculations of ballooning modes of the flute instability are given in Chapter 2. Chapter 3 concerns the analysis of large-scale ballooning modes of the flute instability. Both analytical and numerical results are cited here. Chapter 4 deals with dissipative ballooning modes.

CHAPTER 1

INITIAL EQUATIONS AND METHODS FOR INVESTIGATION OF IDEAL PLASMA STABILITY

1.1. Plasma Equilibrium and Coordinate System

It is more convenient to investigate plasma stability in tokamaks in special curvilinear coordinates a, θ, z in which magnetic surfaces representing a system of nested toruses coincide with coordinate surfaces a = const and the angular variable θ can be chosen in such a way that lines of force in the surface a = const are straight. In this system of coordinates the metric has the form

$$dl^2 = g_{11}da^2 + 2g_{12}dad\theta + g_{22}d\theta^2 + g_{33}dz^2. \qquad (1.1)$$

The metric coefficients g_{ik} occurring in this formula are determined by the form of the cross section of magnetic surfaces. In axisymmetric tokamaks, which will be considered further, g_{ik} do not depend on the longitudinal coordinate z. In this section to simplify the presentation of analytical calculations, we confine ourselves to the case of elliptical magnetic surfaces which differ only slightly from circles. The case of arbitrary ellipticity including triangularity will be discussed in Chapter 2, Section 2.4, in which the criterion for the column stability with noncircular cross section will be deduced.

We take a quasicylindrical coordinate system centered on the magnetic axis of the toroidal column:

$$dl^2 = d\rho^2 + \rho^2 d\omega^2 + (1 - k\rho \cos \omega)^2 dz^2, \qquad (1.2)$$

where ρ is the distance from the magnetic axis, ω is the azimuthal angle, and $k = 1/R$ is the curvature of the magnetic axis.

Magnetic surfaces in tokamaks are shifted relative to the magnetic axis by the value $\xi(a)$ which depends on plasma pressure; therefore, it is natural to change from quasicylindrical coordinates

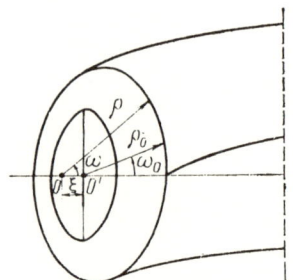

Fig. 1. The relationship between quasicylindrical coordinates ρ, ω and polar coordinates ρ_0, ω_0. The point O is the magnetic axis, the point O' is the center of the magnetic surface a = const.

ρ, ω to polar coordinates ρ_0, ω_0 associated with displaced centers of magnetic surfaces (Fig. 1):

$$\left.\begin{array}{l} \rho \cos \omega = \rho_0 \cos \omega_0 + \xi(a); \\ \rho \sin \omega = \rho_0 \sin \omega_0. \end{array}\right\} \quad (1.3)$$

The equation of magnetic surfaces will be written as

$$\rho_0 = a + \eta(a) \cos 2\theta, \quad (1.4)$$

where $\eta(a)/a \ll 1$.

The azimuthal angle $\omega_0 = \omega_0(a, \theta)$ will be expressed as an expansion

$$\omega_0 = \theta + \lambda(a) \sin \theta + \mu(a) \sin 2\theta, \quad (1.5)$$

where $\lambda(a)$ and $\mu(a)$ are parameters introduced to straighten lines of force.

Substituting these expressions into the formula for the metric (1.2), it is easy to find the initial metric coefficients to the approximation in curvature necessary for further calculations ($\xi/a \sim ka \ll 1$) [8]:

$$\left. \begin{array}{l} g_{11} = 1 + 2\xi' \cos\theta + \left(2\eta' - \dfrac{1}{2} a^2 \lambda'^2 + \xi'\lambda - a\xi'\lambda'\right) \cos 2\theta; \\[4pt] g_{12} = (a^2 \lambda' - a\xi') \sin\theta + \left(a^2 \mu' - 2\eta - a\xi'\lambda + \dfrac{1}{2}\lambda\lambda'\right) \sin 2\theta; \\[4pt] g_{22} = a^2 \left[1 + 2\lambda \cos\theta + \left(2\eta/a + 4\mu + \dfrac{1}{2}\lambda^2\right) \cos 2\theta\right]; \\[4pt] g_{33} = (1 - ka \cos\theta - ka\lambda \sin^2\theta)^2. \end{array} \right\} \quad (1.6)$$

We shall also need the determinant of the metric tensor

$$\sqrt{g} = g_{33}(g_{11} g_{22} - g_{12}^2), \quad (1.7)$$

which determines the volume element on a given magnetic surface:

$$V'(a) = 2\pi R \int_0^{2\pi} \sqrt{g} \, d\theta. \quad (1.8)$$

To second order in curvature $\sqrt{g}$ has the following form:

$$\sqrt{g} = a\left[1 + (\lambda + \xi' - ka)\cos\theta - k\xi - \frac{1}{2} ka\xi' \right.$$
$$\left. + (\eta' + 2\mu + \eta/a + \lambda\xi' - ka\lambda - ka\xi'/2)\cos 2\theta\right]. \quad (1.9)$$

It is clear from the equation of lines of force in curvilinear coordinates a, θ, z

$$d\theta/B^\theta = dz/B^s \quad (1.10)$$

that the lines of force in the coordinates a, θ, z are straight if the ratio of the contravariant components of the magnetic field does not depend on the variable θ (the ratio does not depend on z due to the axial symmetry of the tokamak).

To elucidate the relation between this requirement and the metric coefficients, let us integrate the equation $(4\pi/c)\mathbf{j} = \text{curl } \mathbf{B}$ over the cross section of the torus:

$$\frac{4\pi}{c} \int j \, dS_z = \frac{4\pi}{c} \mathcal{J}(a) = \oint \mathbf{B} \, d\mathbf{l}_\theta = \sqrt{g} \, B^\theta \int_0^{2\pi} \frac{g_{22}}{\sqrt{g}} \, d\theta, \quad (1.11)$$

where $B_i = g_{ik} B^k$ is the covariant component of the magnetic field; the independence of the combination $\sqrt{g} B^\theta$ on θ follows from div $\mathbf{B}$ = $(1/\sqrt{g})(\partial \sqrt{g} B^\theta/\partial \theta) = 0$. Equation (1.11) connects the longitudinal current $\mathcal{J}(a)$ within a given magnetic surface with B^θ. A similar connecnection between the transverse current $I(a)$ and B^S is derived by integrating the same equation over the equatorial cross section of the torus (Fig. 2):

$$\frac{4\pi}{c}\int j\, dS_\theta = -\frac{4\pi}{c} I(a) = \oint B\, dl_z = 2\pi B_S = 2\pi g_{33} B^S. \quad (1.12)$$

From relations (1.11) and (1.12) it follows that when the lines of force are straight in the chosen system of coordinates, the relation $g_{33}/\sqrt{g} \sim B^\theta/B^S$ does not depend on θ. Hence, $\lambda(a)$ is determined by the condition that the coefficient of $\cos\theta$ in the equation for $g_{33}/\sqrt{g}$ vanishes and $\mu(a)$ by the condition that, for $\cos 2\theta$,

$$\left.\begin{array}{l}\lambda = -\xi' - ka; \\ \mu = -\dfrac{1}{2}\left(\eta' + \dfrac{\eta}{a} - \xi'^2 - \dfrac{3}{2} ka\xi' - \dfrac{1}{2} k^2 a^2\right).\end{array}\right\} \quad (1.13)$$

In the system of coordinates with straight lines of force the equation of plasma equilibrium $\nabla p = (1/c)[\mathbf{j}\,\mathbf{B}]$ is reduced to the two equations:

$$pV' = \frac{1}{c}(I'\Phi' - \mathcal{Y}'\chi'); \quad (1.14)$$

$$\chi'\frac{\partial v}{\partial \theta} = cp'(V' - 4\pi^2 R\sqrt{g}), \quad (1.15)$$

whereas the equations for contravariant components of fields and currents have the simple form

$$B^1 = 0;\ B^\theta = \frac{\chi'}{2\pi\sqrt{g}};\ B^S = \frac{\Phi'}{2\pi\sqrt{g}}; \quad (1.16)$$

$$j^1 = 0,\ j^\theta = \frac{I'}{2\pi\sqrt{g}};\ j^S = \frac{\mathcal{Y}' + \partial v/\partial\theta}{2\pi\sqrt{g}}; \quad (1.17)$$

where Φ and χ are the longitudinal and transverse magnetic fluxes, respectively, and $\partial\nu/\partial\theta$ is a periodic function characterizing the polarization current.

For a complete specification of the equilibrium configuration, which will be the basis for the study of stability, it remains to calculate $\xi(a)$, which is the displacement relative to the magnetic axis, and $\eta(a)$, which is the parameter characterizing the ellipticity.

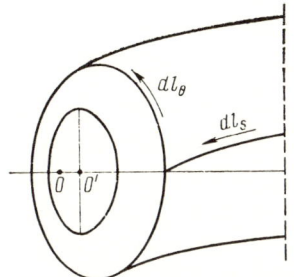

Fig. 2. Meridional dl_θ and equatorial dl_s contours of integration in the torus.

The simplest way to do this is by writing j^S from the equation $(4\pi/c)\mathbf{j} = \operatorname{curl}\mathbf{B}$:

$$\frac{4\pi}{c} j^S = \frac{1}{\sqrt{g}} \left(\frac{\partial B_\theta}{\partial a} - \frac{\partial B_1}{\partial \theta} \right). \tag{1.18}$$

and by substituting (1.18) in the last equation (1.17):

$$\frac{\partial}{\partial a}\left\{\chi'\left[\frac{g_{22}}{\sqrt{g}} - \left(\frac{g_{22}}{\sqrt{g}}\right)_0\right]\right\} - \chi' \frac{\partial}{\partial \theta} \frac{g_{12}}{\sqrt{g}} = \frac{4\pi p'}{\chi'}(V' - 4\pi^2 R\sqrt{g}), \tag{1.19}$$

where $\left(\dfrac{g_{22}}{\sqrt{g}}\right)_0 \equiv \dfrac{1}{2\pi}\displaystyle\int_0^{2\pi} \dfrac{g_{22}}{\sqrt{g}}\,d\theta$.

From Eq. (1.19), collecting all the terms in $\cos\theta$ and $\cos 2\theta$, we obtain equations for ξ and η, respectively:

$$\xi'' + \frac{(aB_J^2)'}{aB_J^2}\xi' = k\left(1 - \frac{8\pi p' a}{B_J^2}\right); \tag{1.20}$$

$$\eta'' + \frac{(aB_J^2)'}{aB_J^2}\eta' - \frac{3\eta}{a^2} = -\frac{12\pi p'}{B_J^2}k\xi' - \frac{3}{2}\left(\frac{B_J'}{B_J} + \frac{1}{a}\right)\xi'^2, \tag{1.21}$$

where $B_J = \mathcal{J}/2\pi a$.

Equations (1.20) and (1.21) prove that displacements of the magnetic surfaces are of first order in curvature, whereas ellipticity appears only when the next order in curvature is taken into account. We clearly mean ellipticity which originates from the plasma pressure rather than the shape of the conducting wall.

Given definite distributions of pressure and current in the plasma, and the shape of the boundary surface or conducting wall, it is possible to solve the system (1.20), (1.21). Thus the equilibrium and the coordinate system with straight lines of force will be completely determined since the "straightening" parameters λ and μ are functions of ξ and η (1.13).

It should be noted that the parameters λ and μ may be chosen otherwise, i.e., to make the system of coordinates orthogonal. For this purpose the metric coefficient g_{12} is set to zero (1.6). In this case λ and μ are expressed through ξ and η in the following way: $a\lambda' = \xi'$, $a^2\mu' = 2\eta + a\xi'\lambda + \lambda\lambda'/2$. However, an orthogonal coordinate system is less convenient for the study of stability than a coordinate system with straight lines of force.

1.2. Method of Normal Mode Equations. Expansion of Equations in a Small Parameter

To obtain the criterion for plasma stability by the normal mode method it is sufficient to consider perturbations to the equilibrium equations. Within the framework of ideal magnetohydrodynamics, the possibility of the formation of a new equilibrium state with a perturbed magnetic field corresponds to instability of the initial state.

The system of equations of magnetostatic equilibrium

$$\nabla p = \frac{1}{c}[\mathbf{j}\,\mathbf{B}]; \qquad (1.22)$$

$$\text{curl }\mathbf{B} = \frac{4\pi}{c}\mathbf{j}; \qquad (1.23)$$

$$\text{div }\mathbf{B} = 0 \qquad (1.24)$$

can be simplified by using the small parameter B_J/B_0 [54]. Therefore, we shall write it in a convenient form for expansion in fields. The current density vector $\mathbf{j}$ is represented as two terms:

$$\mathbf{j} = \mathbf{j}_\perp + \hat{\alpha}\mathbf{B}, \qquad (1.25)$$

where $\mathbf{j}_\perp$ is the component transverse to the magnetic field $\mathbf{B}$, $\hat\alpha = \overline{(\mathbf{jB}/\mathbf{B}^2)}$.

From the condition div $\mathbf{j} = 0$ using (1.24) we obtain

$$\mathbf{B}\nabla\hat\alpha + \mathrm{div}\,\mathbf{j}_\perp = 0, \tag{1.26}$$

where $\mathbf{j}_\perp = \frac{c}{B^2}[\mathbf{B}\nabla p]$, which follows from (1.22).

Now, instead of Eqs. (1.22) and (1.23), we shall use the more convenient equations [55]:

$$\mathbf{B}\nabla\hat\alpha + c\left[\nabla\frac{1}{B^2}\mathbf{B}\right]\nabla p = 0; \tag{1.27}$$

$$\mathbf{B}\nabla p = 0. \tag{1.28}$$

We linearize the system of equations (1.27), (1.28), and (1.24) considering an axisymmetric equilibrium p_0, $\mathbf{B}_0$ as the unperturbed state. Since the longitudinal field in a tokamak is much greater than the field of the current, the perturbation of the longitudinal field can be neglected compared to the perturbation of the transverse field. This enables us from the condition div $\mathbf{B} = 0$ to express the perturbations of transverse fields by one function

$$\widetilde{B}^1 = -\frac{1}{\sqrt g}\frac{\partial\psi}{\partial\theta},\quad \widetilde{B}^\theta = \frac{1}{\sqrt g}\frac{\partial\psi}{\partial a}, \tag{1.29}$$

where $\widetilde{B}^1$ and $\widetilde{B}^\theta \equiv \widetilde{B}^2$ are contravariant components. Using expressions (1.29) we derive

$$\mathbf{B}^0\nabla\widetilde\alpha + \widetilde{\mathbf{B}}\nabla\alpha^0 + \left[\nabla\frac{1}{B_0^2}\mathbf{B}_0\right]\nabla\widetilde p = 0; \tag{1.30}$$

$$\mathbf{B}^0\nabla\widetilde p - \frac{1}{\sqrt g}\frac{dp_0}{da}\frac{\partial\psi}{\partial\theta} = 0, \tag{1.31}$$

where the zero subscript denotes equilibrium values, and the tilde denotes perturbations,

$$\widetilde\alpha = \frac{c}{4\pi}\frac{1}{\sqrt g\,B_0^S}\left[\frac{\partial}{\partial a}\left(\frac{g_{22}}{\sqrt g}\frac{\partial\psi}{\partial a}\right) + \frac{g_{11}}{\sqrt g}\frac{\partial^2\psi}{\partial\theta^2} - 2\frac{g_{12}}{\sqrt g}\frac{\partial^2\psi}{\partial a\,\partial\theta}\right]. \tag{1.32}$$

Instead of the function ψ we can use the function Φ defined by

$$\psi = \frac{\mathbf{B}^0\nabla}{B_0^S}\Phi, \tag{1.33}$$

which permits us from Eq. (1.31) to express the perturbed pressure using Φ:

$$\tilde{p} = \frac{p'_0}{\sqrt{g}\, B_0^S} \frac{\partial \Phi}{\partial \theta}. \qquad (1.34)$$

It is easy to show that the formally introduced function Φ (1.33) when multiplied by a constant is an electrostatic potential. Indeed, in our approximation $\tilde{B}^1$ and $\tilde{B}^\theta$ are derived from one component of the vector potential (1.29), (1.33), namely $\tilde{B}^1 = \frac{1}{\sqrt{g}} \frac{\partial}{\partial \theta} A_S$, $\tilde{B}^\theta = -\frac{1}{\sqrt{g}} \frac{\partial}{\partial a} A_S$, i.e., $A_S = -\frac{\mathbf{B}^0 \nabla}{B_0^S} \Phi$, but exactly this relation between A_S and electrostatic potential follows from the condition of frozen fields. Actually, $\mathbf{E} + \frac{1}{c}[\mathbf{v}\, \mathbf{B}] = 0$, i.e., $E_\parallel \sim (\mathbf{BE}) = 0$, but $\mathbf{E} = -\nabla \varphi - \frac{1}{c} \frac{\partial \mathbf{A}}{\partial t}$; therefore, $-\frac{i\omega}{c}(\mathbf{AB}^0) = \mathbf{B}^0 \nabla \varphi$ or $A_S = \frac{ic}{\omega} \frac{\mathbf{B}^0 \nabla}{B_0^S} \varphi$, which agrees with the preceding relation if we take $\Phi = \frac{ic}{\omega}\varphi$.

We substitute expression (1.34) in Eq. (1.30) and using Eq. (1.32) we obtain the following equation for Φ:

$$L(\Phi) = \frac{\mathbf{B}^0 \nabla}{B_0^S} \sqrt{g}\, \hat{\Delta}_\perp \frac{\mathbf{B}\nabla}{B_0^S} \Phi +$$

$$+ \frac{D\left(\frac{4\pi p'_0}{\sqrt{g}\, B_0^S} \frac{\partial \Phi}{\partial \theta}, \frac{1}{B_0^S}\right)}{D(a,\theta)} + \frac{D\left(\frac{\mathbf{B}^0 \nabla}{B_0^S}\Phi, \frac{4\pi j_0^S}{c B_0^S}\right)}{D(a,\theta)} = 0, \qquad (1.35)$$

where

$$\hat{\Delta}_\perp = \frac{1}{\sqrt{g}}\left(\frac{\partial}{\partial a}\frac{g_{22}}{\sqrt{g}}\frac{\partial}{\partial a} + \frac{g_{11}}{\sqrt{g}}\frac{\partial^2}{\partial \theta^2} - 2\frac{g_{12}}{\sqrt{g}}\frac{\partial^2}{\partial a\, \partial \theta}\right),$$

$$\frac{D(f,g)}{D(a,\theta)} = \frac{\partial f}{\partial a}\frac{\partial g}{\partial \theta} - \frac{\partial f}{\partial \theta}\frac{\partial g}{\partial a}.$$

Equation (1.35) permits us to study the stability of axisymmetric tokamaks with arbitrary cross-sectional shape of magnetic surfaces. Let us consider the physical meaning of its terms. The first term is the result of the perturbation of the magnetic field and is a stabilizing component. The second term describes the ballooning effects produced by the plasma thermal energy; it is proportional to $\varepsilon\beta_J$ and plays a destabilizing role. The last term containing the longitudinal current j_0^S is responsible for the kink instability.

Equation (1.35) enables us to study the limit of plasma stability, i.e., to find stability criteria. For the calculation of growth rates and oscillation frequencies, it is necessary to add an inertial term to the equilibrium equation (1.22):

$$\rho_0 \frac{d\mathbf{v}}{dt} = -\nabla p + \frac{1}{c}[\mathbf{j}\,\mathbf{B}]. \tag{1.36}$$

Using the method discussed above, it is possible to obtain the dispersion relation for small oscillations:

$$\omega^2 \left[\frac{\partial}{\partial a} \frac{g_{22}}{\sqrt{g}\, c_A^2} \frac{\partial \Phi}{\partial a} + \frac{\partial}{\partial \theta} \frac{g_{11}}{\sqrt{g}\, c_A^2} \frac{\partial \Phi}{\partial \theta} \right.$$

$$\left. - 2 \frac{g_{12}}{\sqrt{g}\, c_A^2} \frac{\partial^2 \psi}{\partial a \partial \theta} \right] + L(\Phi) = 0, \tag{1.37}$$

where $c_A^2 = (B_0^S)^2/4\pi\rho_0$ and the operator $L(\Phi)$ is the left-hand side of Eq. (1.35).

For a uniform cylindrical plasma Eq. (1.37) has a simple and obvious form:

$$\frac{\omega^2}{c_A^2} \Delta_\perp \Phi + \hat{k}_\parallel \Delta_\perp \hat{k}_\parallel \Phi + \frac{1}{a} \frac{D\left(\hat{k}_\parallel \Phi, \frac{4\pi}{c} \frac{j_0}{B_0}\right)}{D(a, \theta)} = 0, \tag{1.38}$$

where

$$\Delta_\perp = \frac{1}{a}\frac{\partial}{\partial a} a \frac{\partial}{\partial a} + \frac{1}{a^2}\frac{\partial^2}{\partial \theta^2}, \quad \hat{k}_\parallel = \frac{\mathbf{B}^0 \nabla}{B_0^S}.$$

Here the first two terms describe Alfvén oscillations (operator $\hat{k}_\parallel$ acts along lines of force). The last term is the source of the kink instability.

The expansion of vector normal mode equations (1.36), (1.23), and (1.24) in the field ratio B_J/B_0 led to the scalar equation for Φ

(1.37). The reason is, because of the three types of oscillations described by vector equations, namely, magnetoacoustic, ion-acoustic, and Alfvén, only the most dangerous Alfven oscillations remain. Magnetoacoustic oscillations leading to the compression and rarefaction of lines of force, which is rather difficult in tokamaks due to the strong longitudinal field, are neglected in the approximation considered. Ion-acoustic oscillations which propagate along the magnetic field are singled out as an independent branch $(\omega^2 + c_s^2 \hat{k}_{\|}^2)\tilde{p} = 0$. Alfvén oscillations, in which lines of force bend, can lead to the kink and flute instabilities. Ballooning modes of these instabilities are described by Eq. (1.37).

1.3. The Energy Method. Simplification of the Energy Principle

As is shown in the preceding section, the use of the method of normal modes for the study of plasma stability in tokamaks, even after simplification, leads to a partial differential equation whose solution involves considerable mathematical difficulties. The energy method, which involves the study of plasma perturbations, permits us to evaluate stability by variational methods rather than by using eigenfunctions for the solution of the problem. The variational approach to the problem of stability considerably facilitates the investigation, because choosing trial functions obtained from physical considerations makes it possible to derive the required stability criteria.

As is known, normal mode equations for an ideal plasma can be calculated from a minimum action variational principle [56]:

$$\delta \int \mathcal{L} dt = 0, \qquad (1.39)$$

where $\mathcal{L}$ is the Lagrangian, equal to the difference between the kinetic energy

$$T = \frac{1}{2} \int \rho_0 \left(\frac{\partial \xi}{\partial t}\right)^2 dV \qquad (1.40)$$

and the potential energy [57]:

$$W = \frac{1}{2} \int_{V_i} \left\{ \gamma_0 p_0 (\mathrm{div}\,\xi)^2 + \frac{1}{4\pi}(\mathrm{curl}\,[\xi \mathbf{B}^0])^2 + \xi \nabla p_0 \, \mathrm{div}\,\xi - \right.$$

$$-\frac{1}{4\pi}\left[\boldsymbol{\xi}\text{ curl }\mathbf{B}^0\right]\text{curl}[\boldsymbol{\xi}\mathbf{B}_0]\}\,dV + \frac{1}{8\pi}\int_{V_e}(\text{curl }\mathbf{A})^2\,dV. \quad (1.41)$$

Here the integral over V_i is taken in the plasma region and the integral over V_e is taken in the vacuum, $\boldsymbol{\xi}$ being the displacement of the plasma from the equilibrium state.

The functional (1.39) is written as a function of the vector $\boldsymbol{\xi}$, i.e., as a function of three scalar functions. Using an expansion in fields B_J/B_0, we reduced the normal mode equations to an equation for a scalar function Φ (1.37). This means that the use of an expansion in B_J/B_0 in the energy principle should result in a simplified functional of a single function Φ. The variation of this functional with Φ should give the simplified normal mode equation (1.37). Let us represent the displacement vector $\boldsymbol{\xi}$ in the form

$$\boldsymbol{\xi} = \boldsymbol{\xi}_\perp + \boldsymbol{\xi}_\parallel, \quad (1.42)$$

where $\boldsymbol{\xi}_\parallel = \frac{(\boldsymbol{\xi}\mathbf{B}^0)}{B_0^2}\mathbf{B}_0$, and $\boldsymbol{\xi}_\perp$ is the component of displacement transverse to the magnetic field $\mathbf{B}_0$.

Instead of the contravariant components of displacement $\boldsymbol{\xi}(\xi^1, \xi^\theta, \xi^S)$, it is more appropriate to introduce the following components [58]:

$$\xi_\perp^1 = \xi^1,\ \xi_\perp^\theta = \xi^\theta - \frac{B_0^\theta}{B_0^S}\xi_\parallel^S,\ \xi_\parallel^S = \xi^S + \frac{B_\theta^0}{B_S^0}\xi_\perp^\theta, \quad (1.43)$$

which represent the basic components of longitudinal $\boldsymbol{\xi}_\parallel$ and transverse $\boldsymbol{\xi}_\perp$ displacements: the remaining components $\xi_\parallel^\theta$ and $\xi_\perp^S$ with respect to the fields B_0^θ and B_0^S are less than the basic ones:

$$\xi_\parallel^\theta = \frac{B_0^\theta}{B_0^S}\xi_\parallel^S,\ \xi_\perp^S = -\frac{B_0^\theta}{B_S^0}\xi_\perp^\theta.$$

Note that the expression for the perturbed field $\widetilde{\mathbf{B}} = \text{curl}[\boldsymbol{\xi}\mathbf{B}_0]$ includes only transverse components of displacement; therefore, the components of the perturbed field are functions of $\xi_\perp^1$, $\xi_\perp^\theta$:

$$\widetilde{B}^1 = \frac{1}{\sqrt{g}}\frac{\partial}{\partial\theta}\sqrt{g}\,B_0^\theta\,\xi_\perp^1 + \frac{1}{\sqrt{g}}\frac{\partial}{\partial S}\sqrt{g}\,B_0^S\,\xi_\perp^1 = (\mathbf{B}^0\nabla)\xi_\perp^1; \quad (1.44)$$

$$\tilde{B}^\theta = -\frac{1}{\sqrt{g}} \frac{\partial}{\partial a} B_0^\theta \sqrt{g} \xi_\perp^1 + \frac{1}{\sqrt{g}} \frac{\partial}{\partial S} B_0^S \sqrt{g} \xi_\perp^\theta \frac{B_0^2}{B_0^S B_S^0} ; \quad (1.45)$$

$$\tilde{B}^S = -\frac{1}{\sqrt{g}} \frac{\partial}{\partial a} B_0^S \sqrt{g} \xi_\perp^1 - \frac{1}{\sqrt{g}} \frac{\partial}{\partial \theta} B_0^S \sqrt{g} \xi_\perp^\theta \frac{B_0^2}{B_0^S B_S^0} . \quad (1.46)$$

In addition we shall neglect terms of order of β and B_f^2/B_0^2 as compared to unity ($B_0^2 \simeq B_0^S B_S^0$).

We now rewrite div $\boldsymbol{\xi}$ using $\xi_\perp^1$, $\xi_\perp^\theta$, $\xi_\parallel^S$. It is not difficult to verify that

$$\text{div } \boldsymbol{\xi} \simeq \frac{1}{\sqrt{g}} \frac{\partial}{\partial a} \sqrt{g} \xi_\perp^1 + \frac{1}{\sqrt{g}} \frac{\partial}{\partial \theta} \sqrt{g} \xi_\perp^\theta + \frac{1}{\sqrt{g}} \frac{B^0 \nabla}{B_0^S} \sqrt{g} \xi_\parallel^S$$

$$= \text{div } \boldsymbol{\xi}_\perp + B^0 \nabla \frac{\xi_\parallel^S}{B_0^S} . \quad (1.47)$$

We now show that the third and fourth terms in the potential energy are functions of $\xi_\perp^1$ and $\xi_\perp^\theta$ only; i.e., they do not contain $\xi_\parallel^S$. Using the equilibrium condition $\nabla p_0 = (1/c)[\mathbf{j}^0 \mathbf{B}^0]$ we obtain

$$\boldsymbol{\xi} \nabla p_0 \text{ div } \boldsymbol{\xi}_\parallel + \frac{1}{4\pi} [\text{curl} \mathbf{B}^0 \, \boldsymbol{\xi}_\parallel] \tilde{\mathbf{B}} = \xi_\perp^1 \frac{dp_0}{da} (\mathbf{B}^0 \nabla) \frac{\xi_\parallel^S}{B_0^S}$$

$$+ \frac{\xi_\parallel^S}{B_0^S} \frac{dp_0}{da} (\mathbf{B}^0 \nabla) \xi_\perp^1 = \text{div} \left(\mathbf{B}^0 \xi_\perp^1 \frac{dp_0}{da} \frac{\xi_\parallel^S}{B_0^S} \right) . \quad (1.48)$$

Because of this expression the integral can be transformed into a surface integral, and since the normal component of the equilibrium magnetic field equals zero, it disappears. The terms under consideration take the following form in this case:

$$\xi_\perp^1 \frac{dp_0}{da} \text{ div } \boldsymbol{\xi}_\perp - \frac{\sqrt{g}}{c} [j_0^S(\xi_\perp^\theta \tilde{B}^1 - \xi_\perp^1 \tilde{B}^\theta) + j_0^\theta \xi_\perp^1 \tilde{B}^S]. \quad (1.49)$$

Subject to (1.47) and (1.49), we rewrite expression (1.49):

$$W = \frac{1}{2} \int_{V_i} \left\{ \gamma_0 p_0 (\text{div } \boldsymbol{\xi}_\perp + B^0 \nabla \xi_\parallel^S / B_0^S)^2 + \frac{1}{4\pi} \tilde{B}^2 \right.$$

$$+ (\text{div } \boldsymbol{\xi}_\perp + B^0 \nabla \xi_\parallel^S / B_0^S) \xi_\perp^1 \frac{dp_0}{da} - \frac{\sqrt{g}}{c} [j_0^S (\xi_\perp^\theta \tilde{B}^1 - \xi_\perp^1 \tilde{B}^\theta) +$$

$$+ j_0^\theta \xi_\perp^1 \widetilde{B}^S] \} dV + \frac{1}{8\pi} \int_{V_e} \widetilde{\mathbf{B}} \, dV. \tag{1.50}$$

The kinetic energy is written in terms of components of the displacement simply as

$$T = \frac{1}{2} \int_{V_i} \rho_0 \left[\frac{\partial}{\partial t} \left(\xi_\perp + \frac{\xi_\parallel^S}{B_0^S} \mathbf{B}_0 \right) \right]^2 dV, \tag{1.51}$$

where $dV = \sqrt{g} \, da \, d\theta \, dS$.

We now turn to the variation and minimization of the functional taking into account the assumptions made above. In fact, the result of the variation is already known since the variational equations are simply the equations of motion. Therefore, in performing the variation of the Lagrangian $\delta \int \mathcal{L} dt = 0$ with respect to $\xi_\parallel^S$, we obtain the longitudinal component of the equation of motion:

$$\omega^2 \rho_0 B_0^2 \xi_\parallel^S / B_0^S = \gamma_0 p_0 (\mathbf{B}^0 \nabla) \, \mathrm{div}\, \boldsymbol{\xi}. \tag{1.52}$$

The dependence of perturbations on time is taken as $\exp(-i\omega t)$. Substituting $\xi_\parallel^S / B^S$ from this equation into expression (1.47) for div $\boldsymbol{\xi}$, we can write div $\boldsymbol{\xi}$ in terms of div $\boldsymbol{\xi}_\perp$:

$$\left[1 - \mathbf{B}^0 \nabla \frac{\gamma_0 p_0}{\omega^2 \rho_0 B_0^2} (\mathbf{B}^0 \nabla) \right] \mathrm{div}\, \boldsymbol{\xi} = \mathrm{div}\, \boldsymbol{\xi}_\perp. \tag{1.53}$$

It will be clear from further discussion that div $\boldsymbol{\xi}_\perp$ is of order $\varepsilon = a/R$. Therefore, in the approximation considered, Eq. (1.53), symbolically solved for div $\boldsymbol{\xi}_\perp$, assumes the following form:

$$\mathrm{div}\, \boldsymbol{\xi} = \left(1 - \frac{c_S^2 \hat{k}_\parallel^2}{\omega^2} \right)^{-1} \mathrm{div}\, \boldsymbol{\xi}_\perp, \tag{1.54}$$

where $c_S^2 = \gamma_0 p_0 / \rho_0$.

Naturally, the operator $(1 - c_S^2 \hat{k}_\parallel^2 / \omega^2)^{-1}$ is an integral operator; however, it is possible to avoid the use of an integral operator in Eq. (1.54) in two practically important cases. In the first case, when investigating the stability boundary, i.e., when $\omega^2 \to 0$, we can take div $\boldsymbol{\xi} = 0$, as is apparent from Eq. (1.54). If, however, the frequency $\omega \sim k_\parallel c_A$ is of the same order as that of Alfvén waves ($k_\parallel$ is the longitudinal component of the wave vector **k**, $c_A^2 = B_0^2/4\pi\rho_0$), then $(1 - c_S^2 k_\|^2/\omega^2) \simeq (1 + \gamma_0 \beta) \simeq 1$ and div $\boldsymbol{\xi}$ = div $\boldsymbol{\xi}_\perp$ to terms of order β. This, then, is the minimizing rule for $\xi_\|^S$ in the second case. This minimization is correct even for quite low frequency oscillations of the flute type, when $\omega \sim \varepsilon \beta_J k_\| c_A$. In this case $(1 - c_S^2 k_\|^2/\omega^2) \simeq [1 + (\gamma_0 \beta / \varepsilon^2 \beta_J^2)] \simeq 1$ provided $\beta_J \gg 1/q^2$.

In other words, the procedure for the minimization with respect to $\xi_\|^S$ is to omit $\xi_\|^S$ in the inertial term but to replace div $\boldsymbol{\xi}$ with zero if investigating the stability boundary or with div $\boldsymbol{\xi}_\perp$ if the frequency is not too low.

Further, in order to eliminate one more function, we minimize the functional (1.50) with respect to the function $\xi_\perp^\theta$. The corresponding variational equation is the first contravariant component of the linearized equation curl **B** = $(4\pi/c)$**j**, where **j** is represented as $\mathbf{j} = \mathbf{j}_\| + \mathbf{j}_\perp = \mathbf{B}\,(\mathbf{j}\mathbf{B})/B^2 + (c/B^2)\,[\mathbf{B}\nabla p]$. Here we have neglected the inertial term in $\mathbf{j}_\perp$, since it describes magnetoacoustic waves and is immaterial for perturbations with frequencies $\omega \lesssim k_\| c_A$ discussed below. On the other hand, it is relatively easy to complete the equation and show by direct evaluation that the inertial term is small within the assumptions we have made above.

Thus, the variational equation for $\xi_\perp^\theta$ has the following form:

$$\frac{1}{\sqrt{g}} \left(\frac{\partial}{\partial \theta} \widetilde{B}_S - \frac{\partial}{\partial S} \widetilde{B}_\theta \right) = \frac{4\pi}{c} \left[\frac{\widetilde{B}^1 (\mathbf{j}^0 \mathbf{B}^0)}{B_0^2} + \frac{c}{B_0^2} \frac{1}{\sqrt{g}} \right.$$
$$\left. \times \left(B_\theta^0 \frac{\partial \widetilde{p}}{\partial S} - B_S^0 \frac{\partial \widetilde{p}}{\partial \theta} \right) \right]. \qquad (1.55)$$

Using expressions (1.44)-(1.46) for perturbed fields and the expression $\widetilde{p} = - \xi_\perp^1 dp_0/da - \gamma_0 p_0$ div $\boldsymbol{\xi}$ for perturbed pressure, one can estimate the different terms in Eq. (1.55) and see that if terms of order β and $B_J^2/B_0^2 = \varepsilon^2/q^2$ are neglected compared with unity, the expression will simply reduce to the condition

$$\widetilde{B}_S = 0. \qquad (1.56)$$

This minimization rule with respect to $\xi_\perp^\theta$ has a straightforward physical explanation and signifies that we neglect magnetoacoustic waves. Condition (1.56) makes it possible to introduce the function Φ, which reduces $\widetilde{B}^S$ (1.44) to zero by the following relations:

$$\xi_\perp^1 = -\frac{1}{\sqrt{g}\,B_0^S}\frac{\partial \Phi}{\partial \theta}, \quad \xi_\perp^\theta = \frac{1}{\sqrt{g}\,B_0^S}\frac{\partial \Phi}{\partial a}. \tag{1.57}$$

The field components $\widetilde{B}^1$ and $\widetilde{B}^\theta$ are symmetrically expressed in terms of Φ as

$$\widetilde{B}^1 = -\frac{1}{\sqrt{g}}\frac{\partial}{\partial \theta}\frac{B^0_\nabla}{B_0^S}\Phi, \quad \widetilde{B}^\theta = \frac{1}{\sqrt{g}}\frac{\partial}{\partial a}\frac{B^0_\nabla}{B_0^S}\Phi. \tag{1.58}$$

The function Φ multiplied by a constant is an electrostatic potential as was shown above.

After substitution of the displacement components from Eq. (1.57), the expression div $\xi_\perp$ becomes

$$\operatorname{div}\xi_\perp = \frac{1}{\sqrt{g}}\left(\frac{\partial \Phi}{\partial a}\frac{\partial}{\partial \theta}\frac{1}{B_0^S} - \frac{\partial \Phi}{\partial \theta}\frac{\partial}{\partial a}\frac{1}{B_0^S}\right) = \frac{1}{\sqrt{g}}\frac{D(\Phi, 1/B_0^S)}{D(a, \theta)}. \tag{1.59}$$

After minimization with respect to $\xi_\perp^\theta$ and $\xi_\parallel^S$ all the terms in the expression for W (1.50) and T (1.51) are written in terms of Φ. Now it is not difficult to write final simplified equations for potential and kinetic energies which are accurate up to order ε^2 [59]:

$$W = \frac{1}{2}\int_V \left\{\hat{\gamma} p_0 \left(\frac{1}{\sqrt{g}}\frac{D(\Phi, 1/B_0^S)}{D(a,\theta)}\right)^2 + \frac{1}{4\pi}\left[g_{11}\left(\frac{1}{\sqrt{g}}\frac{\partial}{\partial \theta}\hat{k}_\parallel \Phi\right)^2\right.\right.$$
$$\left.+ g_{22}\left(\frac{1}{\sqrt{g}}\frac{\partial}{\partial a}\hat{k}_\parallel \Phi\right)^2 - 2g_{12}\frac{1}{g}\left(\frac{\partial}{\partial \theta}\hat{k}_\parallel \Phi\right)\left(\frac{\partial}{\partial a}\hat{k}_\parallel \Phi\right)\right]$$
$$\left. -\frac{p_0'}{\sqrt{g}\,B_0^S}\frac{\partial \Phi}{\partial \theta}\frac{1}{\sqrt{g}}\frac{D(\Phi, 1/B_0^S)}{D(a,\theta)} + \frac{j_0^S}{c\sqrt{g}\,B_0^S}\frac{D(\Phi, \hat{k}_\parallel \Phi)}{D(a,\theta)}\right\}dV; \tag{1.60}$$

$$T = \frac{1}{2}\int_V \rho_0\left\{g_{11}\left(\frac{1}{\sqrt{g}\,B_0^S}\frac{\partial \Phi}{\partial \theta}\right)^2 + g_{22}\left(\frac{1}{\sqrt{g}\,B_0^S}\frac{\partial \Phi}{\partial a}\right)^2\right.$$
$$\left. - 2g_{12}\frac{1}{g(B_0^S)^2}\frac{\partial \Phi}{\partial a}\frac{\partial \Phi}{\partial \theta}\right\}dV. \tag{1.61}$$

The term with $\hat{\gamma}$ is to be understood in the sense explained after equation (1.54). The dot in (1.61) means differentiation with respect to time.

The expression obtained for the potential energy is clear enough. We now consider the meaning of its terms. The first term, proportional to $\hat{\gamma}$, includes compression effects. As is apparent, it is a term of order ε^2. In other words, compression can appear only in this order. Note that in a cylinder compression appears by considering terms of order $B_J^2/B_0^2 = \varepsilon^2/q^2$ which we have omitted. The second term in the brackets is the energy of the vacuum and plasma magnetic fields. The use of a common expression for perturbations of the plasma and vacuum magnetic fields was made possible by the formal introduction of the potential Φ for the vacuum in accordance with the equation $A_S = -\frac{B^0 \nabla}{B_0^S} \Phi$, which is equivalent to the introduction in the vacuum of a formal displacement given by $\mathbf{B} = \mathrm{curl}\,[\boldsymbol{\xi} \mathbf{B}^0]$. Of course, only the combination $[\boldsymbol{\xi} \mathbf{B}^0]$, which represents the vector potential $\mathbf{A}$ rather than each term individually, has a physical meaning. This permits us to choose in the vacuum a fictitious field in such a way that lines of force are straight, i.e., $B_0^\theta \sim B_0^S$. Note that this fictitious magnetic field has no relation to the magnetic field in the vacuum.

It should be stressed that all field components are proportional to the operator $\hat{k}_\parallel = \mathbf{B}^0 \nabla / B_0^S$ and thus near the resonant surfaces where $k_\parallel \simeq 0$, the magnetic field does not influence the stability. The third term is proportional to $\varepsilon\, dp_0/da$; i.e., it appears only in toroidal geometry. This term generates ballooning modes of the flute and kink instabilities. The last term describes current effects, and in it the toroidal corrections have only a geometric character. This term is the source of the kink instability.

CHAPTER 2

BALLOONING MODES OF THE FLUTE INSTABILITY

2.1. Method of Equivalent Harmonics

In the study of plasma stability, the potential Φ, which describes oscillations, is usually expanded into a Fourier series:

$$\Phi(a, \theta, z) = \sum_{m,n} \Phi_{mn}(a) \exp\left\{-\left[im\theta - i\frac{n}{R}s\right]\right\}. \qquad (2.1)$$

In tokamaks with axial symmetry, harmonics with respect to S are independent; therefore, only the n-th harmonic need be considered. Harmonics with respect to θ interact with each other because of toroidal curvature which presents a basic technical problem in the calculation of plasma stability in tokamaks.

Plasma stability for low harmonics is usually calculated by numerical methods. As for high harmonics, it is more convenient to study them analytically, since they are strongly localized along the plasma radius which enables us to obtain local criteria for plasma stability for arbitrary equilibrium distributions.

It is clear from the expression for potential energy W (1.60) that the stabilizing effect of the magnetic field is proportional to the square of the operator $\hat{k}_{\parallel} = (\mathbf{B}^0 \nabla)/B_0^S$ whose effect upon Fourier harmonics yields $\hat{k}_{\parallel} = -i(B_0^\theta/B_0^S)[m - nq(a)]$. This means that perturbations strongly extended along the magnetic field and localized near resonance surfaces a_m [$q(a_m) = m/n$] weakly perturb the magnetic field and thus present the major danger to stability. High-order modes of such perturbations are commonly termed flute modes.

For large harmonic number, $m \simeq nq \gg 1$ one can believe that all harmonics with respect to θ are equivalent, i.e., differ but slightly from each other (Fig. 3). Each m-th harmonic in this case "sits" upon its rational surface and can be obtained from any other by a shift along the column radius. Consistent with this is the following choice of Φ_{mn}:

$$\Phi_{mn}(a) \simeq \Phi_n(m - nq(a)). \qquad (2.2)$$

The distance between neighboring rational surfaces can easily be obtained from resonance conditions for the m-th ($m - nq = 0$) and (m + 1)-th ($m + 1 - nq = 0$) modes; $\Delta a = a_{m+1} - a_m \simeq 1/nq'_m = a_m/nqS$. Thus, with increasing shear, resonant surfaces draw together and harmonics localized at neighboring surfaces can overlap.

Functions (2.2) at fixed n differ only in m, which is equivalent to displacement in a; therefore, it is natural to expand these functions as a Fourier integral with respect to the radial variable [$m - nq(a)$]:

$$\Phi_n(m - nq(a)) = \int_{-\infty}^{\infty} F(a, y) \exp\{i[m - nq(a)]y\}\, dy, \qquad (2.3)$$

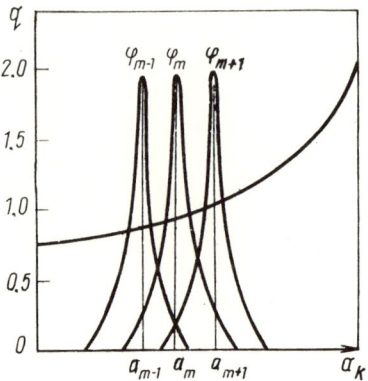

Fig. 3. Azimuthal harmonics of perturbation for $m \gg 1$ localized near neighboring resonant surfaces.

where $F(a, y)$ is independent of m!

It is not difficult to see that the assumption of equivalence of harmonics led to the transformation suggested independently in [19] and [59]. Actually, after inserting (2.3) in (2.1) we obtain

$$\Phi(a, \theta, s) = \sum_m \exp\left\{-im\theta + i\frac{n}{R}s\right\} \int_{-\infty}^{\infty} F(a, y)$$

$$\times \exp\{i[m - nq(a)]y\}\, dy$$

$$= \sum_m \exp\left[-im\theta + i\frac{n}{R}s\right] \int_{-\infty}^{\infty} \exp(imy)\, \hat{\Phi}(a, y)\, dy \quad (2.4)$$

$$= \sum_N F(a, \theta + 2\pi N) \exp\left[-inq(\theta + 2\pi N) + i\frac{n}{R}s\right]. \quad (2.5)$$

These expressions show that the function Φ, periodic in θ, can be represented in the form of an infinite series of nonperiodic functions. This transformation automatically solves the problem that originated in the study of flute perturbations in ordinary space. The crux of the problem can be explained in the following way. The normal mode equation (1.37), which describes the flute instability, in particular, is an equation in partial derivatives with respect to a, θ. The method of its solution to a certain extent is known beforehand. The perturbations should be strongly extended along the lines of force of the magnetic field. In other words, for them $\hat{k}_{\|}\Phi \simeq 0$, and in the system of coordinates with straight lines of force the operator $\hat{k}_{\|} \sim [\partial/\partial\theta - inq(a)]$.

Thus, the condition $\hat{k}_{\|}\Phi \simeq 0$ leads to a solution of the form $\Phi \sim F(a, \theta) \exp(inq\theta)$, where F is a slowly changing function. In the case of large nq, which corresponds to a local approximation, the most important terms arise when differentiating with respect to a of an exponent $\partial\Phi/\partial a \simeq inq'\theta \cdot F \exp(inq\theta)$. The introduction of Φ in the form of $F\exp(inq\theta)$ in the initial differential equation with respect to θ (at nq $\to \infty$) gave both periodic and nonperiodic coefficients (nq'θ). The problem arose concerning the interpretation and solution of such an equation, since the function $F(\theta)$ should be periodic so that the true perturbation Φ was periodic.

Let us show that transformation (2.4) eliminates this problem since now we have to solve an analogous equation in the infinite interval $-\infty < y < \infty$ relative to a Fourier component F(y) which should not be periodic, whereas $\Phi(\theta)$ is naturally a periodic function.

The normal mode equation (1.37) will be written symbolically in the standard form for the problem of eigenvalues:

$$\mathcal{L}(a, \theta)\Phi(a, \theta) = \omega^2 \Phi(a, \theta), \qquad (2.6)$$

where $\mathcal{L}$ is a periodic operator with respect to θ ($0 \le \theta \le 2\pi$) and the function Φ should be periodic in θ and limited in a. After introducing the transformation (2.4) in Eq. (2.6), it becomes clear that any function $\hat{\Phi}(a, y)$ which is a solution of the equation

$$\mathcal{L}(a, y)\hat{\Phi}(a, y) = \omega^2 \hat{\Phi}(a, y) \qquad (2.7)$$

in the interval $-\infty < y < \infty$ generates a periodic function $\Phi(a, \theta)$ — the solution of Eq. (2.6) corresponding to the same eigenvalue. One can prove an opposite statement that a certain solution of Eq. (2.7) actually corresponds to any nonzero periodic solution of Eq. (2.6) [60].

In other words, instead of solving the problem of stability in real space, which requires periodicity, one can solve it in a Fourier space, since it gives the same eigenvalues.

The formal procedure of change is rather simple. Thus it is sufficient in the normal mode equations (1.37) and in the energy principle (1.60) and (1.61) to go over from the function $\hat{\Phi}(a, \theta, S)$ to the function $\hat{\Phi}(a, y, S)$ and in all operators to change the variable θ ($0 \leq \theta \leq 2\pi$) to the variable y in the infinite interval $-\infty < y < \infty$.

It is convenient to represent the function $\hat{\Phi}(a, y, S)$ in the form of an eikonal or "quasi-mode" [61]:

$$\hat{\Phi}(a, y, S) = F(a, y) \exp\left[-inq(a)\,y + i\frac{n}{R}\,S\right]. \qquad (2.8)$$

Here the phase is constant along the magnetic field, but (nq $\gg$ 1) changes very rapidly in directions perpendicular to the field. The amplitude $F(a, y)$ is a slow function compared to the phase and can be calculated by expansion in powers of the parameter 1/nq.

We turn now from Eq. (1.37) to the corresponding equation in Fourier space, in accordance with the procedure described above, and introduce in it $\hat{\Phi}$ in the form (2.8). It is easily seen that as nq $\to \infty$ the main terms (terms of the order 1/nq compared to main ones are neglected) have the following form:

$$\frac{\partial}{\partial y}\left[G(a, y)\frac{\partial F}{\partial y}\right] - \left\{\frac{4\pi p_0' R^2 q^2}{\sqrt{g}\,B_0^S}\left[\frac{\partial}{\partial a}\frac{1}{B_0^S} - \frac{Sy}{a}\frac{\partial}{\partial y}\frac{1}{B_0^S}\right]\right.$$
$$\left. - \omega^2(a)\,\tau_\theta^2\,G(a, y)\right\}F = 0, \qquad (2.9)$$

where

$$G = a\left(\frac{g_{11}}{\sqrt{g}} + \frac{g_{22}}{\sqrt{g}}\frac{S^2 y^2}{a^2} - 2\frac{g_{12}}{\sqrt{g}}\frac{Sy}{a}\right); \quad S = \frac{q'a}{q}.$$

As is apparent from Eq. (2.9) the problem of stability in the approximation considered is reduced to the solution of an ordinary differential equation having both periodic [g_{ik} includes only cos y and sin y (1.6)] and nonperiodic coefficients. The variable a which characterizes magnetic surfaces is a parameter here. This means that oscillations on each magnetic surface are independent and have a local frequency $\omega^2(a)$.

For finding the dependence of the amplitude F on a, i.e., for a complete determination of the radial structure of the perturbations and for finding the true eigenvalue Ω^2, one should take into account the next approximation with respect to $1/nq$. Pertinent detailed calculations were made in [60]. It was shown that the eigenvalue is determined from the following expression:

$$\Omega^2 = \omega^2_{\min}(a)\,[1 + O\,(1/nq)], \qquad (2.10)$$

where the correction $O(1/nq)$ is a positive value and $\omega_{\min}(a)$ is the minimum value of the frequency of radius a. It is apparent from this expression that high modes are most unstable within $nq \to \infty$.

Analogous use of the method of equivalent harmonics, i.e., transformation into a Fourier space, introduction of "quasi-modes," and expansion in $1/nq$, allows us to write the following expressions for the main terms of the potential (1.60) and kinetic energies (1.61) [21]:

$$W = \frac{1}{2}\int_0^{a_k} (nq)^2 \left(\frac{B_0^\theta}{B_0^S}\right)^2 da \int_{-\infty}^{\infty} \left\{ G(a,y)\left(\frac{\partial F}{\partial y}\right)^2 \right.$$
$$\left. + \frac{4\pi p_0' R^2 q^2}{\sqrt{g}\, B_0^S}\left[\frac{\partial}{\partial a}\frac{1}{B_0^S} - \frac{Sy}{a}\frac{\partial}{\partial y}\frac{1}{B_0^S}\right] F^2 \right\} dy; \qquad (2.11)$$

$$T = \frac{1}{2}\int_0^{a_k} (nq)^2 \left(\frac{B_0^\theta}{B_0^S}\right)^2 \tau_\theta^2 \int_{-\infty}^{\infty} G(a,y)(\dot F)^2\, dy. \qquad (2.12)$$

It is easily seen that varying the time integral of the Lagrange function $\mathscr{L} = T - W$ and assuming $\delta \int \mathscr{L} dt = 0$, we find a normal mode equation of the form (2.9).

2.2. Asymptotic Variational Method for Solution of Differential Equations

The investigation of plasma stability to flute perturbations in tokamaks led to an ordinary second-order differential equation having both periodic and nonperiodic coefficients (2.9). The form of the coefficient depends on the form of the magnetic surfaces of the equilibrium configuration studied from the viewpoint of stability. Accurate analytical solutions of such equations are unknown; therefore, we shall consider an approximate asymptotic variational method of solution [33].

For simplicity we shall exemplify this method by a model equation which takes into account all the basic features of Eq. (2.9), but does not overload the calculation with details:

$$\frac{d}{dy}(1 + S^2 y^2) \frac{dF}{dy} + (B \cos y + CSy \sin y - U) F = 0. \qquad (2.13)$$

Here S is the shear, U is the average magnetic well, and quantities B and C characterize the curvature of the lines of force (the ballooning effect).

After substitution of $F = u(1 + S^2y^2)^{-1/2}$ it is convenient to rewrite Eq. (2.13) in the following form:

$$\frac{d^2 u}{dy^2} - V(y) u = 0, \qquad (2.14)$$

where

$$V(y) = \frac{S^2}{(1 + S^2 y^2)^2} + \frac{U - B \cos y - CSy \sin y}{1 + S^2 y^2}. \qquad (2.15)$$

For $S^2 \ll 1$ the potential V(y) is a well, slowly changing in a distance of the order of 1/S with rapid oscillations superposed; i.e., Eq. (2.14) has two scales $y \sim 1/S$ and $y \sim 1$. This permits us to introduce a slow variable $t = Sy$ alongside the variable y, as is usually done in the Van der Pol averaging method. We shall split the potential V(y, t) into slowly and rapidly changing parts:

$$\left.\begin{array}{l} \overline{V} = \dfrac{1}{2\pi} \displaystyle\int_0^{2\pi} V(y, t) \, dy = \dfrac{S^2}{(1 + t^2)^2} + \dfrac{U}{1 + t^2}; \\[1em] \widetilde{V} = V(y, t) - \overline{V} = -\dfrac{B \cos y + CSy \sin y}{1 + t^2}. \end{array}\right\} \qquad (2.16)$$

In the same way we present the solution $u = \overline{u} + \widetilde{u}$.

To find the oscillating part $\widetilde{u}$ at $U \ll 1$ one has to solve the following equation:

$$\frac{d^2}{dy^2} \widetilde{u} + \frac{B \cos y + CSy \sin y}{1 + S^2 y^2} \overline{u} = 0. \qquad (2.17)$$

The solution of this equation to any accuracy in powers of S can easily be obtained by integrating it by parts. In the lowest approximation we have

$$\tilde{u} = \frac{B \cos y + CSy \sin y}{1 + S^2 y^2} \bar{u}. \qquad (2.18)$$

Substituting expression (2.18) into Eq. (2.14) and averaging it over rapid oscillations, one can obtain an equation for the slowly changing part of the solution $\bar{u}$:

$$\frac{d^2 \bar{u}}{dt^2} - \langle V \rangle \bar{u} = 0, \qquad (2.19)$$

where

$$\langle V \rangle = \frac{1}{(1+t^2)^2} + \frac{U}{S^2(1+t^2)} - \frac{B^2 + C^2 t^2}{2S^2(1+t^2)^2}. \qquad (2.20)$$

If we assume that $t^2 \gg 1$ and retain only main terms, we obtain a Suydam type equation with potential $\langle V \rangle = (U - \frac{1}{2}C^2)/S^2 t^2$. It is well known that the absence of eigen solutions of such an equation corresponds to plasma stability [2] and is determined by the condition

$$\frac{1}{4}S^2 + U - \frac{1}{2}C^2 > 0. \qquad (2.21)$$

This condition is the Mercier criterion for the model equation under consideration (2.14). It is quite apparent that in deriving criterion (2.21), only a distant "tail" of the potential well was taken into account. However, it is known [62] that even if condition (2.21) is satisfied, eigensolutions of Eq. (2.19) can exist due to the central well.

In fact, the parameter B does not enter criterion (2.21), though it serves to determine the depth of the central well [see expression (2.20)]; i.e., in obtaining the Mercier criterion, the ballooning effects are not taken into account completely. Therefore, at a sufficiently large B there will be a level in the potential well (2.20), hence in the potential well (2.15) also, despite the fact that condition (2.21) is satisfied and the level in the "tail" of the potential well is absent.

Now we go back to Eq. (2.14), for which we have a solution obtained by asymptotic expansion in S:

$$u = \bar{u}\left(1 + \frac{B\cos y + CSy\sin y}{1 + S^2 y^2}\right). \qquad (2.22)$$

We write the variational principle for Eq. (2.14):

$$W = \int_{-\infty}^{\infty} \left[(u'_y)^2 + V(y)u^2\right] dy \qquad (2.23)$$

and introduce the asymptotic solution (2.22) as a trial function. In this case in (2.22) parameter S cannot be considered a small one, whereas the slowly varying function $\bar{u}$ can be found variationally or chosen from physical considerations.

The asymptotic variational method gives high accuracy. There are two reasons for this. The first reason is that the variational principle itself has a tendency to raise accuracy after the introduction of a trial function which differs slightly from the proper one. The second reason lies in the fact that in this case, as will be shown below, account is taken of exponential terms $\exp(-1/|S|)$ nonanalytical in S which in general cannot be obtained by the averaging method. (These exponential terms take into account effects like quantum-mechanical tunnelling.)

The efficiency of the method in solving equations like (2.14) will be illustrated by a simple case where the Mercier criterion (2.21) is already satisfied, namely, U = 0, C = 0 (the Mercier criterion is $S^2 > 0$ in this instance).

In this case the potential is somewhat simplified:

$$V(y) = \frac{S^2}{(1 + S^2 y^2)^2} - \frac{B\cos y}{1 + S^2 y^2}, \qquad (2.24)$$

and the trial function can be chosen in the following form:

$$u = 1 + \frac{B\cos y}{1 + S^2 y^2}. \qquad (2.25)$$

Here the function $\bar{u}$ is slowly varying and, tending to a constant at the point of marginal stability [62], is taken equal to unity. Upon substitution of (2.25) in the variational principle and integration, it is not difficult to obtain the following stability criterion from the condition W > 0:

$$\frac{1}{4} S^2 - \frac{3}{8} B \exp\left(-\frac{1}{|S|}\right) - \frac{1}{2} B^2 > 0 \qquad (2.26)$$

(here terms of the order S^2 and B^2 as compared to unity are ignored).

One can see from this criterion that, although the Mercier criterion is satisfied, the plasma may be unstable in this model problem due to the ballooning effect.

The potential of Eq. (2.14) has two specific features, namely, periodic structure and behavior at infinity proportional to $1/t^2$. In the case considered at U = 0 and C = 0 attention was given to the rapidly varying part of the trial function which plays here the dominant role. To understand the role of the slowly varying part of the trial function, we turn to Eq. (2.19), but as an example we take a potential which allows an exact analytical solution:

$$\langle V \rangle = \begin{cases} B^* & 0 \leq t \leq 1; \\ -M/t^2 & 1 \leq t < \infty. \end{cases} \qquad (2.27)$$

Here the value B* characterizes the depth of the central part of the well, and the value M regulates the behavior of the potential at infinity.

For this potential the Mercier criterion is M < 1/4. The necessary and sufficient stability criterion which determines the actual boundary of stability can be easily obtained by joining logarithmic derivative solutions at t = 1:

$$\bar{u} = C_1 \begin{cases} \operatorname{ch} B^* t, & B^* > 0 \\ \cos |B^*| t, & B^* < 0 \end{cases} \quad t \leq 1; \qquad (2.28)$$

$$u = C_2 t^{\frac{1}{2}(1 - \sqrt{1 - 4M})}, \qquad t \geq 1. \qquad (2.29)$$

The stability boundary in this case is determined by the following expression:

$$\frac{1}{2}(1-\sqrt{1-4M}) = \begin{cases} \sqrt{B^*}\,\text{tg}\,\sqrt{B^*},\ B^*>0; \\ -\sqrt{|B^*|}\,\text{tg}\sqrt{|B^*|},\ B^*<0. \end{cases} \quad (2.30)$$

Now we shall solve the same task by the variational method. For this purpose we insert in the variational principle (2.23) with potential (2.27) the function $\bar{u} = 1$ as a trial function and from the condition W = 0 we obtain the required boundary of stability relative to the ballooning modes:

$$M = B^*. \quad (2.31)$$

It is obvious that for $B^{*2} \ll 1$ and $M^2 \ll 1$ expression (2.30) goes over into (2.31). This means that at small values of B* and M the slowly varying part of the trial function can be taken as a constant with sufficient accuracy.

For large values of B* and M the actual boundary on the plane of B*, M starts moving away from the straight line M = B*, thus increasing the unstable region, which is clearly seen in Fig. 4. The instability region is hatched in this figure; oblique hatching means violation of the Mercier criterion, the horizontal hatching points to the violation of the criterion for the stability of ballooning modes, and the vertical hatching denotes the region where the criterion for ballooning mode stability is satisfied but the necessary and sufficient stability criterion is violated. Thus Fig. 4 clearly demonstrates the relation between the necessary stability criteria (the Mercier criterion and the criterion for the ballooning modes) and the necessary and sufficient stability criterion.

2.3. Analytical Criterion for the Stability of Ballooning Modes

The differential equation (2.9) with frequency equal to zero and the expression for potential energy (2.11) serve as the basis for obtaining analytical criteria for plasma stability in tokamaks with an arbitrary form of cross section of magnetic surfaces.

First we shall discuss an equilibrium toroidal configuration with circular magnetic surfaces. The metric coefficients of a coordinate system with straight lines of force required for calculation in square-law approximation with respect to curvature have the form [see formulas (1.6), (1.9), and (1.13)]:

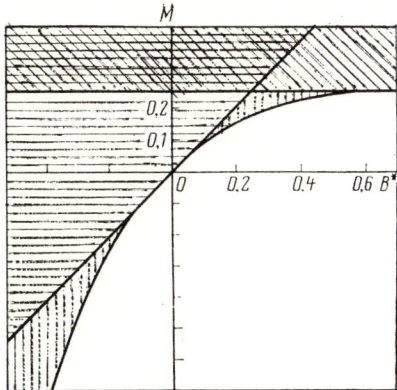

Fig. 4. Stability diagram. The region where the Mercier criterion (M > 1/4) is violated is cross-hatched by sloping lines, horizontal hatching indicates violation of the ballooning mode criterion (M > B), vertical hatching marks the region where both necessary criteria are satisfied whereas the necessary and sufficient stability criterion is violated.

$$\left.\begin{aligned}
g_{11} &= 1 + 2\xi' \cos y; \\
g_{12} &= a(a\lambda' - \xi') \sin y; \\
g_{22} &= a^2(1 + 2\lambda \cos y); \\
g_{33} &= \left(1 - ka \cos y - k\xi + \frac{1}{2} ka\lambda\right)^2; \\
\sqrt{g} &= a\left[1 + (\lambda + \xi' - ka)\cos y - k\xi - \frac{1}{2}ka\xi'\right],
\end{aligned}\right\} \quad (2.32)$$

where $\lambda = \xi' - ka$ and the value $\xi(a)$, which is the displacement of the magnetic surface cross-section center a = const relative to the magnetic axis, is determined from the equilibrium equation (1.20):

$$a\xi'' + (3 - 2S)\xi' = \alpha + \varepsilon. \qquad (2.33)$$

Using a standard substitution, we get rid of the first derivative in Eq. (2.9) $F = uG(a, y)^{-1/2}$ and reduce it to the form

$$d^2u/dy^2 - V(y)u = 0, \qquad (2.34)$$

where

$$V(y) = \frac{1}{G}\left[\frac{1}{2}\frac{\partial^2 G}{\partial y^2} - \frac{1}{4G}\left(\frac{\partial G}{\partial y}\right)^2 \right.$$

$$\left. \frac{4\pi p'_0 R^2 q^2}{\sqrt{g}\, BS_0}\left(\frac{\partial}{\partial a}\frac{1}{BS_0} - \frac{Sy}{a}\frac{\partial}{\partial y}\frac{1}{BS_0}\right)\right]. \qquad (2.35)$$

Using a formal expansion curvature which is valid at $\varepsilon\beta_J \ll 1$ and $\alpha \ll 1$, we write the potential (2.35) in the following way:

$$V(y) = D(t) - A(t)\cos y - E(t)\sin y, \qquad (2.36)$$

where $t = Sy$,

$$D = \frac{S^2}{(1+t^2)^2} + \frac{\alpha}{1+t^2}\left(\varkappa - \frac{1}{2}\alpha\frac{t^2}{1+t^2}\right) + \frac{1}{2}\frac{t^2}{(1+t^2)^2}(\alpha - C_{12})^2$$
$$+ \frac{1}{2}\frac{C_{11}(C_{11}-2t^2\xi')}{(1+t^2)^2} + \frac{\alpha(t^2\xi - C_{11})}{(1+t^2)^2} - \frac{S\alpha(t^2\xi' + C_{11})}{(1+t^2)^2};$$

$$E = t\left[\frac{\alpha - C_{12}}{1+t^2} + \frac{2S\xi'}{1+t^2} + \frac{2S(\xi' + C_{11})}{(1+t^2)^2}\right];$$

$$A = \frac{\alpha - C_{11}}{1+t^2} + \frac{2SC_{12}}{(1+t^2)^2}; \quad C_{11} = \xi' + \varepsilon;$$

$$C_{12} = a\xi'' + \xi' + \varepsilon; \quad \varkappa = \frac{1}{2}(a\xi'' + 3\xi' - \varepsilon) + \varepsilon(1 - q^{-2}).$$

Now we use the asymptotic variational method discussed in the foregoing section. For this purpose we insert in the variational principle (2.23) with potential (2.36) a trial function obtained by asymptotic expansion in S:

$$u = 1 + A\cos y + E\sin y. \qquad (2.37)$$

After integration, from the condition $W > 0$ we derive the required criterion for stability of ballooning modes [33]:

$$\frac{1}{2}S^2 + \alpha\varepsilon(1-q^{-2}) - \frac{3}{2}\alpha\exp\left(-\frac{1}{|S|}\right) - \frac{1}{2}S\alpha^2 > 0. \qquad (2.38)$$

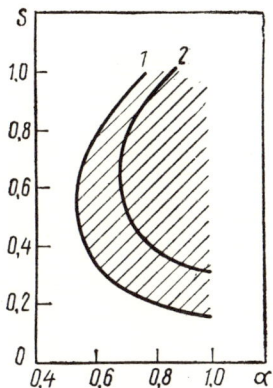

Fig. 5. Stability diagram for $q^2 \gg 1$ for two values of toroidicity: 1) $\varepsilon = 0.1$; 2) $\varepsilon = 0.2$. The region of instability for $\varepsilon = 0.1$ is cross-hatched.

Here terms of the order of S^2 and α^2 compared to unity are omitted.

In this approximation the Mercier criterion has the following form:

$$\frac{1}{4} S^2 + \alpha\varepsilon (1 - q^{-2}) > 0. \qquad (2.39)$$

To obtain the Mercier criterion it is sufficient to study the behavior of the potential (2.36) as $t \to \infty$, as was done in the foregoing section with the model potential. At $q^2(r) > 1$ the Mercier criterion does not set a limit on the pressure gradient α, whereas the criterion for ballooning modes does limit α depending on the value of S. The boundary of plasma stability determined by criterion (2.38) at various values of toroidicity ε and at $q^2 \gg 1$ is depicted in Fig. 5.

The physical nature of the first two terms of the criterion for ballooning modes (2.38) is the same as in the Mercier criterion (2.39). The first term characterizes the stabilizing effect of the shear, hindering the development of a flute perturbation localized near a rational surface and extended along a line of force. The second term consists of two terms: the first stabilizing term has appeared due to the presence in the tokamak of the "geometrical" magnetic well, which is independent of the plasma pressure [15]; and the second $\sim q^{-2}$ destabilizing Suydam term is connected with the curvature of lines of force which is present in a cylinder.

The third term of criterion (2.38), proportional to $\exp(-1/|S|)$, is a destabilizing one and describes the effect of the interaction of perturbations developing at neighboring resonant surfaces. This effect is connected with the broadening of the eigenfunctions when the shear is increased. The nonanalytical dependence on S is due to the fact that perturbations localized near their own surfaces exponentially decrease with distance, the distance between resonance surfaces being $\Delta a = a_{m+1} - a_m \sim 1/|S|$.

The fourth term of criterion (2.38), proportional to $S\alpha^2$, characterizes the ballooning effect associated with shear. This effect plays a significant role, since the usual ballooning effect, α^2 in the approximation considered, is completely annihilated by the deepening of the magnetic well due to the plasma pressure. For the case where the current density decreases with radius ($S = q'a/q > 0$) the ballooning effect associated with the shear is destabilizing, since in this case the rotational transform decreases with increasing radius. This means that there is an increase of the length of perturbation on the low field side of the torus, and an increase of the ballooning part of the perturbation during its displacement along the radius.

Because the criterion for ballooning modes is a local one, it permits us to see the exact place along the radius where the ballooning instability first starts to develop for given profiles of plasma pressure and current density. Figures 6 and 7 show plots of the dependence of critical $\beta* = (8\pi/B^2)(\bar{p}^2)^{1/2}$ on the normalized radius for three profiles of pressure of the form

$$p = p_0 (1 - x^2)^l, \qquad (2.40)$$

where l is a numerical coefficient ($1 \leq l \leq 2$) and $x = a/a_k$ is the normalized radius. Distributions of current density are given by the profile of the safety factor:

$$q = q_0 (1 + \nu x^{2n})^m, \; n, \, m = 1, 2, \qquad (2.41)$$

where $\nu = [(q_a/q_0)^{1/m} - 1]$ characterizes the safety factor at the boundary. The results of calculations of criterion (2.39) for the two

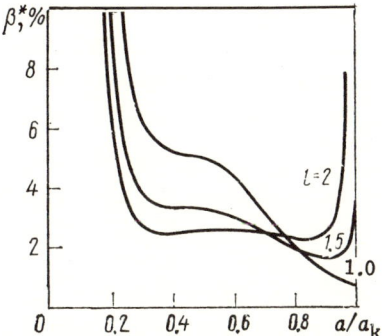

Fig. 6. β^* as a function of normalized radius for a toroidal column ($A = R/a_k = 4.6$) of circular cross section with given current profile [$q_0 = 1$, $q(a_k) = 3.8$; $m = 1$, $n = 2$] for different pressure profiles: 1) $l = 1$; 2) $l = 1.5$; 3) $l = 2$.

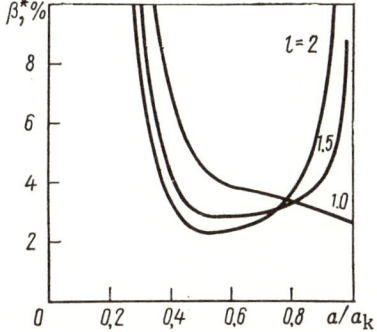

Fig. 7. The same as in Fig. 6 but with a flatter current profile [$q_0 = 1$, $q(a_k) = 1.6$].

current profiles are presented in Figs. 6 and 7. The current profile shown in Fig. 7 is flatter.

The plots show as a function of radius the plasma pressure at which the ballooning instability arises for given distributions of pressure and current density. It is seen that instability first develops at the column periphery leading to a rather small value of the pressure threshold. A comparison of the plots reveals that a flattening of the current density increases the limiting plasma pressure.

Criterion (2.38) has been obtained for circular magnetic surfaces. At the same time, it is well known that a change in the shape of the cross sections of magnetic surfaces influences plasma stability considerably [22]. In this connection the analytical criterion for tokamaks with noncircular magnetic surfaces is of practical interest.

As in Section 1.1, we take coordinate system (1.2) as the starting point for calculations. We denote by ξ the displacement of the center of magnetic surface $\psi = a^2 = $ const relative to the magnetic axis and introduce the coordinate system

$$\left. \begin{array}{l} \rho \cos \omega = \exp(-\eta/2)\, \rho_0 \cos \omega_0 + \xi(a); \\ \rho \sin \omega = \exp(\eta/2)\, \rho_0 \sin \omega_0, \end{array} \right\} \quad (2.42)$$

where $\tanh \eta = (l_z^2 - l_R^2)/(l_z^2 + l_R^2)$ and l_z, l_R are the vertical and horizontal semiaxes of the ellipse, respectively. In this case, in contrast with Section 1.1, we shall not assume the ellipticity of magnetic surfaces to be small; in addition, to take into account the triangularity of magnetic surfaces, we write them in the form

$$\psi = \rho_0^2 + Q\varepsilon\rho_0^2 3\omega_0 = a^2, \quad (2.43)$$

where $\varepsilon = \rho_0/R$ is the toroidicity and Q is the parameter of triangularity. In the noncircular coordinate system this equation has the following form:

$$\psi = (1-e^2)^{-1/2} \{(1 + e \cos 2u)\, \rho^2 \\ + Q\varepsilon\rho^2 \exp(\eta/2)\, [(1-e/2) \cos 3u + (3e/2) \cos u]\} = a^2, \quad (2.44)$$

where

$$e = (K^2 - 1)/(K^2 + 1), \quad K = l_z/l_R .$$

We shall perform calculations by expanding in curvature. We take an elliptical cylinder as a zeroth approximation, and take into account triangularity to the first order in curvature. From Eq. (2.43) at $Q \leq 1$ we obtain

$$\rho_0 = a - (1/2)Q\varepsilon a \cos 3\omega_0. \tag{2.45}$$

We represent the azimuthal variable in the form of an expansion

$$\omega_0 = \theta + \lambda_1 \sin\theta + \lambda_3 \sin 3\theta, \tag{2.46}$$

where the parameters λ_1, λ_3 are found from the condition for straight lines of force.

Inserting formulas (2.42), (2.45), and (2.46) into expression (1.2), we find metric coefficients g_{ik} of the coordinate system with straight lines of force a, θ:

$$\begin{aligned}
g_{11} &= (\operatorname{ch}\eta - \operatorname{sh}\eta \cos 2\omega_0)\rho_0'^2 + \rho_0^2 (\operatorname{ch}\eta + \operatorname{sh}\eta \cos 2\omega_0)\omega_0'^2 \\
&\quad + 2\rho_0 \sin\omega_0 \operatorname{sh}\eta \rho_0' \omega_0' + \xi'^2 + 2\xi'[\rho_0' \exp(-\eta/2)\cos\omega_0 \\
&\quad - \rho_0 \omega_0' \exp(\eta/2)\sin\omega_0]; \\
g_{12} &= (\operatorname{ch}\eta - \operatorname{sh}\eta \cos 2\omega_0)\rho_0' \dot\rho_0 + \rho_0^2 (\operatorname{ch}\eta + \operatorname{sh}\eta \cos 2\omega_0) \\
&\quad \times \omega_0' \dot\omega_0 + \rho_0 (\rho_0' \dot\omega_0 + \rho_0 \omega_0') \operatorname{sh}\eta \sin 2\omega_0 + \xi'[\exp(-\eta/2) \\
&\quad \times \cos\omega_0 \dot\rho_0 - \exp(-\eta/2)\sin\omega_0 \dot\omega_0]; \\
g_{22} &= (\operatorname{ch}\eta - \operatorname{sh}\eta \cos 2\omega_0)\dot\rho_0^2 + \rho_0^2 (\operatorname{ch}\eta + \operatorname{sh}\eta \cos 2\omega_0)\dot\omega_0^2 \\
&\quad - 2\operatorname{sh}\eta \rho_0 \sin 2\omega_0 \dot\rho_0 \dot\omega_0, \\
g_{33} &= (1 - k\exp(-\eta/2)\rho_0 \cos\omega_0 - k\xi)^2,
\end{aligned} \tag{2.47}$$

where the prime means a derivative with respect to a and the dot means a derivative with respect to θ.

Now the problem as in the case of circular magnetic surfaces requires the solution of Eq. (2.34) with potential (2.35) in which the function $G = g_{11}/\sqrt{g} + (g_{22}/\sqrt{g})(S^2 y^2/a^2) - (2g_{12}/\sqrt{g})(Sy/a)$ is set by metric coefficients (2.47). Solving this equation by the asymptotic variational method, we find the required stability criterion for ballooning modes in a tokamak with noncircular cross section [63]:

$$\frac{S^2}{2} + \alpha\varepsilon\left[-\frac{1}{2}V_0'' B_0^2 R \sqrt{1-e^2} - \frac{1}{q^2}\right] - \frac{3}{2}\alpha K^{-1/2}\exp\left(-\frac{1}{|S|}\right)$$
$$- \frac{\alpha^2}{4}(2Sf_1 + e^2 f_2) > 0. \tag{2.48}$$

Here V_0'' is a "geometrical" magnetic well

$$V_0'' = -\frac{2}{B_0^2 R\sqrt{1-e^2}}\left[1 - \frac{3}{2}e\frac{1+e}{2+e} + 6eQ\frac{1-e}{2+e}\right];$$

$$f_1 = 4K^{3/2} \frac{3+K^2}{(1+K^2)(1+3K^2)}, \quad f_2 = \frac{1}{K^2(1+3K^2)} \left(\frac{1+K^2}{1+K}\right)^3. \quad (2.49)$$

Criterion (2.48) differs from criterion (2.38) in the shape of the "geometrical" magnetic well. It is clear from formula (2.49) that when ellipticity is not present (e = 0), the triangularity Q does not change the depth of the magnetic well. However, ellipticity essentially changes the magnetic well even when triangularity is not present. Apparently, the case of the joint action of ellipticity and triangularity (D-shaped form) is most interesting from a practical point of view.

The ellipticity also strongly changes (decreases when the column is extended along the major axis) the effect of the interaction of neighboring perturbations, as well as the ballooning effect proportional to the shear (function f_1). Another destabilizing term proportional to the square of ellipticity emerged in criterion (2.49). This term is the difference of the usual ballooning effect proportional to α^2 and the depth of the magnetic well due to the plasma pressure.

Figure 8 shows the dependences of the plasma pressure threshold calculated with the help of criterion (2.48) for a tokamak with a D-shaped form of magnetic surfaces. The triangularity parameter is fixed (Q = 1.25), whereas the ellipticity e = $(K^2 - 1)/(K^2 + 1)$ changes in the range from zero to 3/5 (the ratio of semiaxis is $1 \leq K \leq 2$). The pressure profile is given by formula (2.40) with $l = 2$ and the current profile is given by (2.41) (curves 1 and 2 correspond to different current profiles at fixed q_0 and q_a). The results of numerical calculations of ballooning modes from [64], represented by dots, are given for comparison.

A detailed analysis of numerical calculations of ballooning modes is given in the next section.

Figure 9 shows the dependences of the plasma pressure threshold on the triangularity parameter at fixed ellipticity (K = 1.65). It is clear that in the case of small triangularity ballooning modes can strongly reduce the pressure threshold for a column cross section extended along the major axis. The violation of the stability criterion occurs near the magnetic axis, which follows from Fig. 10 representing the values of $\beta *$ which induce instability at a

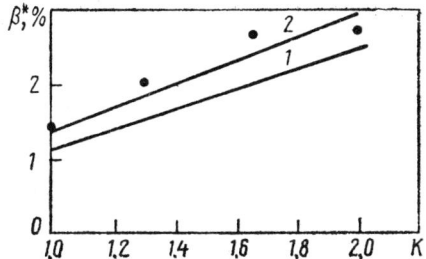

Fig. 8. β^* as a function of ellipticity K for triangularity Q = 1.25 for given profiles of pressure (l = 2) and current. 1) q_0 = 1, $q(a_k)$ = 3.8, m = 1, n = 1; 2) q_0 = 1, $q(a_k)$ = 3.8, m = 2, n = 1. The points are the results of numerical calculations from [64].

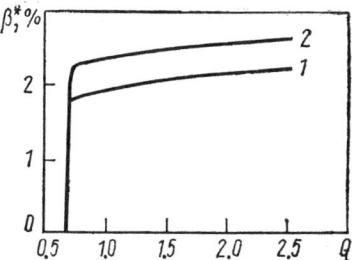

Fig. 9. β^* as a function of triangularity Q for K = 1.65 and the same profiles of current and pressure as in Fig. 8.

given radius for different current profiles. It follows from this figure that flattening of the current profile in the center leads to a widening of the region. The emergence of instability in the central part of the column depends on the decrease of the "geometrical" magnetic well due to ellipticity. This decrease can be compensated by increasing the stability safety factor at the magnetic axis q_0 or

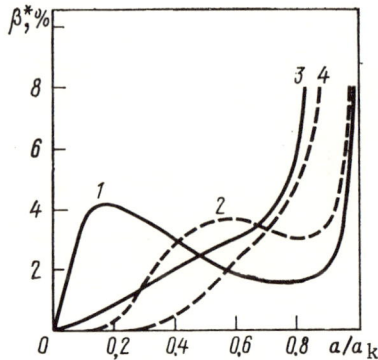

Fig. 10. β^* as a function of the relative radius for a column with elliptical cross section ($K = 1.5$, $Q = 0$) for fixed pressure profile ($l = 2$) at different current profiles: 1) $q_0 = 1$, $q(a_k) = 3.8$, $m = 1$, $n = 1$; 2) $q_0 = 1$, $q(a_k) = 3.8$, $m = 1$, $n = 2$; 3) $q_0 = 1$, $q(a_k) = 1.6$, $m = 1$, $n = 1$; 4) $q_0 = 1$, $q(a_k) = 1.6$, $m = 1$, $n = 2$.

by making a D-shaped form of the cross section. Upon satisfying the condition

$$1 - \frac{1}{q_0^2} + 3\frac{(K^2-1)(4Q-K^2)}{(K^2+1)(3K^2+1)} > 0 \qquad (2.50)$$

the instability in the central part of the column disappears, which is shown in Fig. 11. Now the value of the critical pressure is reduced by the peripheral region of plasma.

In this case the instability essentially depends upon the pressure profile. Figure 12 shows that a parabolic distribution of pressure ($l = 1$) is the most unfavorable since it has the greatest pressure gradient at the column boundary. The increase of the power l in the expression for the pressure profile (2.40) leads to an improvement of stability at the periphery and to a worsening of stability in the middle part of the column which is the result of the de-

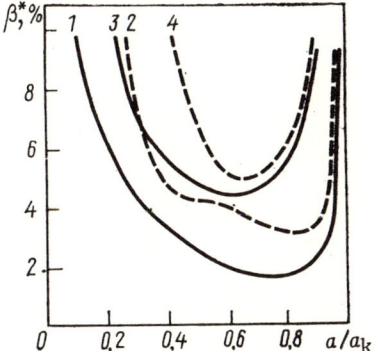

Fig. 11. $\beta*$ as a function of normalized radius for a column with D-shaped cross section (K = 1.5, Q = 1) and the same profiles of current and pressure as in Fig. 10.

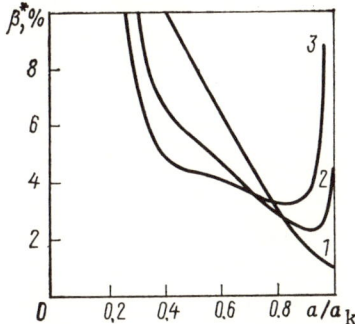

Fig. 12. $\beta*$ as a function of a/a_k for a column with a D-shaped cross section (K = 1.5, Q = 1) and the same profiles of pressure and current as in Fig. 6.

crease and increase, respectively, of the pressure gradient in these regions. The minimum value of $\beta*$ for the entire plasma region may increase or decrease with an increase of l depending on the current profile. A comparison of Figs. 12 and 13 shows that flattening

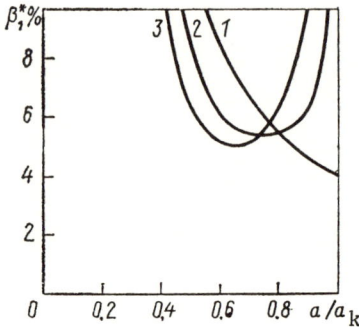

Fig. 13. β^* as a function of a/a_k for a column with a D-shaped cross-section (K = 1.5, Q = 1) and the same profiles of pressure and current as in Fig. 7.

of the current profile, as in the case of circular surfaces, makes the plasma pressure threshold considerably greater for ballooning modes.

2.4. Numerical Calculations of Ballooning Modes

The analytical investigation of ballooning modes made in the foregoing section permits us to comprehend the physical nature of the limitation of the plasma pressure and shows how this limitation can be weakened. A formal expansion for the system curvature has been used in this study, and since the product $\varepsilon \beta_J \sim \alpha$ is the actual small parameter in the problem, the analysis carried out is valid at relatively small pressure. Also, the analytical methods permit us to study the stability asymptotically, accurate only at $nq \to \infty$; i.e., modes with small azimuthal numbers ($m \simeq nq$) have not been investigated analytically. Numerical methods enable us to perform calculations of ballooning modes without the approximations mentioned above.

In this section we shall mainly proceed from the works of the Princeton group, which was the first to discover ballooning modes [29] and made a great contribution to their study [64]. The nu-

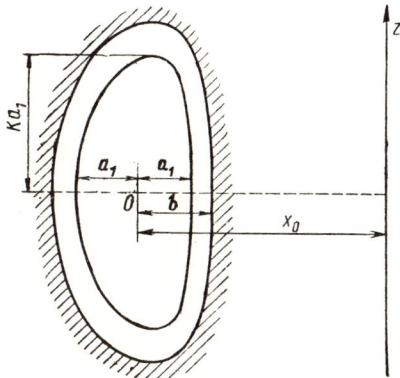

Fig. 14. Geometry of the model used in numerical calculations. a_1 characterizes the size of the plasma and b characterizes the size of the perfectly conducting wall.

merical code PEST [65] was used in these works. The initial calculation concerned an equilibrium with chosen parameters from which a series of equilibria was generated with the same shape of plasma boundary, pressure profile, and safety factor profile but with an increasing value of $\beta*$. In this case the condition of flux conservation was used [66]. Later a study was made of the stability of this series of equilibria.

The shape of the plasma boundary (see Fig. 14) was set in the following way:

$$x = x_0 + a_1 \cos(\theta + \delta \sin \theta); \quad z = K a_1 \sin \theta, \quad (2.51)$$

where x measures distance normal and z parallel to the major axis, x_0 is the position of the boundary surface center, a_1 is the minor radius of the column, b is the minor radius of a perfectly conducting wall, K is the ellipticity parameter, and δ is the triangularity parameter. For comparison of the boundary shape given by Eqs. (2.51) with the shape determined by the analytical expression (2.43) we have the following relations between the parameters: $a_1^2 = e^{-\eta} a^2$, $K = e^{\eta}$, $\delta \approx \frac{3}{2} e^{-\eta/2} a_1 Q/R$.

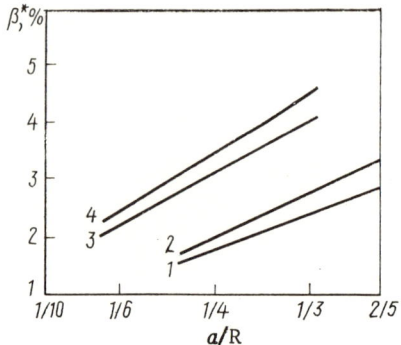

Fig. 15. Critical β^* as a function of torodicity: 1) $n = 3$ mode with a free boundary; 2) $n = 3$ mode with a fixed boundary, both curves for a circular cross section; 3, 4) corresponding curves for a D-shaped cross section ($K = 1.65$, $\delta = 0.25$).

Distributions of the plasma pressure and current density across the column cross section were taken analogously to (2.40) and (2.41). The effect of the current distribution on stability was studied for a fixed safety factor on the magnetic axis q_0 by changing q at the boundary (q_a) or for a fixed q_a by changing q_0.

Low modes with $n = 1$ and $n = 2$ in the case of a free boundary are excited mainly due to the magnetic field energy of the longitudinal current and are the manifestation of the kink instability. The influence of ballooning effects on this instability will be considered in the next chapter. Emphasis will be put on mode $n = 3$, since this value of n is sufficiently great for the mode to be influenced by the ballooning effect even with a free boundary, and at the same time it is sufficiently small to facilitate numerical calculations.

The complicated interrelationship of equilibrium plasma parameters even at moderate values of β^* and their strong effect upon stability make the general analysis quite difficult. For this reason all calculations were made when only one parameter was changed, the rest being fixed.

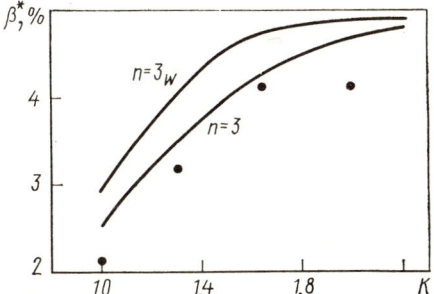

Fig. 16. Critical β^* as a function of ellipticity for a configuration with A = 3, δ = 0.25, l = 2, and 1 < q < 3.8. The points indicate the result of a ballooning mode calculation for n → ∞.

As seen from Fig. 15, the dependence of critical β^* on toroidicity turned out to be linear both for circular and noncircular magnetic surfaces. In both cases calculations were performed for a pressure profile with power l = 2 and for a safety factor changing from 1 on the magnetic axis to 3.8 at the plasma boundary. In the case of D-shaped surfaces the ellipticity K = 1.65 and the triangularity δ = 0.25. The stability was calculated for the n = 3 mode both with a fixed boundary (dotted curves) and with a free plasma boundary (solid curves). It is evident from the curves that the presence of a free plasma boundary for this mode only slightly decreases the critical β^* as compared to a fixed boundary.

Figure 16 shows the dependence of critical β^* on ellipticity. The calculations were performed for a D-shaped cross section which ensures the Mercier stability criterion (δ = 0.25), the aspect ratio being A = R/a = 3, l = 2, and 1 < q < 3.8. The critical β^* increases with increasing distance along the torus symmetry axis K > 1 for the n = 3 mode in the case of both free and fixed boundaries as well as for ballooning modes of the flute instability at n → ∞. It is clear that for K ≃ 1.7 the increase of critical β^* slows down and a further increase of ellipticity is not effective. An analytical investigation of ballooning modes using criterion (2.48) does not indicate a saturation of critical β^* with increasing

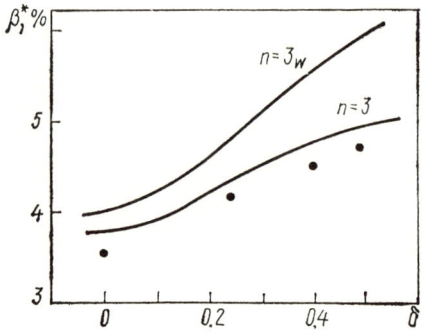

Fig. 17. Critical β^* as a function of triangularity for the configuration with $A = 3$, $K = 1.65$, $l = 2$, and $1 < q < 3.8$.

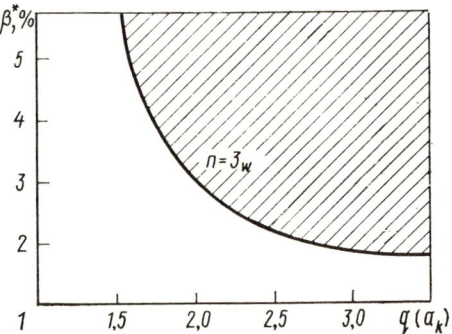

Fig. 18. Critical β^* as a function of $q(a_k)$ for $q_0 = 1$ for the configuration $A = 4.6$, $K = 1.0$, $\delta = 0$, $l = 2$.

ellipticity (Fig. 8), though there is good agreement at relatively small β^*. This difference is explained by the fact that the applicability of analytical calculations is limited, as was mentioned above. Triangularity, as follows from the expression for the magnetic well (2.49), can only influence the stability in the presence of ellipticity; therefore, calculations of the dependence of β^* on δ were performed for $K = 1.65$. Figure 17 shows results for the $n = 3$ mode and for

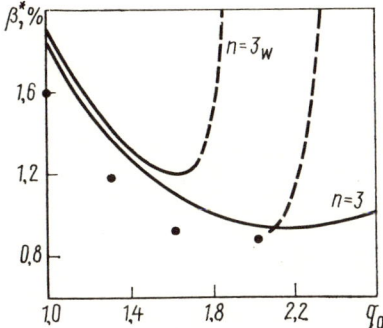

Fig. 19. Critical β^* as a function of q_0 for $q(a_k) = 3.8$ for the configuration $A = 4.6$, $K = 1.0$, $\delta = 0$, $l = 2$. The points indicate the result for $n \to \infty$.

ballooning modes with $A = 3$, $l = 2$, and $1 < q < 3.8$. It is clear that small triangularity is not effective and that only $\delta = 0.5$ (a D-shaped form) significantly improves stability over that of the pure ellipse.

Above we analyzed the effect of purely geometrical factors such as the aspect ratio (A), ellipticity (K), and triangularity (δ) on the plasma stability in a tokamak. This analysis shows that β^* is proportional to the aspect ratio A, a purely elliptical form permitting an approximately twofold increase of β^* as compared with a circle and a D-shaped form permitting a further increase of 1.3 times.

Now we shall consider the effect of the safety factor profile on the value of critical β^*. The result of a calculation of the stability of the $n = 3$ mode for a circular plasma column with $A = 4.6$, $l = 1.4$, and a fixed value $q_0 = 1$ as a function of q_a is depicted in Fig. 18. The critical β^* decreases as q_a increases, which is in agreement with the analytical criterion (2.48), since in this case there is an increase of the destabilizing influence of the ballooning effect associated with shear (the term proportional to $S\alpha^2$) as the shear increases. As $q_a \to 1$, the shear tends to zero and the limitation on β^* disappears in accordance with the stability diagram given in Fig. 5.

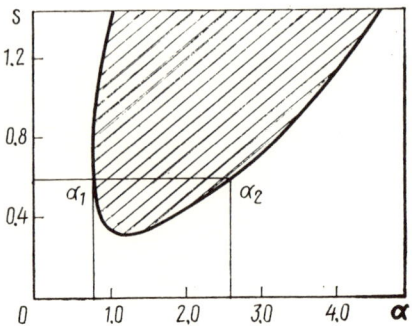

Fig. 20. Stability diagram for a D-shaped column. The stability region is cross-hatched.

Analogous calculations for a circular column with $l = 2$ and a fixed value of $q_a = 3.8$ as a function of q_0 for the $n = 3$ mode, as well as for ballooning modes ($n \to \infty$), are shown in Fig. 19. The sharp increase in critical β^* for small-scale ballooning modes $n \to \infty$ for $q_0 > 2.2$ results from a decrease of the shear and corresponds to the region of small S and large α in the stability diagram (Fig. 5).

Good agreement between the first numerical calculations [29, 30, 64] and the analytical criteria (2.38) and (2.48) is explained by the small value of the critical pressure gradient at which ballooning modes start to develop ($\alpha < 1$). Later it was shown that with increasing pressure, ballooning modes are suppressed and a second stability zone appears [20, 35, 38]. In Fig. 20 from [20] the instability region of ballooning modes is hatched, and it is seen that for given shear, instability exists for a pressure gradient lying in the interval $\alpha_1 < \alpha < \alpha_2$. For an analytical description of the second stability zone one should take into account terms of the order α^4 and $S^2\alpha^2$, which were omitted in deriving criteria (2.38) and (2.48). The question of the accessibility of the second stability zone was discussed in a number of works [75, 76] and can be of practical interest.

Most works calculate plasma stability with a scalar pressure. Achievements in additional heating methods raise questions about the plasma stability with anisotropic pressure. Pressure anisot-

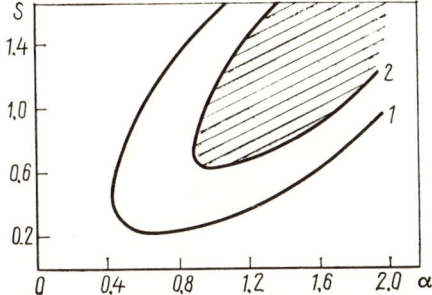

Fig. 21. Stability diagram for a circular column. Curve 1 corresponds to a scalar plasma pressure and curve 2 shows the restricted, cross-hatched region of instability for anisotropic pressure.

ropy caused by perpendicular injection of neutrals was modelled in [77]. The result of the calculation of stability is given in Fig. 21, from which it follows that this method of heating facilitates access to the second stability zone. Curve 1 corresponds to a scalar pressure and curve 2, shifted to a zone of a greater pressure and a greater shear, corresponds to an anisotropic pressure.

The analytical criterion (2.48) and the analysis of numerical calculations carried out above refer to the plasma stability boundary when the frequency of excited oscillations is $\omega^2 = 0$. The dependence of the growth rate $\gamma^2 = -\omega^2$ on different parameters is also of interest since it characterizes the "degree" of instability, i.e., plasma behavior above the stability threshold. Figure 22 from [67] shows the dependence of the normalized growth rate on q_0 for a fixed value of $q_a = 3.3$ and $\beta^* = 8\%$. It is obvious that at $q_0 = 1.73$ the plasma becomes stable for the n = 3 mode with a fixed boundary. It follows from Figs. 21 and 22 that the flattening of the profile due to an increase of q_0 permits a considerable increase of critical β^*.

Up to the present the free plasma boundary has been considered only in the case of a conducting wall at infinity. Now we start

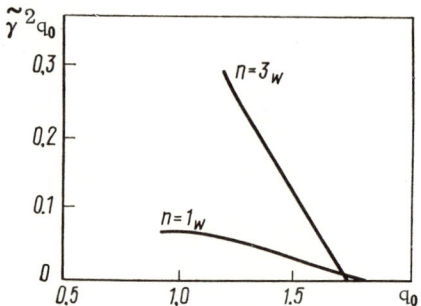

Fig. 22. Parameter $\tilde{\gamma}^2 q_0$ ($\tilde{\gamma}^2 = \gamma^2 a_k^2/C_\theta^2$) as a function of q_0 for $q(a_k) = 3$ for a D-shaped column with $\beta* = 8\%$.

moving the conducting wall nearer to the plasma boundary. It is clear that in this case the growth rate of the mode under consideration will start to decrease and, as soon as the conducting wall coincides with the plasma boundary, it will become equal to the growth rate of the internal mode with a fixed boundary. This approach of the conducting wall is shown in Fig. 23 in which the normalized growth rate is plotted as a function of $\beta*$ for the $n = 3$ mode in a D-shaped tokamak for different values of the ratio b/a. It is seen from the figure that the conducting wall should be placed near the plasma boundary to achieve a noticeable effect.

An interesting result is depicted in Fig. 24, which gives the critical $\beta*$ as a function of the reciprocal of the mode number $1/n$ both for the case of a free boundary ($b/a = \infty$, curve 2) and for the case of a fixed boundary (curve 1). The calculations were performed for a D-shaped tokamak with $A = 3.5$, $l = 1.4$, and $1 < q < 3$. It is clear from the figure that in the case of a fixed plasma boundary the low order modes are not dangerous and that the highest modes set bounds on $\beta*$, the plasma stability relative to the highest modes being independent of the plasma boundary in this case. The result is clear since oscillations with high mode numbers are small-scale. In the case of a free plasma boundary, the limitation on $\beta*$ can be due to low modes and the result in this case essentially depends on the profile of the safety factor q and the value of q_0. To elucidate the cause of the limitation on the plasma pressure

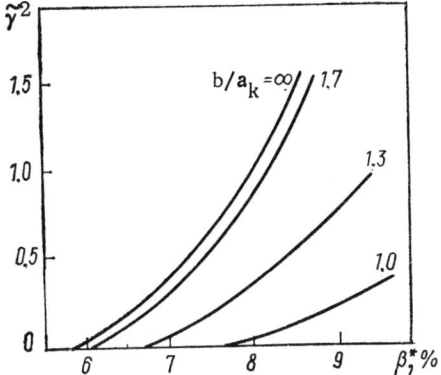

Fig. 23. Normalized growth rate of the n = 3 mode $\tilde{\gamma}^2$ as a function of $\beta*$ for different positions of the conducting wall b/a_k = ∞, 1.7, 1.3, 1.0 for the configuration with A = 3.5, K = 1.65, δ = 0.25, l = 1.4, 1 < q < 3.

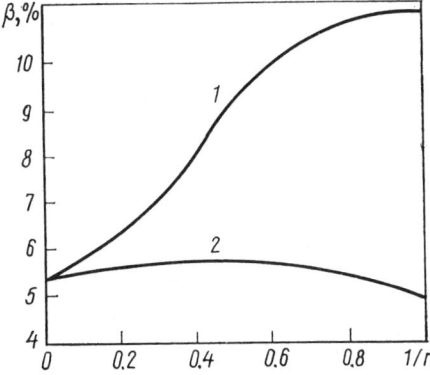

Fig. 24. Critical $\beta*$ as a function of $1/n$ for the configuration with A = 3.5, K = 1.65, δ = 0.25, l = 1.4, and 1 < q < 3. Curve 1 corresponds to a fixed boundary and curve 2 corresponds to a free plasma boundary.

in this case, one should analyze the influence of the ballooning effect on the kink instability, which is done in the next chapter of the survey.

CHAPTER 3

BALLOONING MODES OF THE KINK INSTABILITY

3.1. Theory of the Kink Instability of a Toroidal Column

To elucidate the role of ballooning effects on the kink instability in a tokamak, we shall examine a simple model for the current density distribution across the column cross section which permits us to solve the problem analytically. We assume that the current density across the column cross section is homogeneous and that the toroidal column of circular cross section with radius a_k is placed inside a conducting wall of radius b. First we assume that $b/a_k \to \infty$ and neglect the influence of the conducting wall on stability and later we examine the effects of taking it into account.

It is convenient to solve the problem by a variational method using the perturbation theory with torus curvature $\varepsilon = a_k/R$ as a small parameter. It is well known that for the calculation of corrections of the first order in energy, one should know the eigenfunctions of the cylindrical plasma column. The dispersion relation of a cylindrical plasma column was obtained in Chapter 1 (1.38). For a homogeneous column the equation takes the following form:

$$\left(\frac{\omega^2}{c_A^2} + \hat{k}_\parallel^2\right)\Delta_\perp \Phi = \frac{i\hat{k}_\parallel}{aB_0} \frac{\partial \Phi}{\partial \theta} \frac{4\pi}{c} \frac{dj_0}{da}. \qquad (3.1)$$

The right-hand side of this equation is equal to zero both within the plasma column ($dj_0/da = 0$) and outside in the vacuum ($j_0 = 0$). Therefore, in both plasma and vacuum the eigenfunctions satisfy the equation

$$\Delta_\perp \Phi = 0. \qquad (3.2)$$

Equation (3.1) serves to obtain the boundary condition at $a = a_k$.

We shall consider a cylindrical plasma column with identical faces and length $L = 2\pi R$, where R is the major radius of the equivalent torus. We take the perturbed potential Φ in the form

$$\Phi = \varphi(a)\exp(im\theta - i(\tfrac{n}{R}s)) \tag{3.3}$$

since, due to cylindrical symmetry, radial harmonics are independent.

Within the column we write the solution of Eq. (3.2) in the form

$$\varphi^i = (a/a_k)^m, \tag{3.4}$$

hence, using the condition of potential continuity at $a = a_k$, we obtain for the vacuum:

$$\varphi^e = (a/a_k)^{-m}. \tag{3.5}$$

Integrating Eq. (3.1) over an infinitely thin layer at the boundary between plasma and vacuum, we find

$$\left(k_m^2 + \frac{\gamma^2}{c_A^2}\right)\frac{d\varphi^i}{da} - k_m^2 \frac{d\varphi^e}{da} = \frac{m(4\pi/c)j_0}{aB_0} k_m \varphi^i, \tag{3.6}$$

where

$$k_m = \frac{1}{qR}(m - nq).$$

Substituting (3.4) and (3.5) into (3.6), we obtain the dispersion equation [68]

$$\tilde{\gamma}^2 - 2(m - nq) + 2(m - nq)^2 = 0, \tag{3.7}$$

where

$$\tilde{\gamma}^2 = \gamma^2 a^2/c_\theta^2; \quad c_\theta^2 = B_J^2/4\pi\rho_0.$$

In expression (3.7) for $nq < m$ the second term plays a destabilizing role; it arises from the longitudinal current in a tokamak and is the source of the kink instability. The third term reflects the stabilizing effect of the perturbed magnetic field, and as will be seen later, this component increases as the conducting wall approaches the plasma boundary. Expression (3.7) describes the oscillation spectrum of the cylindrical plasma column in the case of an infinitely distant conducting wall.

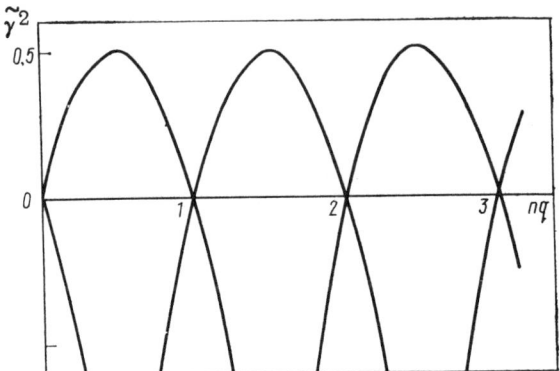

Fig. 25. Oscillation spectrum for a cylindrical column with uniform current without a conducting wall.

This oscillation spectrum is depicted in Fig. 25, the oscillations being unstable for $\tilde{\gamma}^2 > 0$ and stable for $\tilde{\gamma}^2 < 0$. The figure shows that at $\tilde{\gamma}^2 = 0$ different branches of oscillations cross over the stability boundary. Thus, at nq = 2 branches of modes with m = 2 and m = 3 intersect; i.e., the spectrum is degenerate.

It is well known from quantum mechanics and radiophysics that the influence of small perturbations is to remove the degeneracy and this occurs first of all near the points of branch intersection. The role of such a perturbation in the problem under consideration is played by toroidal curvature. Later we show that when toroidicity is taken into account, intersections of the m-th and (m + 1)-th branches of the neighboring modes of oscillations split, as is the case with the zonal structure of the electronic spectrum of solids, and the oscillation spectrum takes the form depicted in Fig. 26. Regions emerge in the spectrum where the propagation of oscillations is impossible, whereas the growth rate ($\tilde{\gamma}^2 > 0$) in the unstable region increases on average.

Indeed, the equation describing the kink instability in a torus (1.37) can be schematically written in the form

$$(\tilde{\gamma}^2 + \hat{L}_0)\Phi(a, \theta) - 2\varepsilon\beta_J \cos\theta \Phi(a, \theta) = 0. \quad (3.8)$$

Here the operator $(\tilde{\gamma}^2 + \hat{L}_0)\Phi(a, \theta) = 0$ characterizes the kink instability in the cylinder, and the term proportional to $\varepsilon\beta_J$ (the

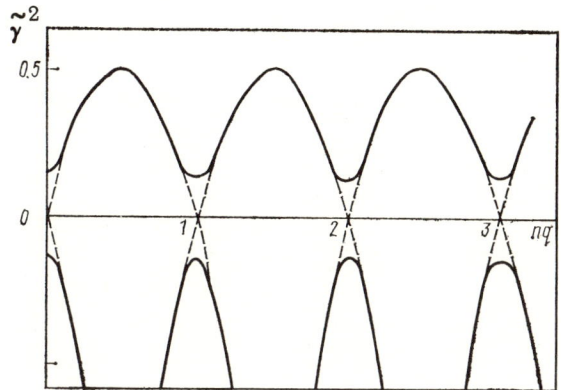

Fig. 26. Oscillation spectrum for a toroidal column with uniform current without a conducting wall.

ballooning effect) has originated due to the bending of the cylinder into the torus.

We shall consider the point of intersection of the m-th and (m + 1)-th branches in Fig. 25. It is obvious that here we deal with a problem having double degeneracy, the solution of which is well known. The correct eigenfunction has the form

$$\Phi(a, \theta) = C_m \chi_m + C_{m+1} \chi_{m+1}, \tag{3.9}$$

where $\chi_m = \varphi_m^0(a) \cos m\theta$, $\varphi_m^0(a)$ is an eigenfunction of the cylindrical problem. We multiply Eqs. (3.8) by χ_m and integrate over a and θ:

$$\Delta\tilde{\gamma}^2 \langle \varphi_m^{0\,2} \rangle C_m = \varepsilon\beta_J \langle \varphi_m^0 \varphi_{m+1}^0 \rangle C_{m+1}. \tag{3.10}$$

Here the difference $\Delta\tilde{\gamma}^2 = \tilde{\gamma}^2 - \tilde{\gamma}_0^2$, is introduced, where $\tilde{\gamma}_0^2$ is the growth rate for the cylinder. The value $\Delta\tilde{\gamma}^2$ characterizes the influence of the ballooning effect. Brackets denote integration with respect to the radius.

In an analogous manner, multiplying Eq. (3.8) by χ_{m+1} and integrating it, we obtain

$$\Delta\tilde{\gamma}^2 \langle \varphi_{m+1}^{0\,2} \rangle C_{m+1} = \varepsilon\beta_J \langle \varphi_m^0 \varphi_{m+1}^0 \rangle C_m. \tag{3.11}$$

In deriving (3.11) we have taken into account that operator $\hat{L}_0$ acting upon the function χ_{m+1} at the point of intersection gives the same value of the growth rate $\tilde{\gamma}_0^2$ as when acting upon χ_m. It follows from Eqs. (3.10) and (3.11) that

$$\Delta\tilde{\gamma}^2 = \pm \varepsilon\beta_J \frac{\langle \varphi_m^0 \varphi_{m+1}^0 \rangle}{\sqrt{\langle \varphi_m^{0\,2}\rangle \langle \varphi_{m+1}^{0\,2}\rangle}}, \qquad (3.12)$$

This is the splitting of the degenerate level $\tilde{\gamma}_0^2 = 0$ caused by "bending" the cylinder into a torus. Equation (3.12) characterizes the maximum splitting at the very point of branch intersection. On moving away from this point, $\Delta\tilde{\gamma}^2 \to 0$ (Fig. 26), or to be more precise, becomes of the order $\varepsilon^2\beta_J^2 \ll 1$, but terms of this order have been neglected.

It is apparent from Eq. (3.12) that the ballooning effect of kink modes with a free boundary is of first order in the small parameter $\varepsilon\beta_J$, and therefore it can be considerably more dangerous than the ballooning effect of small-scale modes of oscillations proportional to $\varepsilon^2\beta_J^2$ and considered in the foregoing section.

For a quantitative calculation of the ballooning modes of the kink instability in the case of a conducting wall, we use the energy principle $\delta \int (T - W)dt = 0$ with simplified expressions for the potential and kinetic energies (1.60) and (1.61). We shall write these expressions retaining only those terms which are required for the calculation of growth rates near the stability boundary:

$$W = \frac{1}{2}\int a\, da\, d\theta\, dS \left\{ \left[\frac{1}{4\pi}\left(\frac{1}{a}\frac{\partial}{\partial \theta}\hat{k}_\parallel \Phi\right)^2 + \frac{1}{4\pi}\left(\frac{\partial}{\partial a}\hat{k}_\parallel \Phi\right)^2 \right] \right.$$
$$\left. - \frac{1}{aB_0}\frac{\partial \Phi}{\partial \theta}\frac{P_0'}{a}\frac{1}{a}\frac{D(\Phi, 1/B_0^S)}{D(a, \theta)} + \frac{j_0^S}{caB_0^S}\frac{D(\Phi, \hat{k}_\parallel \Phi)}{D(a, \theta)} \right\}, \qquad (3.13)$$

$$T = \frac{1}{2}\int a\, da\, d\theta\, dS \left\{ \frac{\rho_0}{B_0^2}\left[\left(\frac{1}{a}\frac{\partial \Phi}{\partial \theta}\right)^2 + \left(\frac{\partial \Phi}{\partial a}\right)^2\right] \right\} \qquad (3.14)$$

Contravariant components of the fields and the current entering Eq. (3.13) in the approximation considered have the following form (see 1.15, 1.16, and 1.17):

$$B_0^S = B_0/g_{33}, \quad B_0^\theta = B_J/\sqrt{g}, \quad j_0^S = \mathcal{J}'/4\pi a \qquad (3.15)$$

where $B_J = \mathcal{J}/2\pi a$, $g_{33} = 1 - 2ka\cos\theta$, $\sqrt{g} = (1 - 2ka\cos\theta)a$. Here use is made of the system of coordinates with straight lines of force where the ratio of contravariant components $B_0^\theta/B_0^S = 1/qR$ does not depend on the angular variable θ. This means that the operator $\hat{k}_\parallel$, when acting upon separate harmonics, gives a constant which simplifies calculations.

We insert a combination of neighboring eigenfunctions of the cylinder as a trial function in the energy principle:

$$\Phi = C_m \varphi_m^0 \exp(im\theta) + C_{m+1} \varphi_{m+1}^0 \exp[i(m+1)\theta] + \text{c.c.}, \quad (3.16)$$

where $\varphi_m^0 \sim a^m$ is within the column and $(a^{-m} - a^m/b^{2m})$ is between the plasma and the conducting wall.

After integration over variables a, θ, S a functional is derived depending on four parameters C_m, C_m^*, C_{m+1}, and C_{m+1}^*. Variation of this functional in C_m^* and C_{m+1}^* gives the following system of equations for the determination of C_m and C_{m+1}:

$$\left.\begin{array}{l}\left[\tilde{\gamma}^2 - 2x_m + \dfrac{2x_m^2}{1-(a/b)^{2m}}\right] C_m + \varepsilon(\beta_J + 1) C_{m+1} = 0; \\[2ex] \left[\tilde{\gamma}^2 - 2x_{m+1} + \dfrac{2x_{m+1}^2}{(1-(a/b)^{2m+2}}\right] C_{m+1} + \varepsilon(\beta_J + 1) \dfrac{m}{m+1} C_m = 0,\end{array}\right\} \quad (3.17)$$

where $x_m = m - nq$.

The dispersion equation

$$\begin{vmatrix} \tilde{\gamma}^2 - 2x_m + \dfrac{2x_m^2}{1-(a/b)^{2m}} & \varepsilon(\beta_J + 1) \\[2ex] \varepsilon(\beta_J + 1)\dfrac{m}{m+1} & \tilde{\gamma}^2 - 2x_{m+1} + \dfrac{2x_{m+1}^2}{1-(a/b)^{2m+2}} \end{vmatrix} = 0 \quad (3.18)$$

is the condition for the solution of this system of equations.

As $\varepsilon \to 0$ it splits into two independent equations for the m-th and (m + 1)-th modes, which differ from Eq. (3.7) by the presence of denominators in the stabilizing terms. These denominators originated when the conducting wall was taken into account, the stabilizing influence of the conducting wall on nondiagonal terms in the approximation considered being insignificant. It is apparent that as the conducting wall approaches the plasma boundary $b \to a$, the kink instability is suppressed. In this case intervals for nq (see Fig. 27) where oscillations do not grow ($\tilde{\gamma}^2 < 0$) appear in the sta-

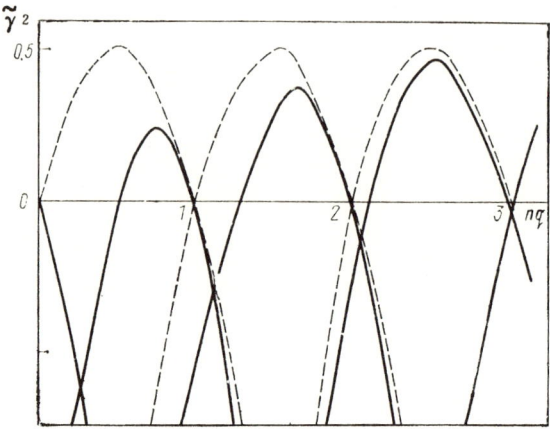

Fig. 27. Oscillation spectrum for a cylindrical column with uniform current with a conducting wall.

bility diagram. These intervals are commonly termed stability gaps. The width of the m-th gap depends on the location of the conducting wall and equals $\Delta nq = (a/b)^{2m}$. It is known that modern tokamaks operate just in these gaps.

At $\varepsilon \neq 0$ it follows from Eq. (3.18) that the width of the stability gap depends on the plasma pressure. Assuming that the width of the gap is sufficiently small, we find an approximate solution of the dispersion equation

$$\tilde{\gamma}^2 \simeq \pm \varepsilon (\beta_J + 1) \sqrt{m/(m+1)} - (a/b)^{2m+2}. \quad (3.19)$$

Apparently, the splitting increases with increasing β_J, whereas the width of the stability gap decreases (see Fig. 29). In other words an increase of the plasma pressure in a tokamak allows the growth of the ballooning modes of the kink instability. We note that the term proportional to β_J in expression (3.19) arises from the term dp_0/da in the potential energy (3.13), the unity having orginated from the current term.

The condition for stable operation of a tokamak in the m-th gap has the form

$$\beta < \beta_c = \frac{\varepsilon}{q^2} \left[\sqrt{\frac{m+1}{m}} \left(\frac{a_h}{b} \right)^{2m+2} - \varepsilon \right]. \quad (3.20)$$

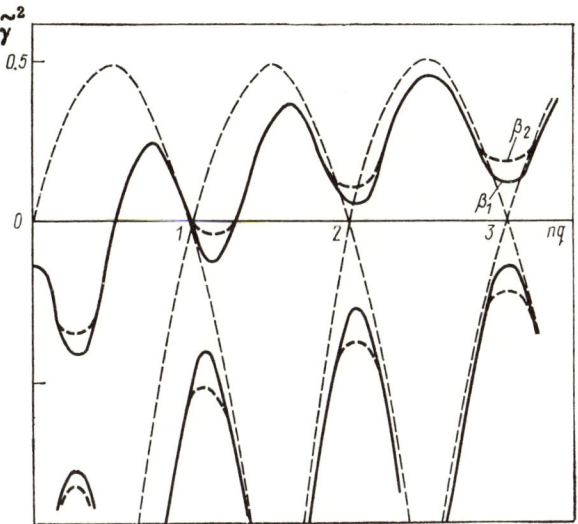

Fig. 28. Oscillation spectrum for a toroidal column with uniform current with a conducting wall. Dotted lines correspond to the splitting at higher pressure $\beta_{J_2} > \beta_{J_1}$.

At a plasma pressure less than critical, determined from condition (3.20), and for any monotonically decreasing current density profile, the plasma column is stable. At a pressure greater than β_c, the question of stability depends on the actual profile since peaking of the current profile has a stabilizing effect [70]. Figure 29 shows the dependence of the normalized growth rate on $nq(a_k)$ for a parabolic current distribution $j = j_0(1 - a^2/a_k^2)$ and for $\varepsilon \beta_J$ equals zero. Comparison of Figs. 27 and 29 shows that peaking of the current profile produces stability gaps analogous to those of a homogeneous current when the conducting wall approaches the plasma boundary. It is clear that for any monotonically decreasing distribution of the current density, an increase in pressure closes the stability gaps, the difference lying in the fact that a conducting wall influences mainly low modes, whereas peaking of the current profile influences high modes.

An analytical investigation of the influence of the ballooning effect on the kink instability in the case of a surface current was

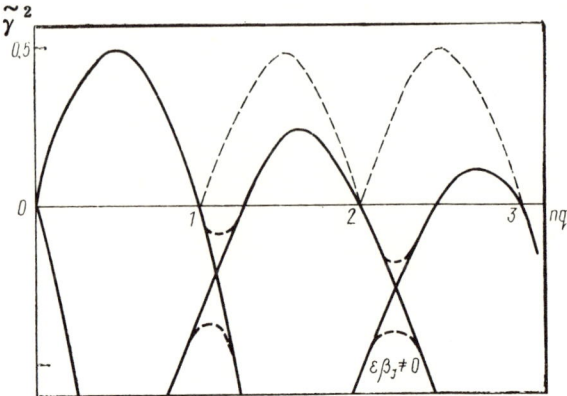

Fig. 29. Oscillation spectrum for a cylindrical column with parabolic current distribution without a conducting wall (continuous curves). Dotted curves show splitting originating from the bending of the cylinder into a torus.

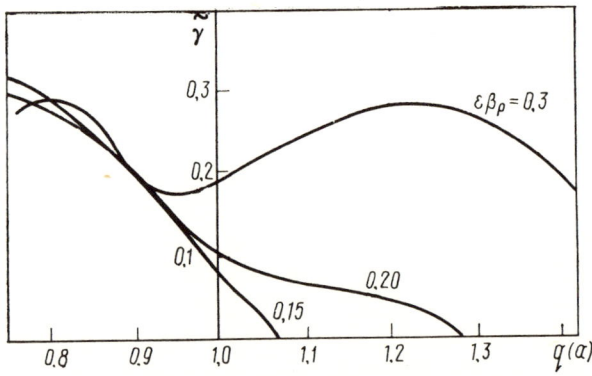

Fig. 30. Normalized growth rate $\tilde{\gamma}$ as a function of $q(a)$ of the n = 1 mode for a skin current distribution with $b/a = 1.6$; 1) $\varepsilon\beta_p = 0.3$; 2) $\varepsilon\beta_p = 0.2$; 3) $\varepsilon\beta_p = 0.15$.

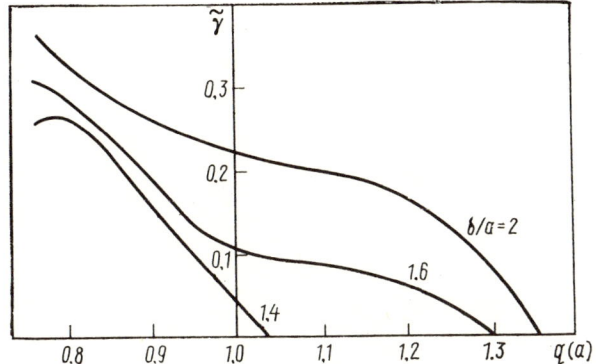

Fig. 31. Normalized growth rate $\tilde{\gamma}$ as a function of $q(a)$ for $\varepsilon \beta_p = 0.2$. 1) $b/a = 2$; 2) $b/a = 1.6$; 3) $b/a = 1.4$.

carried out in [69]. Figure 30 shows the normalized growth rate as a function of $q(a_k)$ for the $n = 1$ mode for different values of the parameter $\varepsilon \beta_J$. It is seen from this figure that at a given position of the conducting wall $b/a_k = 1.6$ an increase in pressure leads to an increase of the growth rate, and at $\varepsilon \beta_J = 0.3$ the stability region disappears altogether. For $b/a < 1.4$ the $n = 1$ mode becomes stable when $q(a_k) < 1$. When the conducting wall is further away, an increase of the safety factor is required for stability.

Though the stability of arbitrary distributions of current density requires specific calculations, qualitatively this effect is easy to comprehend in the framework of the models mentioned above. The value of β_c as a function of the current profile for low mode numbers will be considered numerically in the next section.

3.2. Numerical Calculations for Low Mode Numbers

Calculations of the plasma stability in a tokamak for low mode number oscillations with arbitrary boundary are, in fact, the study of the influence of the ballooning effects on the kink instability, i.e., the study of the ballooning modes of the kink instability. The physi-

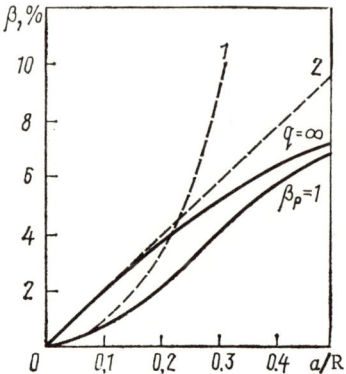

Fig. 32. Critical β as a function of toroidicity for circular surfaces in two cases $q = \infty$ and $\beta_p = 1$. Curves 1 and 2 show the corresponding analytical calculations.

cal nature of these modes has been elucidated in the last section; now we shall consider limitations on the critical pressure originating in tokamaks with arbitrary cross section for specific equilibrium parameter distributions.

We shall start with a model which is simplest for numerical calculations, when the current flows in a skin and the plasma pressure is constant across the column cross section [70].

The equilibrium equation has the following form in this case:

$$\frac{B_p^2(\theta)}{B_0^2} = \frac{8\pi p}{B_0^2} - \frac{(1 - B_i^2/B_0^2)}{(1 + \varepsilon g \cos \theta)^2}, \qquad (3.21)$$

where $B_p(\theta)$ is the poloidal field at the plasma surface, B_i and B_0 are the toroidal fields at the internal and external plasma surfaces at $\theta = \pi/2$, and $g(\theta)$ is a function describing the cross-sectional form of the plasma surface $\rho = R + ag(\theta)$. There are two characteristic regions in parameter space corresponding to low and high plasma pressures. At low pressure $B_i \simeq B_0$ and the pressure is

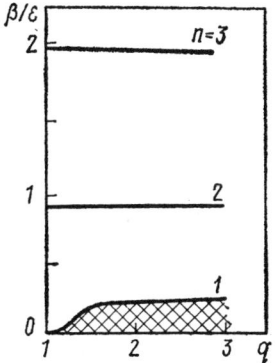

Fig. 33. Parameter β/ε as a function of q for a skin current distribution with circular surfaces for three modes (n = 1, 2, 3). The region of instability for the n = 1 mode is crosshatched.

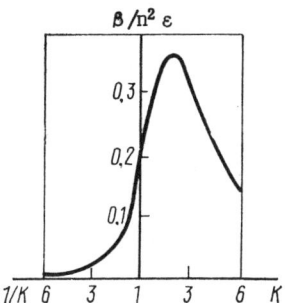

Fig. 34. Parameter $\beta/n^2\varepsilon$ as a function of ellipticity for the n = 1 mode with a skin current distribution.

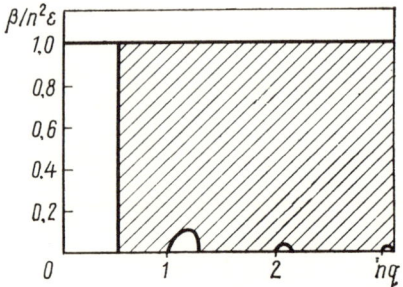

Fig. 35. Parameter $\beta/n^2\varepsilon$ as a function of nq for a quasihomogeneous current with $b/a = 1.4$. The instability region is cross-hatched.

counterbalanced by the poloidal field B_p ($\beta_p \simeq 1$). At high pressure $8\pi p/B_0^2 \simeq (1 - B_i^2/B_0^2)/(1 - \varepsilon)^2$ and with increasing pressure ($\beta_p \simeq R/a$) the separatrix approaches the plasma surface ($q \to \infty$). Figure 32 shows critical β as a function of toroidicity for circular magnetic surfaces in two cases $\beta_p = 1$ and $q = \infty$. For comparison dashed lines are used to show analytical dependences obtained by expansion in ε [71]. The first curve is derived on the assumption $\beta_p \sim 1$, $\beta \sim \varepsilon^2$, and the second on the assumption $\beta_p \sim 1/\varepsilon$, $\beta \sim \varepsilon$. It is seen that numerical calculation yields lower critical β. Calculations were performed for the first three harmonics, the n = 1 mode being the most unstable, which follows from Fig. 33, where the β threshold is shown as a function of the safety factor. For each n the region of stability lies below the curve. At low pressure the n = 1 mode is unstable when q is below the Kruskal–Shafranov value (q = 1) and stable above this value. When the pressure increases, a destabilizing ballooning effect manifests itself, and for $\beta > 0.21\varepsilon$ the n = 1 kink mode is unstable for any value of q.

The stability of the n = 1 mode was also calculated for elliptical magnetic surfaces at high plasma pressure ($q \to \infty$). Figure 34 shows that flattening the cross section in the horizontal direction worsens stability, whereas extending the cross section in the vertical direction (K > 1) yields an optimal value of K = 2.2 at

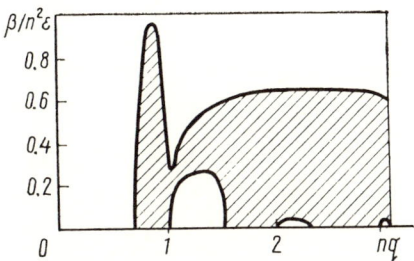

Fig. 36. Parameter $\beta/n^2\varepsilon$ as a function of nq for $b/a = 1.2$.

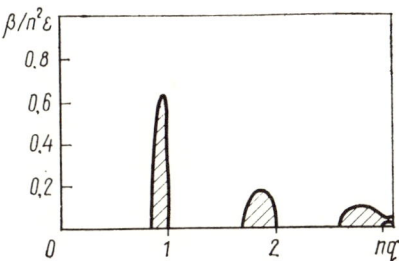

Fig. 37. Parameter $\beta/n^2\varepsilon$ as a function of nq for $b/a = 1.1$.

which $\beta = 0.37\varepsilon$ (in the circle $\beta = 0.21\varepsilon$). A more realistic model with a quasihomogeneous longitudinal current and a conducting wall was treated in [70]. Figure 35 depicts the stability diagram in the plane β, nq for the case when the conducting wall is quite far away from the plasma, $b/a_k = 1.4$. At low plasma pressure the ballooning effect proportional to $\varepsilon\beta_J = \varepsilon^3\beta/q^2$ has little influence on stability, and stability gaps are determined from the equation obtained for the cylindrical column [68]:

$$m - 1 + (a_k/b)^{2m} < nq < m. \qquad (3.22)$$

This result also follows from (3.18) as $\varepsilon \to 0$. When the plasma pressure increases, the stability gaps close, in full agreement with analytical calculations [see former section (3.19)]. When the con-

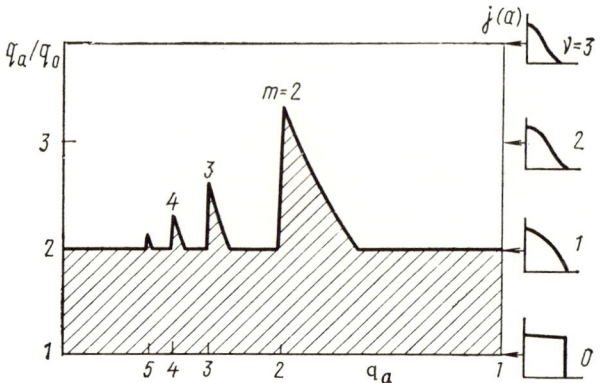

Fig. 38. Instability diagram for low mode numbers $[j(a) = j_0(1 - a^2/a_k^2)^\nu]$.

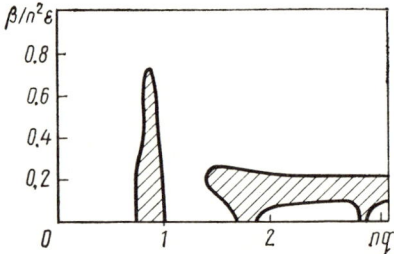

Fig. 39. Parameter $\beta/n^2\varepsilon$ as a function of nq for a bell-shaped current profile ($\nu = 2$) for $b/a = 1.2$. The instability region is cross-hatched.

ducting wall approaches the plasma boundary $b/a_k = 1.2$, the stability gaps widen, and a second stability region appears at high β (see Fig. 36). For even closer approach of the conducting wall to the boundary $b/a_k = 1.1$ (see Fig. 37) the second stability region coalesces with the stability gaps, and it is possible to speak about "gaps" of instability, which disappear when the plasma boundary is fixed.

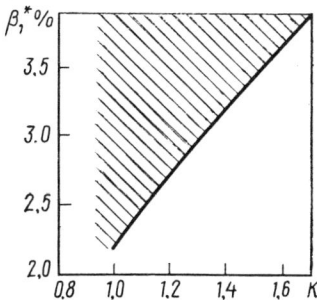

Fig. 40. Critical $\beta*$ for the n = 1 mode as a function of ellipticity for the configuration with A = 3, $\delta = 0.25$, $l = 2$.

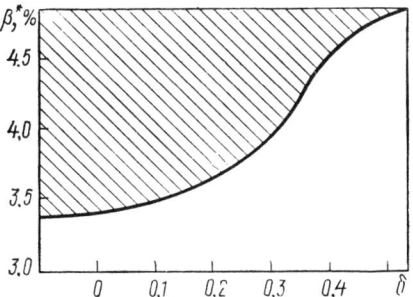

Fig. 41. Critical $\beta *$ as a function of the triangularity for the n = 1 mode for the configuration with A = 3, K = 1.65, $l = 2$, $1 < q < 3$.

The effect of the current profile on the stability of low modes in a tokamak at fixed pressure, calculated in detail in [72], is shown as a stability diagram in the plane q_a/q_0, q_a (Fig. 38) for the modes with n = 1 and m = 2, 3, 4. The current profile was of the form $j = j_0[1 - (a^2/a_k^2)]^\nu$ ($\nu = 0, 1, \ldots$), the conducting wall being absent.

The diagram shows that a peaking of the current profile leads to stabilization of the kink modes.

A comparison of the diagram in Fig. 39 for a bell-shaped current profile with the diagram in Fig. 36 for a quasihomogeneous current for equal distance of the conducting wall from the plasma $b/a_k = 1.2$ permits us to evaluate the effect of a peaking of the current profile on the value of critical β. It is seen from the diagram that for a bell-shaped current profile the second region of stability has merged with the stability gap of the first mode, and the value of the second critical $\beta/\varepsilon n^2$ at which all the modes are stable has dropped from 0.65 to 0.28.

Critical β for the n = 1 mode as a function of the cross-sectional form of the plasma column was calculated in [64] for bell-shaped current profiles and is shown in Figs. 40 and 41. Figure 40 characterizes a purely elliptical plasma column, and Fig. 41 characterizes a plasma column with a D-shaped form of cross section. It is seen to be possible to raise the critical β from 2.1% for a circular cross section up to 3.5% with moderate ellipticity (K = 1.6) and up to 4.9% for a D-shaped form (K = 1.6; δ = 0.5).

CHAPTER 4

INVESTIGATION OF THE DISSIPATIVE PLASMA STABILITY

4.1. Initial Equations and Their Simplification

The plasma stability in the presence of dissipation due to the finite conductivity will be investigated proceeding from the equations of single-fluid magnetohydrodynamics when compressibility is taken into account:

$$\rho \frac{dv}{dt} = -\nabla p + \frac{1}{c}[\mathbf{jB}]; \qquad (4.1)$$

$$\frac{dp}{dt} + \gamma_0 p \operatorname{div} \mathbf{v} = 0, \qquad (4.2)$$

$$\mathbf{j} = \sigma \left(\mathbf{E} + \frac{1}{c}[\mathbf{v}\mathbf{B}] \right), \qquad (4.3)$$

where γ_0 is an adiabatic exponent, and Maxwell equations:

$$\operatorname{curl} \mathbf{B} = \frac{4\pi}{c} \mathbf{j}; \qquad (4.4)$$

$$\operatorname{curl} \mathbf{E} = -\frac{1}{c} \frac{\partial \mathbf{B}}{\partial t}; \qquad (4.5)$$

$$\operatorname{div} \mathbf{B} = 0. \qquad (4.6)$$

Unlike the case of a perfectly conducting plasma ($\sigma = \infty$), when, due to plasma freezing into the magnetic field, only nonpotential perturbations are possible, in plasma with finite conductivity both nonpotential and potential perturbations can develop. Therefore, it is more convenient to introduce the vector potential $\mathbf{A}$ and scalar potential φ:

$$\mathbf{B} = \operatorname{curl} \mathbf{A}; \qquad (4.7)$$

$$\mathbf{E} = -\nabla\varphi - \frac{1}{c} \frac{\partial \mathbf{A}}{\partial t}. \qquad (4.8)$$

The system of equations (4.1)-(4.8) can be simplified by expansion in the field ratio B_J/B_0 [53]. For this purpose, as in the case of an ideal plasma (see Section 1.2), we represent the current density by two components $\mathbf{j} = \mathbf{j}_\perp + \hat{\alpha}\mathbf{B}$. An expression for the transverse current component follows from the equation of motion (4.1):

$$\mathbf{j}_\perp = \frac{c}{B^2} \left([\mathbf{B}\nabla p] + \rho \left[\mathbf{B} \frac{d\mathbf{v}}{dt} \right] \right). \qquad (4.9)$$

The transverse velocity component entering in (4.9) will be determined by Ohm's law (4.3):

$$\mathbf{v}_\perp = -\frac{c[\mathbf{B}\mathbf{E}]}{B^2} - \frac{c^2}{\sigma B^2} \left(\nabla p + \rho \frac{d\mathbf{v}_\perp}{dt} \right). \qquad (4.10)$$

High-frequency magnetoacoustic oscillations in a tokamak are considerably reduced by the strong longitudinal magnetic field, Alfvén and ion-acoustic oscillations being the most dangerous. When considering them, we can neglect transverse inertia and plasma diffusion across the magnetic field, since in high-temperature plasma the resistive skin time $\tau_S = 4\pi\sigma a^2/c^2$ is much greater than the Alfvén time $\tau_A = a/c_A$; i.e., we retain only the first term in expression (4.10).

From the condition of current continuity div $\mathbf{j} = 0$, using (4.9) and (4.10) we obtain a convenient equation for the stability investigation

$$\mathbf{B}\nabla\hat{\alpha} - c\,[\nabla p \mathbf{B}]\nabla \frac{1}{B^2} + \frac{c^2}{B^2}\operatorname{div}\rho\frac{\partial \mathbf{E}}{\partial t} = 0. \tag{4.11}$$

Assuming that lines of force only bend without compressing or rarefying, we shall describe this bending by only one component $\tilde{A}_S$:

$$\tilde{B}^1 = \frac{1}{\sqrt{g}}\frac{\partial \tilde{A}_S}{\partial \theta}, \quad \tilde{B}^\theta = -\frac{1}{\sqrt{g}}\frac{\partial \tilde{A}_S}{\partial a}. \tag{4.12}$$

We shall neglect perturbations of the longitudinal component of the magnetic field since they are of the order of β as compared to transverse components. The linearized equation (4.11) has the following form in this case:

$$\mathbf{B}^0\nabla\tilde{\alpha} + \tilde{\mathbf{B}}\nabla\alpha_0 + c\left[\nabla\frac{1}{B_0^2}\mathbf{B}^0\right]\nabla\tilde{p} - \gamma\frac{c^2}{B_0^2}\operatorname{div}\rho_0\nabla\tilde{\varphi} = 0. \tag{4.13}$$

Here

$$\tilde{\alpha} = -\frac{c}{4\pi}\frac{1}{B_0^S}\tilde{\Delta}_\perp A_S. \tag{4.14}$$

The operator $\hat{\Delta}_\perp$ was introduced during the investigation of an ideal plasma (see 1.35). As will be seen later, we shall not need the explicit form of the metric coefficients in the operator $\hat{\Delta}_\perp$ during the investigation of the dissipative plasma.

The value of $\tilde{\alpha}$ can be found not only from Eq. (4.13) but also from the longitudinal component of Ohm's law (4.3):

$$\tilde{\alpha} = \frac{\sigma B_0^S}{B_0^2}\left(-\frac{\mathbf{B}^0\nabla}{B_0^S}\tilde{\varphi} - \frac{\gamma}{c}\tilde{A}_S\right). \tag{4.15}$$

Equating expressions (4.14) and (4.15), we shall couple $\tilde{A}_S$ and $\tilde{\varphi}$ together:

$$\left(1 - \frac{a^2 \hat{\Lambda}_\perp}{\gamma \tau_S}\right) \widetilde{A}_S = -\frac{c}{\gamma} \frac{\mathbf{B}^0 \nabla}{B_0^S} \widetilde{\varphi}. \tag{4.16}$$

Now we have to express the perturbed pressure through $\widetilde{A}_S$ and $\widetilde{\varphi}$. For this purpose we use Eq. (4.2) after calculating the value div **v** in the zeroth approximation with respect to the curvature:

$$\text{div } \mathbf{v} = -\frac{1}{\gamma p_0} \hat{k}_\| \left(\hat{k}_\| \widetilde{p} + \frac{p_0'}{\sqrt{g} B_0^S} \frac{\partial \widetilde{A}_S}{\partial \theta}\right). \tag{4.17}$$

In deriving this expression we took into account that div $\mathbf{V}_\perp = 0$ in the approximation under consideration, and that the longitudinal velocity $\mathbf{v}_\|$ was determined from the longitudinal component of the equation of motion (4.1). Inserting (4.17) in the linearized equation (4.2) we get

$$\left(1 - \frac{c_S^2}{\gamma^2} \hat{k}_\|^2\right) \widetilde{p} = \frac{p_0'}{\sqrt{g} B_0^S} \left(\frac{c}{\gamma} \frac{\partial \widetilde{\varphi}}{\partial \theta} + \frac{c_S^2}{\gamma^2} \hat{k}_\| \frac{\partial \widetilde{A}_S}{\partial \theta}\right), \tag{4.18}$$

where $c_S^2 = \gamma_0 p_0 / \rho_0$.

To have a clearer idea of the physical nature of the oscillations we shall consider the case of small-scale flute perturbations. Solving Eq. (4.16) for $\widetilde{A}_S$, it is possible to obtain symbolically from Eqs. (4.13) and (4.18):

$$\frac{c^2 \hat{k}_\|}{4\pi} \hat{\Lambda}_\perp \frac{\hat{k}_\|}{D} \widetilde{\varphi} + c \left[\nabla \frac{1}{B_0^2} \mathbf{B}^0\right] \nabla \widetilde{p} - \gamma \frac{c^2}{B_0^2} \text{div } \rho_0 \nabla \widetilde{\varphi} = 0; \tag{4.19}$$

$$\left(1 - \frac{c_S^2}{\gamma^2} \hat{k}_\|^2\right) \widetilde{p} = \frac{p_0'}{\sqrt{g} B_0^S} \frac{c}{\gamma} \frac{\partial}{\partial \theta} \left[1 - \frac{c_S^2 \hat{k}_\|}{\gamma^2} \frac{\hat{k}_\|}{D}\right] \widetilde{\varphi}, \tag{4.20}$$

where

$$D = 1 - a^2 \hat{\Lambda}_\perp / \gamma \tau_S, \quad \hat{k}_\| = \mathbf{B}^0 \nabla / B_0^S.$$

It is seen from these equations that for a cylindrical column [the second term in (4.19) is equal to zero], they describe Alfvén and ion-acoustic oscillations independently. In a tokamak these branches of oscillations interact with each other.

This system of the three equations (4.13), (4.16), and (4.18) is considerably simpler than the initial system (4.1)-(4.8), but it is still rather complex since it consists of equations in partial derivatives. For further simplifications we shall use the method of equivalent harmonics, described in Section 2.1. For large azimuthal mode numbers ($m \simeq nq \gg 1$) we shall express all perturbed values in the form

$$\tilde{\varphi}(a, \theta, s) = \sum_m \exp\left[im\theta + i\frac{n}{R}z\right] \int_{-\infty}^{\infty} \varphi(y) \exp\{-i[m-nq(a)y]\}\,dy \quad (4.21)$$

and obtain a system of two ordinary differential equations for the Fourier transforms $\varphi(y)$ and $p(y)$ [53]:

$$\frac{d}{dy}\frac{G(a,y)}{(1+G/\Gamma)}\frac{d\varphi}{dy} - \gamma^2\tau_\theta^2(1+S^2y^2)\varphi + \frac{1}{2}\alpha R B_0^S$$
$$\times \left(\frac{\partial}{\partial a}\frac{1}{B_0^S} - \frac{Sy}{a}\frac{\partial}{\partial y}\frac{1}{B_0^S}\right)\frac{\gamma}{inq}\frac{\sqrt{g}}{c}\frac{B_0^S}{p_0'}p(y) = 0; \quad (4.22)$$

$$\left(1 - \frac{1}{\gamma^2\tau_c^2}\frac{d^2}{dy^2}\right)p = \frac{inq}{\gamma}\frac{c}{\sqrt{g}}\frac{p_0'}{B_0^S}\left[1 - \frac{1}{\gamma^2\tau_c^2}\frac{d}{dy}\frac{1}{(1+G/\Gamma)}\frac{d}{dy}\right]\varphi, \quad (4.23)$$

where $S = q'a/q$, $\alpha = -8\pi p_0' Rq^2/B_0^2$, $\tau_c = Rq/c_S$,

$\tau_\theta = Rq/c_A$, $\Gamma = \gamma\tau_S/n^2q^2$, $G = a\left(\dfrac{g_{11}}{\sqrt{g}} + \dfrac{g_{22}}{\sqrt{g}}\dfrac{S^2y^2}{a^2} - 2\dfrac{g_{12}}{\sqrt{g}}\dfrac{Sy}{a}\right)$.

The system of Eqs. (4.21) and (4.22) permit us to investigate ideal and dissipative ballooning modes of the flute instability in a tokamak. In the case of a perfectly conducting plasma ($\tau_S \to \infty$ and $\Gamma \gg G$) this system is reduced to one equation for φ [see Eq. (2.9)]. The solution of this equation and correspondingly the condition for plasma stability in the framework of ideal magnetohydrodynamics, as is seen from Sections 2.3 and 2.4, depends considerably on the cross-sectional form of the magnetic surfaces.

In the case of an imperfectly conducting plasma ($\Gamma \ll G$) the system of Eqs. (4.22) and (4.23) assumes a simpler form:

$$\Gamma\frac{d^2\varphi}{dy^2} - \gamma^2\tau_\theta^2(1+S^2y^2)\varphi +$$

$$+ \frac{1}{2} R B_0^S \left(\frac{\partial}{\partial a} \frac{1}{B_0^S} - \frac{Sy}{a} \frac{\partial}{\partial y} \frac{1}{B_0^S} \right) \hat{p} = 0; \qquad (4.24)$$

$$\left(1 - \frac{1}{\gamma^2 \tau_c^2} \frac{d^2}{dy^2}\right) \hat{p} = \left(1 - \frac{\Gamma}{\gamma^2 \tau_c^2} \frac{d}{dy} \frac{1}{G} \frac{d}{dy}\right) \varphi, \qquad (4.25)$$

where

$$\hat{p} = \frac{\gamma}{inq} \frac{\sqrt{g} B_0^S}{p_0'} p(y), \quad G \simeq 1 + S^2 y^2.$$

These equations depend to a much lesser degree on specific geometry. It can be shown that they have two different scales, viz., $y \sim 1$ and $y \gg 1$, which allows us to use the Van der Pol averaging method.

For circular magnetic surfaces the averaged equations are written as follows:

$$\Gamma \frac{d^2 \bar{\varphi}}{dy^2} - \frac{\Gamma^2}{N^2} (1 - S^2 y^2) \bar{\varphi} - \alpha V_0 \bar{p}$$

$$+ \frac{\alpha^2}{2} \frac{(1 + S^2 y^2 + M^2/\Gamma)}{\Gamma (1 + M^2/\Gamma^2) [1 + \Gamma (1 + S^2 y^2)/N^2]} \bar{p} = 0; \qquad (4.26)$$

$$\left(1 - \frac{M^2}{\Gamma^2} \frac{d^2}{dy^2}\right) \bar{p} = \left(1 - \frac{M^2}{\Gamma} \frac{d}{dy} \frac{1}{1 + S^2 y^2} \frac{d}{dy}\right) \bar{\varphi}. \qquad (4.27)$$

Here $M = \frac{\tau_S}{\tau_c} \frac{1}{n^2 q^2}$, $N = \frac{\tau_S}{\tau_\theta} \frac{1}{n^2 q^2}$ are parameters characterizing the ratio of the resistive skin time to the sound and Alfvén times, respectively; $M^2 = \gamma_0 \beta N^2$; $V_0 = \left[\frac{R}{2} \frac{\partial}{\partial a} (g_{33})_0 - \frac{\varepsilon}{q^2}\right]$ characterizes the depth of the magnetic well in a tokamak.

A characteristic feature of these equations, arising from the finite conductivity, is that they are not conjugated. This means that the growth rate Γ is not necessarily real; i.e., oscillatory perturbations can appear in the spectrum [73]. The physical nature of this nonconjugacy is clear; oscillatory instabilities associated with ion sound can develop alongside the purely growing Alfvén instability.

Ballooning modes in a tokamak for $nq \gg 1$ represent perturbations which are localized along the column radius and have a correspondingly large characteristic y in Fourier space. For $y^2 \gg M^2/\Gamma^2$, or for $\gamma^2 \gg k_\parallel^2 c_S^2$ in standard notation, Eqs. (4.26) and (4.27) are reduced to one equation:

$$\Gamma\bar{\varphi}'' - \left\{ \frac{\Gamma^2}{N^2}(1 + S^2 y^2) + \alpha V_0 \right.$$

$$\left. - \frac{\alpha^2}{2} \frac{(1 + S^2 y^2 + M^2/\Gamma)}{\Gamma(1 + M^2/\Gamma^2)[1 + \Gamma(1 + S^2 y^2)/N^2]} \right\} \bar{\varphi} = 0. \quad (4.28)$$

The condition $\gamma^2 \gg k_\parallel^2 c_S^2$ means that the average perturbation of the electrostatic potential has no time to equalize along the lines of force due to finite sound velocity.

Thus, the problem of plasma stability to dissipative ballooning modes is reduced to the solution of Eq. (4.28).

4.2. Threshold-Free Dissipative Ballooning Modes

Now we turn to the analysis and solution of Eq. (4.28):

$$\bar{\varphi}'' - V(\Gamma, y) \bar{\varphi} = 0. \quad (4.29)$$

The potential in this equation has a complicated dependence upon the growth rate Γ:

$$V = \frac{\alpha V_0}{\Gamma} + \frac{\Gamma}{N^2}(1 + S^2 y^2) - \frac{\alpha^2(1 + S^2 y^2 + M^2/\Gamma)}{2\Gamma^2(1 + M^2/\Gamma^2)[1 + \Gamma(1 + S^2 y^2/N^2)]}. \quad (4.30)$$

All the terms in this expression have apparent physical sense: the first term characterizes the stabilizing effect of the magnetic well, the second term takes into account plasma inertia, and the third term describes the destabilizing ballooning effect. The condition for the occurrence of the first minimum in the potential well (4.30) determines the maximum possible growth rate of dissipative ballooning modes.

It can be shown that the solution of Eq. (4.29) with potential (4.30) depends essentially on the values of the ratios S/S_k ($S_k = \alpha^{1/2}/V_0^{1/2}N$), α/α_k (α_k is determined from the condition $\alpha = 2V_0$), and β/β_k ($\beta_k = \alpha^{4/3}N^{2/3}$). In high-temperature plasma $N \gg 1$.

Shear of the order of unity and much greater than S_k ($S_k \ll S \sim 1$) is of practical interest. First we separately consider two limiting cases of high $\beta \gg \beta_k$ plasma pressure ($\beta \lesssim \beta_k$).

In the case of high plasma pressure when the perturbation of the magnetic field is considerable and oscillations are nonpotential, Eq. (4.29) represents an oscillator equation which permits us to write at once the expression for the growth rate:

$$\Gamma = \begin{cases} \dfrac{N^{2/3}}{\alpha}\left(\dfrac{\alpha^2}{2} - \alpha V_0\right)^{2/3}, & \alpha \geqslant \alpha_h; \\ 0, & \alpha \leqslant \alpha_h. \end{cases} \quad (4.31)$$

This growth rate describes the nonpotential gravitational-dissipative instability [49, 74] which is unstable above a threshold value of the relative pressure gradient $\alpha = -8\pi p_0' R q^2 / B_0^2$.

It is interesting to compare the condition for dissipative plasma stability, which follows from (4.31),

$$\alpha V_0 - \alpha^2/2 > 0, \quad (4.32)$$

with ideal plasma stability criteria: the Mercier criterion

$$\frac{1}{4}S^2 + \alpha V_0 - \frac{\alpha^2}{2} > 0 \quad (4.33)$$

and the ballooning mode criterion

$$\frac{1}{2}S^2 - \frac{1}{2}S\alpha^2 - \frac{3}{2}\alpha \exp\left(-\frac{1}{S}\right) + \alpha V_0 - \frac{\alpha^2}{2} > 0. \quad (4.34)$$

It follows from these expressions that the condition for dissipative plasma stability (4.32), which does not include shear, is always more rigorous than the Mercier criterion, but it can be less rigorous than the ballooning mode criterion. Hence, it is clear that nonpotential gravitational-dissipative instability is not the most dangerous dissipative instability.

To check the analysis we shall consider the case of low plasma pressure $\beta < \beta_k$ and $\alpha < \alpha_k$ when the oscillations are purely potential. In this case the potential well (4.30) has a minimum which is shifted relative to the origin of coordinates. The approximation of this well near its minimum for small potential changes permits us to obtain the following equation for the growth rate [53]:

$$\Gamma^3 + \gamma_0 \beta N^2 \Gamma = \tfrac{1}{2}\alpha^2 N^2. \qquad (4.35)$$

This equation has one real root:

$$\Gamma_1 = u + v, \qquad (4.36)$$

here

$$u = \left(\frac{1}{4}\alpha^2 N^2 + \sqrt{\frac{1}{16}\alpha^4 N^4 + \frac{\gamma_0^3}{27}\beta^3 N^3}\right)^{1/3},$$

$$v = \left(\frac{1}{4}\alpha^2 N^2 - \sqrt{\frac{1}{16}\alpha^4 N^4 + \frac{\gamma_0^3}{27}\beta^3 N^3}\right)^{1/3},$$

and two complex conjugate roots:

$$\Gamma_{2,3} = -\frac{1}{2}\Gamma_1 \pm i\frac{\sqrt{3}}{2}(u-v). \qquad (4.37)$$

The first root describes threshold-free dissipative ballooning modes which are Alfvén perturbations unstable for any small pressure gradient with a growth rate

$$\gamma \sim \frac{1}{\tau_S}\left(\frac{\tau_S}{\tau_\theta}\right)^{2/3}. \qquad (4.38)$$

This growth rate is much greater than the inverse skin time, since in high-temperature plasma $\tau_S/\tau_\theta \gg 1$. As the plasma pressure increases, the growth rate of the ballooning modes decreases. The other two roots correspond to the branches of ion-acoustic modes. Increasing the pressure causes an increase in the frequency of these modes and a decrease in the damping.

In order to get a result valid for any value of the parameters β and α, i.e., to trace the change from potential to nonpotential perturbations, we solve Eq. (4.29) by the variational method. For this purpose we write the functional corresponding to Eq. (4.29):

$$\mathcal{L} = \int_{-\infty}^{\infty} [\varphi'^2 + V(\Gamma, y)\varphi^2]dy, \qquad (4.39)$$

and insert in it as the trial function a function of the form

$$\varphi = 1/(y^2 + \lambda^2), \qquad (4.40)$$

where λ is a variational parameter.

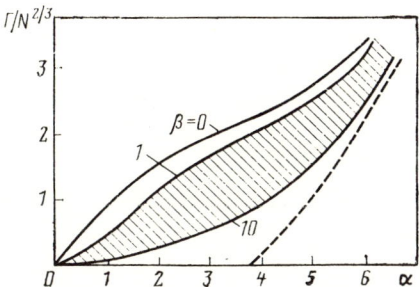

Fig. 42. Normalized threshold-free growth rate $\Gamma/N^{2/3}$ as a function of α for different values of $\tilde{\beta} = \beta\gamma_0 N^{2/3}$: 1) $\tilde{\beta} = 0$; 2) $\tilde{\beta} = 1$; 3) $\tilde{\beta} = 10$. The region corresponding to a high-temperature plasma is cross-hatched. The dotted line shows the growth rate of the nonpotential gravitational-dissipative instability.

Upon accomplishing the necessary integration and variation, from the conditions $\mathscr{L} = 0$ and $d\mathscr{L}/d\lambda = 0$ it is not difficult to obtain a system of equations for the determination of the growth rate and the parameter λ:

$$(\Gamma^3 + \gamma_0 \beta N^2 \Gamma)\left[1 - \frac{N^2(\alpha^2 - 2\alpha V_0)}{4S^2 \Gamma^2 \lambda^2}\right] = \frac{1}{2}\alpha^2 N^2. \qquad (4.41)$$

$$\frac{N^2}{2\Gamma^2 \lambda^2}(\alpha^2 - 2\alpha V_0) + \frac{S^3 \Gamma^{1/2} \lambda}{N} = \frac{N^2}{\Gamma \lambda^4}. \qquad (4.42)$$

From these expressions one can get both limiting cases considered above.

Figure 42 gives the growth rate of the dissipative ballooning modes as a function of the plasma pressure gradient α calculated according to formulas (4.41) and (4.42) for different values of β. Curve 1 corresponds to β equal to zero or to a fully compressible liquid $\gamma_0 = 0$, curve 2 corresponds to $\tilde{\beta} = \beta\gamma_0 N^{2/3} = 1$, and curve 3 to $\tilde{\beta} = 10$. The region corresponding to a high-temperature plasma lies between curves 2 and 3. The dashed curve showing a threshold represents the growth rate of a nonpotential gravitational-dissipative instability. In obtaining this growth rate, we assumed that the sound velocity is infinite. The ion sound in this case equalized perturbations of the electrostatic potential and pressure along the lines of force and for $\alpha < \alpha_k$ the instability did not develop. The picture is different when the sound velocity is taken into account: perturbations of potential and pressure do not have time to equalize along the lines of force and the ballooning instability, which develops due to Alfvén perturbations, can exist at any pressure gradient. The growth rate of these dissipative threshold-free ballooning modes decreases with increasing plasma pressure, as seen in Fig. 42.

CONCLUSIONS

The present survey has treated the influence of the ballooning effects on plasma stability in a tokamak within the framework of the linear theory.

Analytical and numerical calculations permit us to comprehend the basic mechanisms leading to ballooning mode instability. The critical pressure to destabilize ballooning modes of both flute and kink instabilities turns out to be rather small (for a tokamak with circular cross section of magnetic surfaces $\beta_c = 2\text{-}3\%$). If experiments prove these instabilities to be dangerous, the theory shows that critical β can be somewhat raised by choosing the cross-sectional form of the magnetic surfaces (D-shaped form) and by profiling the density across the column cross section. Thus, it is quite easy to obtain $\beta_c \simeq 5\text{-}6\%$.

When dissipation is taken into account, ballooning modes are unstable at arbitrary pressure. However, in this case the growth rate and localization region turn out to be considerably smaller than in the ideal case.

It is difficult to say how dangerous ballooning modes can be experimentally without studying the nonlinear stage of instability development.

Only from this nonlinear investigation (as well as, naturally, from the experiment) can one obtain information on plasma losses and macroscopic effects (diffusion, thermal conductivity) which can result from the development of instability. So far only preliminary results have been obtained from the nonlinear theory. Therefore, we will mainly confine ourselves to a qualitative discussion of possible effects.

We shall start with the dissipative instability. It follows from the theory that its growth rate and localization are much less than the corresponding values for the ideal case. Therefore, one can assume that it will manifest itself more weakly.

If one uses dimensional estimates for the transport coefficients $\chi \sim D \sim \gamma/k_\perp^2$, then using the results of Section 4.2, it is not difficult to obtain $\chi \sim a^2\alpha/\tau_S \sim c^2\alpha/4\pi\sigma$ for the dissipative instability. This is similar to the pseudoclassical expression and approximates what is observed experimentally.

It is more difficult to determine what will occur on passing the critical β for ideal modes. Some information on this question can be obtained from recent numerical calculations [78] for large-scale tearing modes in a torus. It was shown in these calculations that overlapping of magnetic islands corresponding to different modes and stochastization of magnetic lines of force takes place in the nonlinear stage. These effects occur both in the cylinder and in the torus. In the torus, however, instability develops somewhat more rapidly (ballooning effect!) and leads to a more complete stochastization of magnetic lines of force. For large-scale modes stochastization can be of secondary importance since the column touches the wall due to the development of the disruptive instability much earlier than phenomena connected with stochastization manifest themselves. The situation can be different for the ballooning modes of the flute instability. Considering the fact that small-scale ballooning modes develop first, one can assume that such dramatic phenomena as disruption are not likely to occur and stochastization of lines of force can then be the main effect. This phenomenon, as was repeatedly shown [79, 81], leads in the first place to increased electronic thermal conductivity (diffusion is not excluded either), which can be evaluated by the Ohkawa type formula: $\chi \sim (c^2/\omega_p^2)(v_e/qR)$.

Transport coefficients may be greater than this value if they are affected by plasma convection as well as stochastization.

If the processes actually develop in this way then the energy losses from the column will rise sharply as β exceeds β_c and the plasma pressure will stabilize somewhere near $\beta \sim \beta_c$.

REFERENCES

1. International Tokamak Reactor: Zero Phase. IAEA, Vienna (1979).
2. B. Suydam, Proc. 2 U.N. Int. Conf., PVAE Geneva, 31, 157 (1958).
3. C. Mercier, Nucl. Fusion, 1, 47 (1960).
4. H. D. Furth, I. Killeen, M. Rosenbluth, and B. Coppi, Conference on Plasma Physics and Controlled Nuclear Fusion Research, Vienna, Vol. 1 (1966), p. 127.
5. R. M. Kulsrud, ibid., 1, 103 (1966).
6. L. S. Solov'ev and V. D. Shafranov, ibid., 1, 169 (1966).
7. B. B. Kadomtsev and O. P. Pogutse, Dokl. Akad. Nauk SSSR, 170, 811 (1966).
8. V. D. Shafranov and E. I. Yurchenko, Zh. Eksp. Teor. Fiz., 53, 1157 (1967).
9. A. A. Ware and F. A. Haas, Phys. Fluids, 9, 956 (1966).
10. V. D. Shafranov and E. I. Yurchenko, Plasma Physics and Controlled Nuclear Fusion Research, Vienna, Vol. 2 (1971), p. 519.
11. A. B. Mikhailovskii, Nucl. Fusion, 14, 483 (1974).
12. J. M. Green and J. L. Johnson, Phys. Fluids, 5, 510 (1962).
13. M. Bineau, Conference on Plasma Physics and Controlled Nuclear Fusion Research, Salzburg (1961), p. 35.
14. L. S. Solov'ev, Zh. Eksp. Teor. Fiz., 53, 626 (1967).
15. V. D. Shafranov and E. I. Yurchenko, Nucl. Fusion, 8, 329 (1968).
16. C. Mercier, Nucl. Fusion, 4, 213 (1964).
17. A. B. Mikhailovskii, Zh. Eksp. Teor. Fiz., 64, 536 (1973).
18. A. B. Mikhailovskii and V. D. Shafranov, Zh. Eksp. Teor. Fiz., 66, 190 (1974).
19. J. W. Connor, R. J. Hastie, and J. B. Taylor, Phys. Rev. Lett., 40, 396 (1978).
20. C. Mercier, Plasma Physics and Controlled Nuclear Fusion Research, Innsbruck, Vol. 1 (1979), p. 701.

21. O. P. Pogutse and E. I. Yurchenko, Fiz. Plazmy, 5, 786 (1979).
22. L. S. Solov'ev, V. D. Shafranov, and E. I. Yurchenko, Plasma Physics and Controlled Nuclear Fusion Research, Vienna (1969), p. 197.
23. G. Küppers and M. Tasso, Z. Naturforsch., 27a, No. 1, 23 (1972).
24. Ya. N. Yavlinskiy, Nucl. Fusion, 13, 951 (1973).
25. M. Okamoto, M. Wakatani, and T. Amano, Nucl. Fusion, 15, 225 (1975).
26. R. M. O. Galvao, Nucl. Fusion, 15, 785 (1975).
27. B. Coppi, et al., Phys. Fluids, 15, 2405 (1972).
28. G. Laval, E. K. Maschke, and R. Pellat, Phys. Rev. Lett., 24, 1229 (1970).
29. A. M. M. Todd, M. S. Chance, et al., Phys. Rev. Lett., 38, 826 (1977).
30. G. Bateman and Y.-K. Peng, Phys. Rev. Lett., 38, 829 (1977).
31. B. Coppi, Phys. Rev. Lett., 39, 339 (1977).
32. D. Dobrott, D. B. Nelson, et al., Phys. Rev. Lett., 39, 943 (1977).
33. O. P. Pogutse and E. I. Yurchenko, Pis'ma Zh. Eksp. Teor. Fiz., 28, 344 (1978).
34. M. S. Chance, R. L. Dewar, et al., Plasma Physics and Controlled Nuclear Fusion Research, Innsbruck, Vol. 1 (1979), p. 677.
35. J. D. Jukes, ibid., 733 (1979).
36. A. Sykes, M. F. Turner, et al., ibid., 625 (1979).
37. J. W. Connor, R. J. Hastie, and J. B. Taylor, ibid., 674 (1979).
38. L. E. Zakharov, ibid., 689 (1979).
39. R. A. Dory, D. P. Berger, et al., ibid., 579 (1979).
40. F. Pegoroto and T. Schep, ibid., 507 (1979).
41. G. Bateman, Phys. Fluids, 14, 1506 (1971).
42. J. D. Freidberg and F. A. Haas, Phys. Fluids, 16, 1909 (1973).
43. J. P. Freidberg, J. P. Goedbloed, et al., Plasma Physics and Controlled Nuclear Fusion Research, Tokyo, Vol. 1 (1975), p. 505.
44. O. P. Pogutse and E. I. Yurchenko, PPPL-tr-122 (1977).
45. V. D. Shafranov, Zh. Eksp. Teor. Fiz., 40, 241 (1970).
46. A. M. M. Todd, J. Manickam, et al., PPPL-1470 (1978).
47. P. Furth, J. Killeen, and M. Rosenbluth, Phys. Fluids, 6, 459 (1963).

48. J. L. Johnson and J. M. Greene, Plasma Phys., 9, 611 (1967).
49. A. H. Glasser, J. M. Greene, and J. L. Johnson, Phys. Fluids, 18, 875 (1975).
50. A. B. Mikhailovskii, Nucl. Fusion, 15, 95 (1975).
51. A. B. Mikhailovskii, Fiz. Plasmy, 4, 1226 (1978).
52. G. Bateman and D. B. Nelson, Phys. Rev. Lett., 41, 805 (1978).
53. O. P. Pogutse and E. I. Yurchenko, Pis'ma Zh. Éksp. Teor. Fiz., 31, 479 (1980).
54. B. B. Kadomtsev and O. P. Pogutse, Zh. Éksp. Teor. Fiz., 65, 575 (1973).
55. B. B. Kadomtsev and O. P. Pogutse, in: Reviews of Plasma Physics, Vol. 5, Consultants Bureau (1977).
56. B. B. Kadomtsev, in: Reviews of Plasma Physics, Vol. 2, Consultants Bureau (1966).
57. I. B. Bernstein, E. A. Frieman, et al., Proc. R. Soc. London, A244, 17 (1958).
58. O. P. Pogutse and E. I. Yurchenko, Nucl. Fusion, 18, 1629 (1978).
59. Y. C. Lee and J. W. Van Dam, Proceedings of the High Beta Tokamak Theory Workshop, Varenna (1977), p. 93.
60. J. W. Connor, R. J. Hastie, and J. B. Taylor, Proc. R. Soc., London A365, 1 (1979).
61. K. V. Roberts and J. B. Taylor, Phys. Fluids, 8, 315 (1965).
62. L. D. Landau, Quantum Mechanics [in Russian], Fizmatgizdat (1963), p. 146.
63. O. P. Pogutse, N. V. Chudin, and E. I. Yurchenko, Fiz. Plazmy, 6, 621 (1980).
64. A. M. M. Todd, J. Manickam, Okabayashi, et al., Nucl. Fusion, 19, 743 (1979).
65. R. C. Grimm, J. M. Greene, and J. L. Johnson, in: Methods in Computational Physics, Vol. 16, Academic Press (1976), p. 253.
66. J. F. Clarke and D. J. Sigmar, Phys. Rev. Lett., 38, 70 (1977).
67. R. Grüber, R. Schreiber, et al., Plasma Physics and Controlled Nuclear Fusion Research, Vienna, Vol. 1 (1975), p. 593.
68. V. D. Shafranov, Zh. Eksp. Teor. Fiz., 15, 175 (1970).
69. M. Wakatani, J. Phys. Soc. Jpn., 38, 1481 (1975).
70. J. P. Friedberg, J. P. Goedbloed, et al., Plasma Physics and Controlled Nuclear Fusion Research, Vienna, Vol. 1 (1975), p. 505.

71. J. P. Friedberg and F. A. Haas, Phys. Fluids, $\underline{16}$, 1909 (1973).
72. J. A. Wesson and A. Sykes, Plasma Physics and Controlled Nuclear Fusion Research, Vienna, Vol. 1 (1975), p. 449.
73. G. Z. Ghershuni and E. M. Zhukhovitskii, Convective Stability of a Compressible Fluid [in Russian], Fizmatgiz (1972), p. 174.
74. A. B. Mikhailovskii, Plasma Instabilities in Magnetic Traps [in Russian], Atomizdat, Moscow (1978), p. 126.
75. B. Coppi, Plasma Physics and Controlled Nuclear Fusion Research, Brussels, IAEA, CN B-3 (1980).
76. H. R. Strauss, W. Park, et al., Nucl. Fusion, $\underline{20}$, 638 (1980).
77. P. J. Fielding and F. A. Haas, Plasma Physics and Controlled Nuclear Fusion Research, Innsbruck, Vol. 1 (1979). p. 630.
78. B. Carreras, H. R. Hicks, and D. K. Lee, ORNL/TM-7281 (1980).
79. A. B. Rechester and M. Rosenbluth, Phys. Rev. Lett., $\underline{40}$, 38 (1978).
80. B. B. Kadomtsev and O. P. Pogutse, Plasma Physics and Controlled Nuclear Fusion Research, Innsbruck, Vol. 1, 649 (1979).
81. V. V. Parail and O. P. Pogutse, Plasma Physics and Controlled Nuclear Fusion Research, Brussels, IAEA CN B13-2 (1980).
82. A.H. Glasser, Proceedings of the High Beta Tokamak Theory Workshop, Varenna (1977), p. 55.

EQUILIBRIUM OF CURRENT-CARRYING PLASMAS IN TOROIDAL CONFIGURATIONS

L. E. Zakharov and V. D. Shafranov

INTRODUCTION

The theory of plasma equilibrium in magnetic configurations has been reviewed in papers included in Volumes 2 and 5 of "Reviews of Plasma Physics." These papers were written at a time when the reality of the existence of high-temperature plasmas was not quite evident. During the past years essential progress has been achieved in controlled nuclear fusion research. Toroidal tokamak-type devices have become widespread in the world. The achievable tokamak plasma parameters are now close to those required for a fusion reactor [3]. Accordingly, the number of scientists who need to know the fundamentals of tokamak plasma confinement theory and, in particular, of equilibrium theory has considerably expanded. The aim of the present paper is to give an account of the main information on equilibrium theory that has been accumulated to the present time and which is necessary first of all for scientists dealing with tokamak plasma confinement studies.

CHAPTER 1

EQUILIBRIUM EQUATIONS FOR TOROIDAL

PLASMAS

The basis of plasma equilibrium theory is the following simple system of vector equations:

$$\nabla p = \frac{1}{c}\,[\mathbf{jB}]; \qquad (1.1)$$

$$\operatorname{curl}\mathbf{B} = \frac{4\pi}{c}\,\mathbf{j}; \qquad (1.2)$$

$$\operatorname{div}\mathbf{B} = 0. \qquad (1.3)$$

Here p(r) is the gas-kinetic pressure of the high-temperature plasma; $\mathbf{j}(\mathbf{r})$ and $\mathbf{B}(\mathbf{r})$ are the plasma current density and the magnetic field strength. The first equation expresses the fact that in equilibrium the bulk force $\mathbf{F}_p = -\nabla p$ acting on an element of plasma and directed from the center to the plasma boundary is compensated by the ampere force $(1/c)[\mathbf{jB}]$ in every volume element.

It should be pointed out that the typically "hydrodynamic" form of the equilibrium equations sometimes leads to an incorrect point of view about the region of validity of this equation. For example, some authors state that the system (1.1)-(1.3) corresponds to the ideal MHD model of a plasma. This would imply that the system (1.1)-(1.3) is incompatible with a dissipative expansion (plasma diffusion) even at high collision frequency. Sometimes Eq. (1.1) is believed to be invalid in the collisionless case, i.e., when the mean free path of a particle exceeds the length of the torus, so that the effects of the neoclassical transport theory, developed by Galeev and Sagdeev [4], are essential. Unfortunately, in Volume 7 of "Reviews of Plasma Physics" an incorrect statement was made for the case of collisionless plasma that an additional factor $1 - \alpha\sqrt{\varepsilon}$ ($\varepsilon = a/R$ is the inverse aspect ratio of the plasma torus, $\alpha \sim 1.5\text{-}2$ is a numerical coefficient which depends on the ratio of the electron and ion temperatures) was required in one of the integral relations resulting from the equilibrium equation (that which gives the diamagnetic signal).

Therefore it seems necessary to describe the conditions of validity of the MHD equilibrium equation (1.1) for a real nonideal plasma before we start discussing the main topic of this review. Actually, one can find this treated in papers on neoclassical transport (see, e.g., an excellent review by Hinton and Hazeltine [5]) and some readers may consider it as superfluous. However, taking into account the practical importance of the consequences of equilibrium theory we believe it will be useful to devote a few pages of the present paper to this problem.

Anticipating the extensive discussion below, let us first state the following conditions, which are sufficient for the applicability of Eq. (1.1):

1. The kinetic energy of the directional plasma motion should be less than the thermal energy

$$\rho v^2 \ll p. \tag{1.4}$$

This condition arises from the neglect of the inertia term. Drift and diffusive plasma motions are admissible. They do not contradict Eq. (1.1) since their velocity is smaller than the thermal velocity.

2. The gas-kinetic pressure of plasma particles should be nearly isotropic. Undoubtedly, the deviation from isotropy is small in the hydrodynamic limit when the mean free path of charged particles along the magnetic field line is considerably smaller than the length of the system $\lambda \ll L$. But also in a high-temperature rarefied plasma, when the opposite condition is fulfilled ($\lambda \gg L$), the velocity distribution of the particles relaxes to an isotropic Maxwellian provided the path travelled by a particle during a characteristic macroscopic time t is sufficiently large ($v_T t \gg \lambda$). This condition can be satisfied because toroidal systems are closed (the magnetic field line does not leave the volume of plasma confinement).

So, the second necessary condition for the validity of Eq. (1.1) is that the macroscopic processes are rather slow, i.e.,

$$t \gg \tau_{ee}, \tau_{ii}, \tag{1.5}$$

where τ_{ee} and τ_{ii} are the electron-electron and ion-ion collision times.

3. Because the Larmor radius $r_\perp(v)$ of charged particles is finite, the local distribution function of an inhomogeneous plasma deviates from a Maxwellian by the value $\delta f \sim r_\perp f'(r)$ even in cylindrical geometry (here, r is the cross-sectional radius of a toroidal magnetic surface). This leads to a deviation from isotropic pressure of order $\delta p \sim p r_\perp / r$. This deviation is somewhat larger in toroidal geometry, since the excursion of charged particles from the magnetic surfaces is of order $\Delta(v) \cong r_\perp q$ for passing particles and $\Delta(v) = r_\perp q \varepsilon^{-1/2}$ for trapped particles. Here, q is the

"rotational number" of magnetic field lines (the "safety factor" for tokamaks). This results in a relative deviation of the velocity distribution from a local Maxwellian, $\delta f / f \approx \Delta(v)/a$. The anisotropy of the plasma gas-kinetic pressure associated with this deviation is $\delta p \sim p r_\perp q / a$. Thus, for Eq. (1.1) to be applicable the following condition for the particle Larmor radius should be fulfilled:

$$r_\perp / a \ll 1, \quad r_\perp q / a \ll 1. \tag{1.6}$$

Now, we shall present a derivation of the equilibrium equation (1.1) which clarifies why this has a hydrodynamical form even for rarefied plasmas. Let us use a simple representation of the motion of a charged particle in a magnetic field. Provided (1.6) applies the plasma can be considered as a gas of "Larmor circles" produced by the charged particles rotating in a magnetic field.

The Larmor circles undergo a thermal motion along magnetic field lines and slowly drift across the magnetic field **B**. Therefore, from the point of view of gas-kinetic theory we have a gas characterized by a one-dimensional random motion (along the magnetic field lines). Correspondingly, Larmor circles only produce a longitudinal pressure. Thus, to describe the plasma as a whole, it is sufficient to have only a one-dimensional Euler equation, i.e., a balance of the forces acting along **B**. This equation should be applied together with the Maxwell equation (1.2), in which the current density should be expressed in terms of the plasma parameters. First, we will derive this equation and then we will construct a one-dimensional equilibrium equation.

Let us denote by $v_\parallel$ and $v_\perp$ the components of the velocity **v** of the particle with charge e and mass m parallel and perpendicular to the vector **B** and by $\mathbf{v}_d$ the drift velocity of the Larmor circle center. By analogy with the equation for the average current density in electrodynamics of continuous media, the flux of particles of a single species (i.e., having the same charge e and mass m) can be represented as a sum of two terms, one being associated with the transport of Larmor circles and the other with the circulating flow:

$$\mathbf{\Gamma} \equiv n \langle \mathbf{v} \rangle = n \langle \mathbf{v}_d \rangle + \operatorname{curl} \mathbf{K}; \tag{1.7}$$

$$\mathbf{K} = \frac{1}{2} n \langle [\mathbf{rv}] \rangle. \tag{1.8}$$

Here, n is the particle density of a particular species, **r** is the radius vector of a particle, and the angular brackets denote averaging over the velocity distribution function. The vector **K** may be called the density of the kinematic moment. The mechanical moment of a unit volume element is obviously $2m\mathbf{K}$, and the magnetic moment is

$$\mathbf{M} = \frac{e}{c}\mathbf{K}. \tag{1.9}$$

Note that the circulating flow curl **K** has been dropped from the continuity equation

$$\partial n/\partial t + \operatorname{div} \mathbf{\Gamma} = 0 \tag{1.10}$$

due to the identity div curl = 0. Thus, the density variation is associated only with the flux of Larmor circles.

Next, let us derive an explicit expression for the components of the flux $\mathbf{\Gamma}$.

The velocity of the guiding centers is determined by the equation

$$\mathbf{v}_d = v_\| \mathbf{B}/B + \frac{c}{|\mathbf{B}|^2}[\mathbf{EB}] - \frac{c}{e|\mathbf{B}|^2}[\mathbf{R}_1 \mathbf{B}]$$
$$+ \frac{mc(2v_\|^2 + v_\perp^2)}{4e|\mathbf{B}|^4}[\mathbf{B}\nabla|\mathbf{B}|^2] + \frac{mcv_\|^2}{e|\mathbf{B}|^4}[\mathbf{B}[\operatorname{curl}\mathbf{B}\cdot\mathbf{B}]]. \tag{1.11}$$

Here, **E** is the electric field and $\mathbf{R}_1$ is the mean friction force acting on a charged particle. Let us introduce the longitudinal and transverse pressure of the particular species of particles:

$$p_\| = mn\langle v_\|^2 \rangle, \quad p_\perp = \frac{1}{2}mn\langle v_\perp^2 \rangle. \tag{1.12}$$

Then

$$n\langle \mathbf{v}_d \rangle = n\langle v_\| \rangle \mathbf{B}/B + \frac{cn}{|\mathbf{B}|^2}[\mathbf{EB}] - \frac{cn}{e|\mathbf{B}|^2}[\mathbf{R}_1 \mathbf{B}]$$
$$+ \frac{c(p_\| + p_\perp)}{2e|\mathbf{B}|^4}[\mathbf{B}\nabla|\mathbf{B}|^2] + \frac{4\pi p_\|}{e|\mathbf{B}|^2}\mathbf{j}_\perp. \tag{1.13}$$

Using the equation

$$\operatorname{curl}\mathbf{B} = \frac{4\pi}{c}\mathbf{j} \tag{1.14}$$

we have expressed the last term in Eq. (1.12) in terms of the component of electric current density $j_\perp = j - B\,(jB)/|B|^2$ transverse to the vector **B**.

Now we calculate the kinematic moment of a single Larmor circle $K_\perp = \frac{1}{2}[rv]$. Here **v** should be interpreted as the transverse velocity component only, since the longitudinal motion is explicitly included in the expression for v_d. For **r** one can take the radial vector $r_\perp$ with the origin at the center of the Larmor circle. Then [6]

$$v_\perp = [\omega r_\perp], \quad \omega = -eB/mc \tag{1.15}$$

gives $K_\perp = r_\perp^2 \omega / 2$. Upon averaging we get

$$K = -\frac{cp_\perp}{e|B|^2} B; \tag{1.16}$$

$$\text{curl } K = \frac{c\,[B\nabla p]}{e|B|^2} - \frac{cp_\perp}{e|B|^4}[B\nabla|B|^2] - \frac{4\pi p_\perp}{|B|^2} j. \tag{1.17}$$

Adding Eqs. (1.13) and (1.17) we find the current density of the particles of a given species

$$\Gamma = n\langle v_\| \rangle B/B + \frac{cn}{|B|^2}[EB] - \frac{cn}{e|B|^2}[R_1 B] + \frac{c\,[B\nabla p]}{e|B|^2}$$
$$+ \frac{c(p_\| - p_\perp)}{2e|B|^4}[B\nabla|B|^2] + \frac{4\pi}{e|B|^2}(p_\| - p_\perp)j_\perp - \frac{4\pi p_\perp}{e|B|^2} j_\|. \tag{1.18}$$

The presence of trapped particles in a toroidal system does not change this expression. In fact, expression (1.18) includes both the motion of Larmor circles along the vector **B** and in the transverse direction. Thus, in terms of Larmor circles the "banana" current associated with trapped particles is included in the first term of the expression for Γ, and one should not add it to Eq. (1.18). If a plasma is considered as a mixture of Larmor circles and a "banana" gas, then Eq. (1.18) would represent the flux of passing particles only. The flux associated with trapped particles should be represented in this case by an equation of the type (1.7), where the first term would denote the flux of "banana" centers and **K** their kinematic moment. The latter would include besides the kinematic moment of the Larmor circles also the proper banana kinematic moment which would give some contribution to the parallel component $\Gamma_\parallel$ of the flux. In Eq. (1.18) the current associated with the trapped particles is taken into account by averaging the first term using the corresponding distribution function. The neoclassical effect is present in the fifth term and is associated with the pressure anisotropy $p_\parallel - p_\perp \neq 0$.

Now, we can write the expression for the current density

$$\mathbf{j} = \Sigma e \mathbf{\Gamma}. \tag{1.19}$$

Here, all charge species are summed. In this procedure one should keep in mind that the friction force is an internal force and that the quasineutrality condition is satisfied; i.e., electrical interaction between charges is also considered to be an internal force:

$$\Sigma n \mathbf{R}_1 = 0, \quad \Sigma en \mathbf{E} = \mathbf{E} \Sigma en = 0. \tag{1.20}$$

Denoting in the present section the total plasma pressure by P

$$P_\perp = \Sigma p_\perp, \quad P_\parallel = \Sigma p_\parallel, \tag{1.21}$$

we get

$$\mathbf{j} = \mathbf{j}_\parallel + \mathbf{j}_\perp; \tag{1.22}$$

$$\mathbf{j}_\parallel = \frac{\mathbf{B}}{B} \frac{\Sigma en <v_\parallel>}{1 + 4\pi P_\perp/|\mathbf{B}|^2} ; \tag{1.23}$$

$$\mathbf{j}_\perp = \frac{c[\mathbf{B}\nabla P_\perp]/|\mathbf{B}|^2 + c(P_\parallel - P_\perp)[\mathbf{B}\nabla |\mathbf{B}|^2]/2|\mathbf{B}|^4}{1 - 4\pi(P_\parallel - P_\perp)/|\mathbf{B}|^2} . \tag{1.24}$$

Now we shall denote the one-dimensional equilibrium equation for the gas of Larmor circles. The Larmor circles move inside magnetic tubes. Thus, it is sufficient to consider the condition for equilibrium of a gas within an element of the tube of length dl, volume dV, magnetic flux $d\Phi$, and cross-sectional areas $dS_1 = d\Phi/B_1$, $dS_2 = d\Phi/B_2$. The difference of the hydrostatic forces

$$(p_\parallel dS)_2 - (p_\parallel dS)_1 = d\Phi dl \frac{\partial}{\partial l}\left(\frac{p_\parallel}{B}\right) = dVB \frac{\partial}{\partial l}\left(\frac{p_\parallel}{B}\right) \tag{1.25}$$

is due to the surrounding gas and acts on the ends of the tube. It should be balanced by the volume force

$$n(R_{1\parallel} + eE_\parallel + \langle M_\perp \rangle \nabla B) dV = (nR_{1\parallel} + enE_\parallel - p_\perp \partial B/\partial l) dV. \tag{1.26}$$

Here, $M_1 = -mv_\perp^2 \mathbf{B}/2B$ is the magnetic moment of one circle. Equating these forces we get

$$-\nabla_\| p_\| + \frac{p_\| - p_\perp}{B} \nabla_\| B + enE_\| + nR_{1\|} = 0 \tag{1.27}$$

for particles of a given species and

$$-\nabla_\| P_\| + \frac{P_\| - P_\perp}{B} \nabla_\| B = 0 \tag{1.28}$$

for the plasma as a whole. The system of equations (1.24), (1.28) is equivalent to the equilibrium condition for the conducting fluid with an anisotropic pressure tensor $P_{\alpha\beta} = P_\| B_\alpha B_\beta / |\mathbf{B}|^2 + P_\perp \times (\delta_{\alpha\beta} - B_\alpha B_\beta / |\mathbf{B}|^2)$ (see, e.g., [1, 7-10]).

When $P_\| = P_\perp = P$, we get from Eqs. (1.28), (1.22), and (1.24)

$$\mathbf{B} \nabla P = 0; \tag{1.29}$$

$$\mathbf{j} = \mathbf{j}_\| + c \frac{[\mathbf{B}\nabla P]}{|\mathbf{B}|^2}, \tag{1.30}$$

which is equivalent to the equilibrium equation (1.1).

Now, assume that there exists a slight pressure anisotropy

$$(P_\| - P_\perp)/P_\perp = \delta. \tag{1.31}$$

Then, the maximum correction to $\mathbf{j}_\perp = c[\mathbf{B}\nabla P]/|\mathbf{B}|^2$ in a torus will be of order $\delta \Delta B/B \sim \delta \varepsilon$ as one can see from Eq. (1.24), i.e., it is actually negligible. Similarly, the fifth term of Eq. (1.18) for the flux Γ of particles of a particular species is also small in comparison with the fourth term, which is associated with the drift of particles along the magnetic surfaces. It can be comparable with the normal component of the second and third terms of Γ which are associated with the radial particle flux (diffusion). This component is smaller than the drift flux by a factor of $\omega\tau$. This is the reason for taking the "neoclassical" effect (pressure anisotropy) into account in Γ for the collisionless regime. It is also necessary to take into account a neoclassical effect in Ohm's law, which expresses $\mathbf{j}_\|$ in terms of the electric field. This relation results from the equilibrium equation (1.27) in which $\mathbf{j}_\|$ is included via $R_{1\|}$. Multiplying (1.27) by $\mathbf{B}$ and averaging over a magnetic layer gives

$$n \langle (e\mathbf{E} + \mathbf{R}_1) \mathbf{B} \rangle_B + \left\langle \frac{p_{||} - p_\perp}{2|\mathbf{B}|^2} \mathbf{B}_\nabla |\mathbf{B}|^2 \right\rangle_B = 0. \quad (1.32)$$

Here, the angular brackets with a subscript B mean averaging over a layer between two neighboring toroidal magnetic surfaces. Equations (1.18) and (1.32) coincide with the equations which result from taking moments of the distribution function [5, 11]. The geometric coefficients $\nabla_\perp \ln B$ and $\nabla_{||} \ln B$, entering into the neoclassical terms of Eqs. (1.18) and (1.32), appear to be independent, but they are really related to each other in the case of axial or helical symmetry since each can be expressed in terms of a derivative with respect to the poloidal coordinate.

It is evident that the presence of an additional source of plasma current — the Galeev–Sagdeev or "bootstrap" current — associated with trapped particles in Eq. (1.32) for $\mathbf{j}_{||}$ does not affect the equilibrium equation (1.1), since $\mathbf{j}_{||}$ does not enter it. In particular, the expression used for tokamak experiments to relate the diamagnetic signal to the plasma pressure is not changed.

Thus, we see that the equilibrium equations (1.1)-(1.3) represent not a model of ideal magnetohydrodynamics, but rather the macroscopic equations for the description of a real plasma in a strong magnetic field ($r_\perp/a \ll 1$ is the condition for the validity of the drift approximation) with slowly varying macroscopic parameters (t $\gg \tau_{ee}$, τ_{ii}: weak anisotropy) and in the absence of macroscopic velocities close to the speed of sound (the neglect of inertial drift).

Under such conditions Eqs. (1.1)-(1.3) describe an equilibrium plasma configuration where various processes such as drift motion, diffusion, oscillations, etc., may develop without destroying the configuration itself. The plasma equilibrium equations are insensitive to a number of internal phenomena because they reflect the general law of conservation of momentum of the macroscopic system. The processes associated with the fact that a real plasma is not ideal do not change the momentum of the system, so that the form of the plasma equilibrium equations is not affected.

CHAPTER 2

GENERAL RELATIONS FOR EQUILIBRIUM PLASMA CONFIGURATIONS

2.1. Straight Plasma Column with a Circular Cross Section

The simplest equilibrium plasma configuration is a straight plasma column with a circular cross section. Introduce a cylindrical coordinate system ρ, θ, s with the axis coinciding with that of the column. Due to the cylindrical symmetry ($\partial/\partial\theta = 0$, $\partial/\partial s = 0$) we get $B_\rho = 0$ from Eq. (1.3) and the only components of the magnetic field are B_θ and B_s. Equation (1.2) relates the poloidal and longitudinal field components B_θ and B_s to the longitudinal and poloidal components j_s and j_θ of the current density:

$$\frac{1}{\rho}\frac{\partial}{\partial\rho}\rho B_\theta = \frac{4\pi}{c}j_s, \quad \frac{\partial B_s}{\partial\rho} = -\frac{4\pi}{c}j_\theta, \qquad (2.1)$$

and Eq. (1.1) describes the pressure balance

$$dp/d\rho = \frac{1}{c}(j_\theta B_s - j_s B_\theta). \qquad (2.2)$$

One can see from Eq. (2.2) that the pressure gradient is compensated by the interaction of the longitudinal current with the poloidal field and of the poloidal current with the longitudinal field. Outside the plasma where $\mathbf{j} = 0$ the poloidal field is of the form

$$B_\theta(\rho) = B_\theta(\rho_0)\rho_0/\rho \qquad (2.3)$$

(ρ_0 is the radius of the current channel) and coincides with the field of a straight current-carrying filament, while the longitudinal field is constant:

$$B_s = B_{se} = \text{const.} \qquad (2.4)$$

The poloidal field in this configuration is produced only by the plasma current and poloidal fields of external sources are absent. The longitudinal field B_{se} is produced by external windings.

In the absence of the poloidal field the configuration represents a so-called θ pinch. There is no longitudinal plasma current and pressure balance reduces to the integral relationship

$$p(\rho) + B_s^2(\rho)/8\pi = \text{const.} \tag{2.5}$$

In the absence of a longitudinal field the configuration is a z pinch in which the plasma is confined by the plasma current only. When we multiply Eq. (2.2) by ρ^2 and substitute j_s from Eq. (2.1) and then integrate Eq. (2.2) over the cross section of the column with $B_s = 0$ we get the integral relation

$$2c^2 \int p\, dS / \mathcal{J}^2 = 1, \tag{2.6}$$

where $\mathcal{J}$ is the total plasma current.

The quantity on the left-hand side of Eq. (2.6) is invariant and does not depend on the choice of the area of integration provided that it covers the whole cross section of the plasma column. This quantity is also important for toroidal systems where it is denoted by $\beta_\mathcal{J}$ (the current beta).

A stabilized pinch ($B_\theta \approx B_s$) and a tokamak ($B_s \gg B_\theta$) are systems employing both longitudinal field and the field of the plasma current for confinement. The pressure balance equation for a column of circular cross section can be written in the following integral form [12]:

$$8\pi \langle p \rangle = \frac{4\mathcal{J}^2}{c^2 \rho_0^2} + B_{se}^2 - \langle B_s^2 \rangle, \tag{2.7}$$

where one uses the same procedure as that used in the derivation of (2.6). Here $\langle\ \rangle$ denotes averaging over the plasma cross section. Denoting by $\mu_\mathcal{J}$ the invariant quantity

$$\mu_\mathcal{J} = \frac{c^2 \int [B_{se}^2 - B_s^2(\rho)]\, dS}{4\pi \mathcal{J}^2}, \tag{2.8}$$

which characterizes the plasma diamagnetism, we get from (2.7) the following relationship for the integral quantities $\beta_\mathcal{J}$, $\mu_\mathcal{J}$:

$$\beta_\mathcal{J} = 1 + \mu_\mathcal{J}. \tag{2.9}$$

This is valid for a column with a circular cross section.

In the configuration under consideration the equilibrium is determined by the radial distribution of two variables. In systems with a longitudinal field these variables are the radial profiles of the pressure $p(\rho)$ and the longitudinal current $j_s(\rho)$, which are determined by transport processes in the plasma. In this situation the poloidal current j_θ results from plasma diffusion across the magnetic field and adjusts to a level that guarantees pressure balance (2.2). Thus, in systems with a longitudinal field the pressure balance is reached automatically due to adjustment of the j_θ current and there is no problem of equilibrium along the minor radius, while Eqs. (2.2) or (2.9) can be used for diagnostic purposes. In this case, depending on the value of β_χ the plasma can be either paramagnetic ($\beta_\chi < 1$, $\mu_\chi < 0$, $\langle B_s^2 \rangle > B_{se}^2$) or diamagnetic when the pressure of the plasma is increased ($\beta_\chi > 1, \mu_\chi > 0$, $\langle B_s^2 \rangle < B_{se}^2$).

In z and θ pinch systems there is no free parameter in the equilibrium relations (2.6) and (2.5). For this reason the equilibrium conditions cannot be satisfied and inertia processes play an essential role in the formation of the column which accounts for the pulsed character of these systems.

2.2. Axisymmetric Configurations

Axisymmetric configurations are currently the most important ones since they represent the simplest closed configurations. In the present section we describe the basic relationships without analyzing the equilibrium conditions in detail.

We denote the cylindrical coordinates by r, ζ, z. Sometimes, instead of an azimuthal angle ζ, we use a longitudinal coordinate $s = r\zeta$. In particular, we denote the vector longitudinal components (along s) by an index s.

The magnetic field and the current density in the plasma are the sum of poloidal and toroidal components

$$\mathbf{B} = \mathbf{B}_p + \mathbf{B}_s, \quad \mathbf{B}_p = B_r e_r + B_z e_z, \quad \mathbf{B}_s = B_s e_s; \quad (2.10)$$

$$\mathbf{j} = \mathbf{j}_p + \mathbf{j}_s, \quad \mathbf{j}_p = j_r e_r + j_z e_z, \quad \mathbf{j}_s = j_s e_s. \quad (2.11)$$

From the conditions div $\mathbf{B} = 0$ and div $\mathbf{j} = 0$ and due to the symmetry of the configuration ($\partial/\partial s = 0$) it follows that the poloidal components of the field and current may be expressed in terms of scalar functions $\Psi(r, z)$ and $F(r, z)$:

$$\mathbf{B}_p = \frac{1}{2\pi r}[\nabla \Psi e_s], \quad B_r = -\frac{1}{2\pi r}\frac{\partial \Psi}{\partial z}, \quad B_z = \frac{1}{2\pi r}\frac{\partial \Psi}{\partial r}; \quad (2.12)$$

$$j_p = \frac{1}{2\pi r}[\nabla F \mathbf{e}_s], \quad j_r = -\frac{1}{2\pi r}\frac{\partial F}{\partial z}, \quad j_z = \frac{1}{2\pi r}\frac{\partial F}{\partial r}. \tag{2.13}$$

The flux function $\Psi(r, z)$ and the current function $F(r, z)$ are determined apart from additive constants which may be chosen such that Ψ and F represent the total flux and current within a circular contour $r = \text{const}$, $z = \text{const}$. They are related to the integral of the vector potential A and the magnetic field B around this contour:

$$\Psi = \oint \mathbf{A} d\mathbf{l}_s = 2\pi r A_s; \tag{2.14}$$

$$\frac{4\pi}{c}F = \oint \mathbf{B} d\mathbf{l}_s = 2\pi r B_s. \tag{2.15}$$

The function Ψ that is defined in this way contains in particular the flux of the transformer which serves to maintain the longitudinal current in the plasma column and which does not directly affect the equilibrium. The function F includes the total current in the windings producing the longitudinal field (Fig. 1). This definition is convenient for describing the electrical properties of the device since the longitudinal electric field E_s is directly related to Ψ:

$$\text{curl } \mathbf{E} = -\frac{1}{c}\frac{\partial \mathbf{B}}{\partial t}; \tag{2.16}$$

$$2\pi r E_s = -\frac{1}{c}\frac{\partial \Psi}{\partial t}. \tag{2.17}$$

Frequently, definitions are used in which the poloidal flux and the current I are computed starting from the magnetic axis:

$$\chi(r, z) = \Psi(r_0, z_0) - \Psi(r, z); \tag{2.18}$$

$$I(r, z) = F(r_0, z_0) - F(r, z), \tag{2.19}$$

where r_0 and z_0 are the coordinates of the magnetic axis. The flux χ and the current I are convenient to use when one considers only the interior of a plasma column, as for example in stability problems.

Up to this point in this section, we have used only the symmetry conditions. If we take into account the equilibrium equation, the first two consequences of this are the following: the magnetic surfaces $\Psi(r, z) = \text{const}$ and the current surfaces $F(r, z) = \text{const}$ coincide and they also represent surfaces of constant pres-

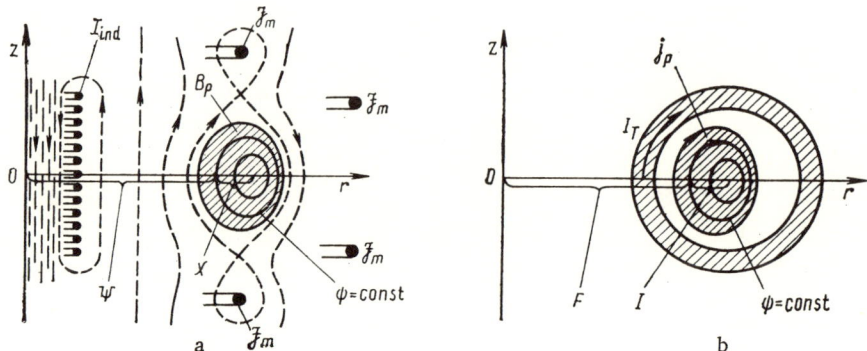

Fig. 1. Poloidal fluxes Ψ, χ and poloidal currents F, I in an equilibrium configuration: a) Ψ is the flux between the symmetry axis and the magnetic surface, including the flux of the transformer with a current I_{ind}, the external part of the fluxes both of the plasma current and the external currents Jm; χ is the flux between the magnetic surface and the magnetic axis. b) F is the poloidal current between the symmetry axis and the magnetic surface, including the current I_T of the toroidal field coils; I is the poloidal current of the plasma between the magnetic surface and the magnetic axis.

sure, p(r, z) = const. This follows from the relations $\mathbf{B} \cdot \nabla p = 0$ and $\mathbf{j} \cdot \nabla p = 0$, which are consequences of Eq. (1.1). Thus, one may consider $p = p(\Psi)$, $F = F(\Psi)$. The third consequence of the equilibrium equation is the relation between the longitudinal current density j_s in an equilibrium configuration and the functions $p(\Psi)$ and $F(\Psi)$:

$$j_s = 2\pi r \left(c \frac{dp}{d\Psi} + \frac{4\pi}{c} \frac{F}{4\pi^2 r^2} \frac{dF}{d\Psi} \right). \qquad (2.20)$$

Thus, the equilibrium conditions simply restrict the possible current distribution j_s in a plasma and instead of a two-dimensional current distribution $j_s(r, z)$ in fact a one-dimensional one is obtained, depending on two profiles $p'(\Psi)$ and $FF'(\Psi)$. These restrictions can be used for diagnostic purposes as will be seen later.

If the current distribution j_s is given, one can write the equation for the flux function $\Psi = 2\pi r A_s$ [13-16]:

$$\Delta^* \Psi = r^2 \operatorname{div} \frac{\nabla \Psi}{r^2} = \frac{\partial^2 \Psi}{\partial r^2} - \frac{1}{r} \frac{\partial \Psi}{\partial r} + \frac{\partial^2 \Psi}{\partial z^2} = -\frac{8\pi^2}{c} r j_s$$

$$= -4\pi^2 \frac{4\pi}{c} \left(cr^2 p' + \frac{4\pi}{c} \frac{FF'}{4\pi^2} \right). \tag{2.21}$$

The factors $4\pi/c$ and c are written separately here to simplify transformation to other systems of units. Equation (2.21), which determines the equilibrium configuration, is simply a magnetostatic equation with a nonlinear longitudinal current distribution.

Outside the plasma where the pressure and the poloidal plasma current are absent, $p' = 0$, $F' = 0$, the right-hand side of Eq. (2.21) vanishes. When solving equilibrium problems, it is convenient to separate the region occupied by the plasma from the outer region of the vacuum magnetic field. The boundary conditions connecting these two regions are: constancy of $\Psi(r, z)$ on the plasma boundary Γ:

$$\Psi_i(r, z)|_\Gamma = \Psi_e(r, z)|_\Gamma \tag{2.22}$$

and the condition of pressure balance across the boundary of the column:

$$(8\pi p_i + B_i^2)_\Gamma = (B_e^2)_\Gamma \tag{2.23}$$

or

$$\left\{ 8\pi p_i + \left(\frac{1}{2\pi r} \frac{\partial \Psi}{\partial n} \right)^2 + B_{si}^2 \right\}_\Gamma = \left\{ \left(\frac{1}{2\pi r} \frac{\partial \Psi}{\partial n} \right)^2 + B_{se}^2 \right\}_\Gamma . \tag{2.24}$$

The subscripts i and e label the inner and outer sides of the plasma boundary, respectively. The form of Eq. (2.24) for the pressure balance permits us to take account of plasma surface currents. When the latter are absent, Eq. (2.24) simply reduces to the continuity of the normal derivative of the flux function:

$$\frac{\partial \Psi_i}{\partial n}\bigg|_\Gamma = \frac{\partial \Psi_e}{\partial n}\bigg|_\Gamma . \tag{2.25}$$

Here we also present the general relations for a straight plasma column which can be considered as the limit of an axisymmetric configuration at $R \to \infty$ ($R = r - x$). Instead of the functions Ψ and F it is convenient to use directly the longitudinal components of the vector potential A_s and the magnetic field B_s which are constant on a magnetic surface, giving $p = p(A_s)$, $B_s = B_s(A_s)$. For a rectilinear configuration we have

$$\mathbf{B} = \mathbf{B}_p + \mathbf{B}_s = [\nabla A_s \, \mathbf{e}_s] + B_s \, \mathbf{e}_s, \quad B_x = \frac{\partial A_s}{\partial y}, \quad B_y = -\frac{\partial A_s}{\partial x}; \quad (2.26)$$

$$\mathbf{j} = \mathbf{j}_p + \mathbf{j}_s = \frac{c}{4\pi}[\nabla B_s \, \mathbf{e}_s] + j_s \, \mathbf{e}_s, \quad \frac{4\pi}{c} j_x = \frac{\partial B_s}{\partial y},$$

$$\frac{4\pi}{c} j_y = -\frac{\partial B_s}{\partial x}; \quad (2.27)$$

and for the longitudinal current density we have the following relation:

$$j_s = c \frac{d}{dA_s}\left(p + \frac{B_s^2}{8\pi}\right), \quad (2.28)$$

resulting from the equilibrium equation. Equation (2.21) for toroidal geometry is replaced by

$$\Delta A_s = \frac{\partial^2 A_s}{\partial x^2} + \frac{\partial^2 A_s}{\partial y^2} = -\frac{4\pi}{c} j_s(A_s) = -4\pi p'(A_s) - B_s B_s'(A_s). \quad (2.29)$$

The important characteristic of an equilibrium configuration, which to a large extent determines its stability, is the value of the rotational transform $\mu(\Psi) = \iota/2\pi$ (ι is the rotational transform angle) or its inverse $q(\Psi) = 1/\mu(\Psi)$, which is called the safety factor. For an axisymmetric configuration

$$q = \oint \frac{B_s \, dl}{2\pi r B_p} = \frac{2F(\Psi)}{2\pi c} \oint \frac{dl_p}{r^2 B_p}. \quad (2.30)$$

For a straight plasma column

$$q = \frac{B_s}{2\pi R} \oint \frac{dl_p}{B_p}, \quad (2.31)$$

where R is the major radius of an equivalent torus. The integration in Eqs. (2.30) and (2.31) is performed around the contour of the cross section of the magnetic surface.

Note that the longitudinal magnetic field, which determines the value q, is included in Eqs. (2.21) and (2.29) only as a derivative of either $dF^2/d\Psi$ or dB_s^2/dA_s. Thus, one and the same equilibrium configuration represents a whole class of equilibria with various values of longitudinal field and q value but with fixed values of the derivatives $dF^2/d\Psi$ or dB_s^2/dA_s. This property, allowing one to vary the q parameter without necessitating additional equilibrium calculations, is used in numerical stability studies [17].

The important difference between axisymmetric toroidal and rectilinear equilibria is that it is impossible to realize the former without a longitudinal current. As follows from (2.20), the current density j_s cannot vanish everywhere due to the additional dependence on the major radius r. The reason for this is the ballooning effect. The total force associated with the plasma pressure is in the direction of increasing major radius because of the difference in values on the outer (remote from the symmetry axis) and inner (nearest to the axis) parts of the toroidal surface. Another reason is the global force directed along the major radius and acting on the poloidal currents in the toroidal magnetic field. These forces must be balanced by the interaction of the longitudinal current and the poloidal field.

With the help of the equilibrium equation one may conclude that an equilibrium of simply nested tori is impossible in the absence of a net total plasma current: $\mathcal{J} = 0$. For a configuration with a single magnetic axis this is obvious since for $\mathcal{J} = 0$ the poloidal field on the plasma surface would change sign. But this is incompatible with the closure of magnetic surfaces inside the plasma volume. Thus, the conclusion can be drawn that for a torus there are critical values $\beta_{\mathcal{J},cr}$ and $\mu_{\mathcal{J},cr}$ which cannot be exceeded. Since the toroidal effects already appear in Eq. (2.20) to the first order in the inverse aspect ratio expansion, the maximum achievable $\beta_{\mathcal{J},cr}$ and $\mu_{\mathcal{J},cr}$ should be of the order of the aspect ratio

$$\beta_{\mathcal{J},cr} \approx R/a, \quad \mu_{\mathcal{J},cr} \approx R/a, \qquad (2.32)$$

where R and a are the characteristic major and minor radii of the configuration.

2.3. Configurations with Helical Symmetry

In the case of helical symmetry, all the physical values characterizing a configuration depend solely on two variables: ρ, $\theta =$

$\omega - \varkappa s$, where ρ, ω, s are the cylindrical coordinates related to the axis of the system. The parameter $\varkappa$ determines the longitudinal period of the system $L = 2\pi/\varkappa$. Here, as for any periodic system, any other length which is a multiple of the smallest period may also play the role of the period of the system. The period becomes physically defined when the system is bent into a torus and hence becomes closed. Such a system can often be considered as helically symmetric when the toroidal effects are treated only as corrections. The circumference $2\pi R$ of an equivalent torus is equal to an integer n times the minimum period. In the cross section s = const one can also define a minimum period with respect to the angle θ, which is equal to $2\pi/m$. Based on the concept of an equivalent torus the helical symmetry can be characterized by the radius R and the integers m and n. Then, the parameter $\varkappa$ will be

$$\varkappa = 2\pi/L = n/mR. \tag{2.33}$$

In the case of helical symmetry

$$\varkappa \partial/\partial \omega = -\partial/\partial s, \tag{2.34}$$

as before, the relationships div $\mathbf{B} = 0$ and div $\mathbf{j} = 0$ permit us to introduce the flux function $\chi^*(\rho, \theta)$ and the current function $I^*(\rho, \theta)$. It is convenient to introduce the following renormalized "poloidal" field and "poloidal" current [18-20]:

$$\left. \begin{array}{l} \mathbf{B}^* \cong B_\rho \, \mathbf{e}_\rho + (B_\omega - \varkappa \rho B_s) \, \mathbf{e}_\omega, \\ \mathbf{j}^* \equiv j_\rho \, \mathbf{e}_\rho + (j_\omega - \varkappa \rho j_s) \, \mathbf{e}_\omega, \end{array} \right\} \tag{2.35}$$

which simplifies the expression of the magnetic field and current density in terms of χ^* and I^*. Then,

$$\mathbf{B}^* = [\mathbf{e}_s \, \nabla \chi^*], \quad \mathbf{j}^* = [\mathbf{e}_s \, \nabla I^*]. \tag{2.36}$$

When defined in such a way, the $\chi^*(\rho, \theta)$ and $I^*(\rho, \theta)$ functions represent the magnetic flux and total current through a helical surface $\theta = \omega - \varkappa s = $ const between $\rho = 0$ and the current value of ρ per unit length in s. One can also introduce a current $F^*(\rho, \theta)$ through the surface between $\rho = \infty$ and the current value of ρ, so that

$$I^*(\rho, \theta) = F^*(0, \theta) - F^*(\rho, \theta). \tag{2.37}$$

The magnetic flux function Ψ^* between $\rho = \infty$ and the current value of ρ in a straight system with helical symmetry, strictly speaking, cannot be defined because of its divergence at $\rho \to \infty$. However, when the system is closed and forms a torus such a flux acquires a meaning and can be used in calculations. In this case

$$\chi^*(\rho, \theta) = \Psi^*(0, \theta) - \Psi^*(\rho, \theta). \tag{2.38}$$

The functions Ψ^*, F^*, and p are surface functions: $p = p(\Psi^*)$, $F = F(\psi^*)$. Thus, for configurations with helical symmetry we have

$$\mathbf{B}^* = [\nabla \Psi^* \mathbf{e}_s], \quad B_\rho = \frac{1}{\rho}\frac{\partial \Psi^*}{\partial \theta}, \quad B_\omega - \varkappa \rho B_s = -\frac{\partial \Psi^*}{\partial \rho}; \tag{2.39}$$

$$\mathbf{j}^* = [\nabla F^* \mathbf{e}_s], \quad j_\rho = \frac{1}{\rho}\frac{\partial F^*}{\partial \theta}, \quad j_\omega - \varkappa \rho j_s = -\frac{\partial F^*}{\partial \rho}; \tag{2.40}$$

$$A_s + \varkappa \rho A_\omega = \Psi^*; \tag{2.41}$$

$$B_s + \varkappa \rho B_\omega = \frac{4\pi}{c} F^*; \tag{2.42}$$

$$j_s + \varkappa \rho j_\omega = c(1 + \varkappa^2 \rho^2)\frac{dp}{d\Psi^*} + \frac{4\pi}{c} F^* \frac{dF^*}{d\Psi^*}. \tag{2.43}$$

The equation for the function $\Psi^*(\rho, \theta)$ determining the magnetic surfaces, $\Psi^*(\rho, \theta) = \text{const}$, has the following form:

$$\frac{1}{\rho}\frac{\partial}{\partial \rho}\left(\frac{\rho}{1 + \varkappa^2 \rho^2}\frac{\partial \Psi^*}{\partial \rho}\right) + \frac{1}{\rho^2}\frac{\partial^2 \Psi^*}{\partial \theta^2} = -4\pi p'$$
$$+ \frac{4\pi}{c}\frac{2\varkappa F^*}{(1 + \varkappa^2 \rho^2)^2} - \left(\frac{4\pi}{c}\right)^2 \frac{F^* F'^*}{1 + \varkappa^2 \rho^2}. \tag{2.44}$$

In the case of helical symmetry the long-wavelength limit, when $\varkappa a \ll 1$ (a is a characteristic transverse size), is the simplest and most important one. With the additional conditions $B_s \gg B_\omega$, B_ρ and $\beta = 8\pi p/B_s^2 \ll 1$, all relationships for helical symmetry become equivalent to those for configurations with translational symmetry, where $\mathbf{B}_p^*$ and $\mathbf{j}_p^*$ play the role of poloidal field $\mathbf{B}_p$ and current density $\mathbf{j}_p$, respectively, while the longitudinal current j_s is substituted by the current $j_s - j_B$, where

$$j_B = 2\varkappa F^* \approx \frac{c}{4\pi} 2\varkappa B_s. \tag{2.45}$$

Equation (2.44) transforms into an analog of the translational one (2.29):

$$\Delta \Psi^* = \frac{1}{\rho} \frac{\partial}{\partial \rho} \rho \frac{\partial \Psi^*}{\partial \rho} + \frac{1}{\rho^2} \frac{\partial^2 \Psi}{\partial \theta^2} = -\frac{4\pi}{c} [j_s(\Psi^*) - j_B]$$
$$= -[4\pi p'(\Psi^*) + B_s B_s'(\Psi^*)] + 2\varkappa B_s. \quad (2.46)$$

The principal difference between Eqs. (2.29) and (2.46) is that the right-hand side of the latter does not vanish outside the plasma column because of the presence of the fictitious current j_B which acts as a source of flux for the longitudinal field through the helical surface. It has been proven that this analogy between a long-wavelength helical column and a rectilinear one also holds for helical MHD plasma streams where inertia is important [20]. For a plasma with high electrical conductivity in the presence of a strong longitudinal field $B_s \gg B_p$ an additional constraint on the motion appears: div $\mathbf{v} = 0$, where $\mathbf{v}$ is the poloidal plasma velocity, since the flow cannot compress the longitudinal magnetic field.

The equation for helical equilibria (2.46) is useful from at least two points of view. First of all, in the absence of a longitudinal current, it describes stellarator-type equilibrium configurations. Secondly, it is useful for studies of the consequences of the development of the helical instability in tokamak-type devices.

2.4. Quasicylindrical Description of Equilibrium Configurations

For every equilibrium configuration the magnetic flux and current surfaces also coincide with the surfaces of equal pressure since $\mathbf{B} \cdot \nabla p = 0$, $\mathbf{j} \cdot \nabla p = 0$. If these surfaces are finite, they form a system of nested toroids because of the general properties of vector fields. The innermost one degenerates in a closed line, viz., the magnetic axis.

Actually, there could be a number of magnetic axes in a configuration where each of these establishes its own system of nested toroids. Different simply connected systems are separated from each other by separatrixes which can represent a very complicated geometry. It is desirable for plasma confinement purposes to have a configuration with a minimum number of magnetic axes because the smoothing of the plasma pressure and temperature result in a decrease of plasma energy losses across the magnetic field.

The conclusion that any equilibrium configuration consists of a system of nested toroids allows the description of the equilibrium of each of these systems by methods which were used for symmetrical configurations and even for a straight column with a circular cross section. Before discussing this type of description we make a general remark about nonsymmetrical equilibria.

In the presence of symmetry, the topology of magnetic configurations does not depend on equilibrium conditions and is determined only by the property that the magnetic field is divergence free. It allows us to introduce the flux function ψ that determines the existence of magnetic surfaces. For the general case it is possible to talk about magnetic surfaces if it is known a priori that the equilibrium conditions are fulfilled. The question of whether an equilibrium could be provided in a given nonsymmetrical external field remains open for the present. Two approaches are used to solve this question. The first uses the fact that a system of nested toroidal magnetic surfaces can be established around certain closed magnetic lines even for a vacuum magnetic field. This is the basic concept of stellarators. Here the problem consists in how a finite-pressure plasma affects this configuration and how large the critical value of the equilibrium pressure is. The second approach consists in studying the effects of asymmetrical perturbations on an equilibrium with a certain type of symmetry. The perturbations can produce a splitting of the configuration and establish a new system of nested toroids — "the magnetic islands." At present, the widespread opinion is that if the amplitude of the perturbations is large enough to create overlapping of the different types of magnetic islands then the configuration becomes stochastic and the property of plasma confinement is lost [21, 22]. However, it is also a legitimate assumption that the plasma may prevent the stochastization of the lines of force and thus maintain the toroidal structure. It is also of importance that in configurations with magnetic islands, regions with closed toroidal surfaces can exist which enclose the regions with wandering magnetic lines. The question of the destruction of magnetic configurations belongs to the theory of stability of equilibrium configurations and we will not deal with it here.

Now, let us consider a simple configuration with only one magnetic axis and show how the equilibrium can be described in an analogous form to that of straight cylinders with circular cross section [23, 24].

It is convenient to introduce a label a for the magnetic surfaces which is analogous to a "radial" coordinate. Besides a, one should have two cyclic variables θ and ζ which are the poloidal and toroidal cyclic coordinates ranging from 0 to 2π going once around the torus in the corresponding directions. The choice of this "natural" coordinate system a, θ, ζ is arbitrary and the natural coordinates should be specified so as to facilitate the description of the problem under consideration. The variable a often implies a well-defined quantity such as the volume V contained inside the toroidal surface. Henceforth, without losing generality, we shall assume that a is analogous to the minor radius of the torus which transforms into ρ for a straight cylinder. On the magnetic axis $a = 0$.

The physical variables of a toroidal configuration are the following: the plasma pressure $p(a)$, the magnetic fluxes $\Psi(a)$ and $\Phi(a)$, and the currents $F(a)$ and $\mathcal{J}(a)$. The flux $\Psi(a)$ is taken to be the poloidal flux through the middle of the torus (the contour a = const, θ = const). If, as in the procedure described in Section 2.2, we define Ψ_0 to be the flux enclosing the magnetic axis, the difference

$$\chi(a) = \Psi_0 - \Psi(a) \tag{2.47}$$

is the flux contained between the magnetic axis and the toroidal surface a = const $[\chi(0) = 0]$. The poloidal current $F(a)$ through the middle of the torus and the partial current $I(a)$

$$I(a) = F_0 - F(a) \tag{2.48}$$

between the magnetic axis and toroidal surface are defined in a similar way. The longitudinal flux $\Phi(a)$ and the current $\mathcal{J}(a)$ are the flux and the current through the cross section of the toroidal surface (the contour a = const, ζ = const).

Consider the relation between the components of the magnetic field and integral variables. Using Eq. (A.42) of the Appendix, the fluxes $\chi(a)$ through the contour θ = const and $\Phi(a)$ through the contour ζ = const can be written in the form

$$\chi(a) = \int \mathbf{B} d\mathbf{S}_2 = \int_0^a da \int_0^{2\pi} B^2 \sqrt{g}\, d\zeta; \tag{2.49}$$

$$\Phi(a) = \int \mathbf{B} dS_3 = \int_0^a da \int_0^{2\pi} B^3 \sqrt{g}\, d\theta, \qquad (2.50)$$

where B^2 and B^3 are the contravariant components of the field: $B^i = \mathbf{B}\nabla x^i$ ($x^i = a, \theta, \zeta$); $\sqrt{g} = |\nabla a \, [\nabla \theta \nabla \zeta]|^{-1}$, so that $dV = \sqrt{g}\, da\, d\theta\, d\zeta$.

Taking into account that $B^1 = \mathbf{B}\nabla a = 0$ we get the following expressions for the contravariant components of the magnetic field:

$$\begin{aligned} j^i &= \left\{ 0, \; \frac{I' - \partial v/\partial \zeta}{2\pi \sqrt{g}}, \; \frac{\mathscr{F}' + \partial v/\partial \theta}{2\pi \sqrt{g}} \right\} \\ &= \left\{ 0, \; -\frac{F' + \partial v/\partial \zeta}{2\pi \sqrt{g}}, \; \frac{\mathscr{F}' + \partial v/\partial \theta}{2\pi \sqrt{g}} \right\}. \end{aligned} \qquad (2.51)$$

Similarly, for the current density we have

$$\begin{aligned} B^i &= \left\{ 0, \; \frac{\chi' - \partial \eta/\partial \zeta}{2\pi \sqrt{g}}, \; \frac{\Phi' + \partial \eta/\partial \theta}{2\pi \sqrt{g}} \right\} \\ &= \left\{ 0, \; -\frac{\Psi' + \partial \eta/\partial \zeta}{2\pi \sqrt{g}}, \; \frac{\Phi' + \partial \eta/\partial \theta}{2\pi \sqrt{g}} \right\}. \end{aligned} \qquad (2.52)$$

In Eqs. (2.51) and (2.52) η and v are some periodic functions of θ and ζ.

Employing the relations $\mathbf{B} = \text{curl } \mathbf{A}$ and $\text{curl } \mathbf{B} = (4\pi/c)\mathbf{j}$ one can find the expressions for the covariant components of the vector potential A_i and the magnetic field B_i:

$$A_i = \frac{1}{2\pi} \{-\eta, \Phi, \Psi\}, \qquad (2.53)$$

$$B_i = \frac{4\pi}{c} \frac{1}{2\pi} \left\{ \frac{\partial \varphi}{\partial a} - v, \; \mathscr{F} + \frac{\partial \varphi}{\partial \theta}, \; F + \frac{\partial \varphi}{\partial \zeta} \right\}. \qquad (2.54)$$

The periodic function φ is the curlfree part of the magnetic field $\mathbf{B}$. In expression (2.53) for the vector potential the similar function can be omitted since an arbitrary vector gradient can be added to $\mathbf{A}$.

Now, the equilibrium equation (1.1) may be written in the form

$$4\pi^2 c p' \sqrt{g} = -(F' + \partial v/\partial \zeta)(\Phi' + \partial \eta/\partial \theta)$$
$$+ (\mathscr{F}' + \partial v/\partial \theta)(\Psi' + \partial \eta/\partial \zeta). \qquad (2.55)$$

Maxwell's equation curl **B** = $(4\pi/c)$**j**, which has been used already to obtain Eq. (2.54), is simply reduced to the relation between the co- and contravariant components of the magnetic field

$$\frac{4\pi}{c}\left(-\nu + \frac{\partial\varphi}{\partial a}\right) = -\frac{g_{12}}{\sqrt{g}}\left(\Psi' + \frac{\partial\eta}{\partial\zeta}\right) + \frac{g_{13}}{\sqrt{g}}\left(\Phi' + \frac{\partial\eta}{\partial\theta}\right); \quad (2.56)$$

$$\frac{4\pi}{c}\left(\mathscr{J} + \frac{\partial\varphi}{\partial\theta}\right) = -\frac{g_{22}}{\sqrt{g}}\left(\Psi' + \frac{\partial\eta}{\partial\zeta}\right) + \frac{g_{23}}{\sqrt{g}}\left(\Phi' + \frac{\partial\eta}{\partial\theta}\right); \quad (2.57)$$

$$\frac{4\pi}{c}\left(F + \frac{\partial\varphi}{\partial\zeta}\right) = -\frac{g_{23}}{\sqrt{g}}\left(\Psi + \frac{\partial\eta}{\partial\zeta}\right) + \frac{g_{33}}{\sqrt{g}}\left(\Phi' + \frac{\partial\eta}{\partial\theta}\right). \quad (2.58)$$

Let us consider each term separately for the case of a straight cylindrical plasma column with a circular cross section. Here, a has the meaning of minor radius ρ, while $\zeta = s/R$, where R is the major radius of an equivalent toroid. In this case

$$g_{11} = 1, \; g_{22} = a^2, \; g_{33} = R^2, \; \sqrt{g} = aR, \; g_{ik} = 0, \; i \neq k. \quad (2.59)$$

In addition,

$$\Phi' = 2\pi a B_s, \; \Psi' = -2\pi R B_\theta. \quad (2.60)$$

Equation (2.55) is reduced to the equation for pressure balance along the minor radius (2.2), while Eqs. (2.57) and (2.58) are reduced to the relationship between the magnetic fields B_θ and B_s and the current densities (2.1).

Consider now the general case of axial symmetry: $\partial/\partial\zeta = 0$, $g_{13} = g_{23} = 0$, $g_{33} = r^2$ and follow the transition to the conventional description of the equilibrium (Section 2.2). Eliminating the function φ from Eq. (2.57) by means of Eq. (2.56) the system of equations is reduced to the following:

$$4\pi^2 cp'\sqrt{g} = -F'\left(\Phi' + \frac{\partial\eta}{\partial\theta}\right) + \left(\mathscr{J}' + \frac{\partial\nu}{\partial\theta}\right)\Psi'; \quad (2.61)$$

$$\frac{4\pi}{c}\left(\mathscr{J}' + \frac{\partial\nu}{\partial\theta}\right) = -\frac{\partial}{\partial a}\frac{g_{22}}{\sqrt{g}}\Psi' + \frac{\partial}{\partial\theta}\frac{g_{12}}{\sqrt{g}}\Psi'; \quad (2.62)$$

$$\frac{4\pi}{c}F = \frac{g_{33}}{\sqrt{g}}\left(\Phi' + \frac{\partial\eta}{\partial\theta}\right). \quad (2.63)$$

Equation (2.61) can be used to express the current density j_s in a form similar to Eq. (2.20):

$$j_s = \sqrt{g_{33}}\,\tilde{j}^3 = \frac{r(\mathcal{J}' + \partial v/\partial\theta)}{2\pi\sqrt{g}} = 2\pi r\left[c\frac{p'}{\Psi'} + \frac{4\pi}{c}\frac{FF'}{4\pi^2 r^2 \Psi'}\right]. \quad (2.64)$$

Substituting $\mathcal{J}' + \partial v/\partial\theta$ from Eq. (2.61) into (2.62) we obtain the analog of Eq. (2.21) for the poloidal flux ψ written in natural coordinates:

$$\frac{1}{\sqrt{g}}\left[\frac{\partial}{\partial a}\frac{g_{22}}{\sqrt{g}}\Psi' - \frac{\partial}{\partial\theta}\frac{g_{12}}{\sqrt{g}}\Psi'\right] = -\frac{4\pi}{c}4\pi^2\left[c\frac{p'}{\Psi'} + \frac{4\pi}{c}\frac{FF'}{4\pi^2 r^2 \Psi'}\right]. \quad (2.65)$$

These equations may be written in quasicylindrical form by introducing the following average values:

$$\bar{B}_\theta = -\Psi'/2\pi R;\quad \bar{B}_s = \Phi'/2\pi a; \quad (2.66)$$

$$\bar{j}_s = \mathcal{J}'/2\pi a;\quad \bar{j}_\theta = -F'/2\pi R, \quad (2.67)$$

where R is some characteristic major radius, for example that of the magnetic axis. Then we have

$$p'\left\langle\frac{\sqrt{g}}{aR}\right\rangle = \frac{1}{c}(\bar{j}_\theta \bar{B}_s - \bar{j}_s \bar{B}_\theta); \quad (2.68)$$

$$\frac{4\pi}{c}\bar{j}_s = \frac{1}{a}\frac{d}{da}\left\langle\frac{g_{22}}{\sqrt{g}}\frac{R}{a}\right\rangle a\bar{B}_\theta; \quad (2.69)$$

$$\frac{4\pi}{c}\bar{j}_\theta = -\frac{d}{da}\frac{\bar{B}_s}{\left\langle\frac{\sqrt{g}}{g_{33}}\frac{R}{a}\right\rangle}. \quad (2.70)$$

The coefficients averaged over the angular variables, like $\langle\sqrt{g}/aR\rangle$ entering these equations, should be determined from the solution of the two-dimensional equilibrium equation (2.65) which, after extraction of the averaged part, becomes equivalent to the following:

$$\Psi'\frac{\partial}{\partial a}\left(\frac{\tilde{g}_{22}}{\sqrt{g}}\right)\Psi' - \Psi'^2\frac{\partial}{\partial\theta}\frac{g_{12}}{\sqrt{g}} = \frac{4\pi}{c}c4\pi^2 p'\times$$

$$\times \left[\langle \sqrt{g} \rangle \frac{(\widetilde{\sqrt{g}}/g_{33})}{\langle \sqrt{g}/g_{33} \rangle} - (\widetilde{\sqrt{g}}) \right] + \Psi' \frac{(\widetilde{\sqrt{g}}/g_{33})}{\langle \sqrt{g}/g_{33} \rangle} \frac{d}{da} \left\langle \frac{g_{22}}{\sqrt{g}} \right\rangle \Psi'. \quad (2.71)$$

Here, parentheses $(\widetilde{\sqrt{g}})$ denote the oscillatory part: $(\widetilde{\sqrt{g}}) = \sqrt{g} - \langle \sqrt{g} \rangle$. For a straight cylinder with a circular cross section all coefficients in the angular brackets of Eqs. (2.68)-(2.70) transform to unity. For a toroidal plasma with a circular cross section, corrections $\sim a^2/R^2$, $a^2/R^2 \beta_{\mathcal{J}}^2$ appear in them.

It will be shown below how to use the equations written in natural coordinates for solving equilibrium problems.

2.5. Maintaining Field. The Principle of a Virtual Casing

For a straight plasma with a circular cross section the equilibrium configuration is determined by the radial distribution of the pressure and the longitudinal current. The application of external poloidal fields alters the shape of the cross section. For the general case the equilibrium configuration is determined by the profile of the pressure $p(a)$ and the longitudinal current $\mathcal{J}(a)$ and the magnetic field $\mathbf{B}_{ext}$ produced by external sources. We shall call the field $\mathbf{B}_{ext}$ the maintaining field. The basic problem in equilibrium theory is to find the spatial distributions of the magnetic fields, the currents, and the pressure when the maintaining field $\mathbf{B}_{ext}$ and $p(a)$, $\mathcal{J}(a)$ or any other two equivalent profiles [for example, $p(a)$, $FF'(a)$ or $p(a)$, $\Psi'(a)$] are given.

However, even for the simplest cases the problem formulated in this way can only be solved numerically, since the plasma boundary separating the current-carrying region from the vacuum field is unknown until the problem is already solved. Therefore, it is customary to divide the problem into three parts:

1. Determination of the configuration for given $p(a)$ and $\mathcal{J}(a)$ inside a fixed plasma boundary whose shape is prescribed by arguments of stability, machine design, etc.

2. Determination of the vacuum field outside the plasma by matching to an internal solution. Next, the extraction of the maintaining field $\mathbf{B}_{ext}$ from the total vacuum field and determination of external currents to provide the equilibrium.

3. Correction of the boundary taking into account real restrictions on the number of external currents, their distribution and amplitude. The simulation of the operation of the system of maintaining fields.

The second problem is a purely magnetostatic one. It is not conventional, however, because two boundary conditions are imposed at the plasma surface: the normal component of magnetic field B_n is required to be zero and the tangential component B_τ is to be matched to an internal solution by the pressure balance equation. Since the vacuum magnetic field is described by a scalar potential φ_B: $\mathbf{B} = \nabla \varphi_B$, the Cauchy problem for the Laplace equation is obtained. A numerical solution is difficult to find, since small-scale perturbations of the boundary conditions result in strong distortions of the external solution, which requires some regularization procedure.

However, to find the maintaining field there is a universal method called the principle of virtual casing, which allows us to do this without solving the external magnetostatic problem [25]. Assume that the plasma is surrounded by a thin closed superconducting casing coinciding with the plasma boundary. The equilibrium would be assumed. In this case the maintaining field would be produced by image currents in the casing. These currents can be easily determined since there are no magnetic fields outside the superconducting casing. It follows that

$$\frac{4\pi}{c} \mathbf{i} = [\mathbf{Bn}]. \qquad (2.72)$$

This allows the explicit calculation of the distribution of the surface current $\mathbf{i}$ in the casing if the magnetic field $\mathbf{B}$ on the plasma boundary is known from the internal solution. In Eq. (2.72) $\mathbf{n}$ is the unit external normal vector. The principle of virtual casing states that the magnetic field of the surface current $\mathbf{i}$ of Eq. (2.72) inside the plasma coincides with the maintaining field required for equilibrium, while outside the plasma it is equal in magnitude and opposite in sign to the self-field of the plasma current.

In particular, the principle of virtual casing solves the general magnetostatic problem by replacing a volume current distribution by a surrounding surface current. To do this it is sufficient to produce some enclosing magnetic surface (maybe through external sources), to calculate the magnetic field **B** on it, and to distribute a surface current **i** over this surface according to Eq. (2.72). Note that if this magnetic surface is formed by the internal current distribution only (external sources are absent), then Eq. (2.72) gives the current distribution on a superconducting toroidal surface coinciding with the chosen magnetic surface.

2.6. Integral Relations for a Toroidal Plasma Column

In this section we shall consider integral relations which enable one to relate some internal plasma parameters to the magnetic fields outside the plasma [26, 27].

It will be convenient to represent the total magnetic field of an equilibrium configuration by the following sum:

$$\mathbf{B} = \mathbf{B}_1 + \mathbf{B}_2, \qquad (2.73)$$

where $\mathbf{B}_2$ is a vortex-free field within the region under consideration:

$$\text{curl } \mathbf{B}_2 = 0. \qquad (2.74)$$

Of course, $\mathbf{B}_2$ is not unique. The following cases are of interest: a) $\mathbf{B}_2 = 0$, i.e., the total field is included in $\mathbf{B}_1$; b) $\mathbf{B}_2$ is equal to the maintaining field produced by external sources, i.e., $\mathbf{B}_2 = \mathbf{B}_{ext}$ (in this case $\mathbf{B}_1$ is the field produced by the plasma current); c) $\mathbf{B}_2$ is the static part of the external field, which is not detected by usual probe measurements.

Multiply the equilibrium equation (1.1) by an arbitrary vector **Q**. Using the vector identity

$$\mathbf{Q}[\text{curl } \mathbf{B} \cdot \mathbf{B}] = \text{div}\left[(\mathbf{QB})\mathbf{B} - \frac{B^2}{2}\mathbf{Q}\right] + \frac{B^2}{2}\text{div }\mathbf{Q} - \mathbf{B}(\mathbf{B}\nabla)\mathbf{Q} \qquad (2.75)$$

and integrating over an arbitrary volume V with a surface S we obtain

$$\int_V \left[\left(p + \frac{B_1^2}{8\pi}\right)\text{div }\mathbf{Q} - \frac{\mathbf{B}_1(\mathbf{B}_1\nabla)\mathbf{Q}}{4\pi}\right]dV = \oint_S \left[\left(p + \frac{B_1^2}{8\pi}\right)(\mathbf{Qn}) - \right.$$

$$-\frac{(QB_1)(B_1 n)}{4\pi}\Big]dS - \frac{1}{c}\int_V Q[jB_2]\,dV, \tag{2.76}$$

where **n** is the external unit vector normal to S.

If the polar vector $\mathbf{R} = r\mathbf{e}_r + z\mathbf{e}_z$ is taken as **Q**, and $\mathbf{B}_2 = 0$, then this general integral condition for the equilibrium is reduced to a virial theorem

$$\int_V \left(3p + \frac{B^2}{8\pi}\right)dV - \oint_S \left[\left(p + \frac{B^2}{8\pi}\right)(\mathbf{Rn}) - \frac{(\mathbf{RB})(\mathbf{Bn})}{4\pi}\right]dS. \tag{2.77}$$

From Eq. (2.77) the impossibility of plasma equilibrium provided by a self-magnetic field only (in the absence of a maintaining field) follows. If the magnetic field **B** were a plasma self-field then the integral on the right-hand side would reduce to zero for a volume of integration extending up to infinity, since the self-field of a localized source is $B_{pl} \sim 1/R^3$. Thus, Eq. (2.77) would be invalid [1].

To obtain some other consequences of Eq. (2.76) we restrict ourselves to the case of axial symmetry, $\partial/\partial\zeta = 0$. It is convenient to separate the poloidal and toroidal fields:

$$\mathbf{B} = \mathbf{B}_p + \mathbf{B}_s. \tag{2.78}$$

We assume that the surface S is outside the plasma, where curl $\mathbf{B} = 0$, and that the vectors **Q** and $\mathbf{B}_2$ are purely poloidal (no toroidal $\mathbf{B}_2$ component):

$$\mathbf{Q} = Q_r\mathbf{e}_r + Q_z\mathbf{e}_z, \quad \mathbf{B}_2 = B_{2r}\mathbf{e}_r + B_{2z}\mathbf{e}_z. \tag{2.79}$$

Noting that the external longitudinal magnetic field varies in accordance with

$$B_{se} = \text{const}/r, \tag{2.80}$$

we obtain the following auxiliary relations:

$$\oint_S B_{se}^2 (\mathbf{Qn})\,dS = \int_V B_{se}^2 \left(\text{div } \mathbf{Q} - 2\frac{Qe_r}{r}\right)dV; \tag{2.81}$$

$$\mathbf{B}_1(\mathbf{B}_1 \nabla)\mathbf{Q} = \mathbf{B}_{pl}(\mathbf{B}_{pl}\nabla)\mathbf{Q} + B_s^2 \frac{Qe_r}{r}. \tag{2.82}$$

Using these relations we rewrite the integral equilibrium equation (2.76) in such a way that the longitudinal field enters in the form of the difference $B_s^2 - B_{se}^2$ and only in the left-hand side of the equation. In this connection, we omit the subscript p from the poloidal field; so instead of Eq. (2.76) we will have

$$\int_V \left[\left(p + \frac{B_1^2}{8\pi}\right) \operatorname{div} \mathbf{Q} + \frac{B_s^2 - B_{se}^2}{8\pi}\left(\operatorname{div} \mathbf{Q} - 2\frac{Q e_r}{r}\right) - \frac{\mathbf{B}_1(\mathbf{B}_1 \nabla)\mathbf{Q}}{4\pi}\right] dV$$

$$= \oint_S \left[\left(p + \frac{B_1^2}{8\pi}\right)(\mathbf{Qn}) - \frac{(\mathbf{QB}_1)(\mathbf{B}_1 \mathbf{n})}{4\pi}\right] dS - \frac{1}{c}\int_V \mathbf{j}[\mathbf{B}_2 \mathbf{Q}] dV. \quad (2.83)$$

We introduce the notation

$$P = p + (B_s^2 - B_{se}^2)/8\pi; \quad (2.84)$$

$$T = Bp + B_1^2/8\pi - (B_s^2 - B_{se}^2)/8\pi. \quad (2.85)$$

Then, the integrand on the left-hand side of Eq. (2.83) can be written in the form

$$[\ldots] = P\left(\frac{\partial Q_r}{\partial r} + \frac{\partial Q_z}{\partial z}\right) + \frac{T}{r} Q_r + \frac{B_{1r}^2 - B_{1z}^2}{8\pi}\left(\frac{\partial Q_z}{\partial z} - \frac{\partial Q_r}{\partial r}\right)$$

$$- \frac{B_{1r} B_{1z}}{4\pi}\left(\frac{\partial Q_r}{\partial z} + \frac{\partial Q_z}{\partial r}\right). \quad (2.86)$$

Now we choose the vector $\mathbf{Q}$ such that the two last terms are reduced to zero. For convenience we introduce the complex variable

$$u = r - R + iz = \rho \exp(i\omega), \quad (2.87)$$

where R is some characteristic major radius.

If we choose the vector $\mathbf{Q}$ such that

$$Q_r = F(u), \quad Q_z = -iF(u), \quad (2.88)$$

where F is an analytic function of the complex variable u, then the two last terms of Eq. (2.86) become equal to zero. In particular, taking

$$F(u) = u^{m+1}, \quad (2.89)$$

we obtain

$$\int_V \left[2(m+1) P u^m + \frac{T}{r} u^{m+1}\right] dV = \oint_S \left[\left(p + \frac{B_1^2}{8\pi}\right)(n_r - in_z) - \right.$$

$$-\frac{(\mathbf{B}_1 \mathbf{n})(B_{1r}-B_{1z})}{4\pi}\right] u^{m+1}\, dS - \frac{1}{c}\int_V \mathbf{j}\,[\mathbf{B}_2\,\mathbf{Q}]\, dV. \tag{2.90}$$

In what follows we shall distinguish two approximations: the x approximation (large aspect ratio):

$$|r - R| \ll R, \tag{2.91}$$

and the ρ approximation (low toroidicity):

$$\rho \ll R. \tag{2.92}$$

The second restriction is evidently more stringent since it implies that the transverse dimensions of the cross section are small not only in the direction of the major radius, but also in the direction of the z axis. Therefore, the second approximation is unsuitable for strongly elongated systems with $z \approx R$.

Let us consider relation (2.90) for $m = -1$. Let $\mathbf{B}_2$ be the maintaining field of the external currents and $\mathbf{B}_1$ the field of the plasma current. If the volume of integration tends to infinity, the surface integral vanishes and we obtain the condition

$$\frac{1}{c}\int_V j_s B_{\text{ext},z}\, dV = -\int_V \frac{T}{r}\, dV, \tag{2.93}$$

or

$$\frac{1}{c}\int j_s B_{\text{ext},z}\, r\, dS_\zeta = -\int \left[p + \frac{B_{\text{pl}}^2}{8\pi} - \frac{B_s^2 - B_{sl}^2}{8\pi} \right] dS_\zeta. \tag{2.94}$$

The integral on the left-hand side represents the force (directed along the major radius) of interaction between the longitudinal current and the maintaining field. This force provides the equilibrium along the major radius. From Eq. (2.94) it is seen that it should compensate the ballooning force $\int p\, dS_\zeta$, the force of the interaction between the plasma poloidal current and the longitudinal field $\int \frac{B_{se}^2 - B_s^2}{8\pi}\, dS_\zeta$ and, also, the electromechanical force due to the action of the self-field of the plasma loop $\int \frac{B_{\text{pl}}^2}{8\pi}\, dS_\zeta$.

In the ρ approximation it may be assumed that the r dependence enters T in the form of the difference $r - R$ so that $dT/dr = -dT/dR$.

Noting that $T/r = \operatorname{div} \mathbf{T} e_r - dT/dr$ in the ρ approximation one obtains

$$\int \frac{T}{r} dV = -\int \frac{\partial T}{\partial r} dV \approx \frac{\partial}{\partial R} \int T dV. \qquad (2.95)$$

Here, we have used the condition that T vanishes at infinity. Further, we introduce the following definitions for the average transverse maintaining field and for the inductance of the plasma column:

$$B_\perp = -\frac{1}{2\pi R \mathcal{J}} \int j_s B_{\mathrm{ext},z} dV; \qquad (2.96)$$

$$L_{\mathrm{pl}} = \frac{c^2 \int B_{\mathrm{pl}}^2 dV}{4\pi \mathcal{J}^2} \qquad (2.97)$$

($\mathcal{J}$ is the total current in the plasma column). Then, in the ρ approximation Eq. (2.93) gives the expression for the transverse component of the maintaining field

$$B_\perp = -B_{z,\mathrm{ext}} = \frac{\mathcal{J}}{2cR}\left[\beta_\mathcal{J} + \mu_\mathcal{J} + \frac{\partial L_{\mathrm{pl}}}{2\pi \partial R}\right]. \qquad (2.98)$$

For $m = 0$, Eq. (2.90) gives the equilibrium integral condition along the minor radius

$$\int_V PdV + \frac{R_T-R}{2R_T}\int TdV = \frac{1}{2}\oint_S \left[\frac{B_{\tau 1}^2 - B_{n1}^2}{8\pi}\ (\mathbf{n}e_\rho)\right.$$
$$\left. -\frac{B_{\tau 1} B_{n1}}{4\pi}(\mathbf{\tau} e_\rho)\right]\rho dS - \frac{1}{2c}\int_V j_s [B_{z2}(r-R) - B_{r2}z]dV, \qquad (2.99)$$

where R_T is the "mean radius" of the toroidal configuration bounded by a chosen surface

$$R_T = \int TrdS_\zeta / \int TdS_\zeta; \qquad (2.100)$$

B_τ and B are the components of the poloidal magnetic field tangential and normal to S. Since the term with the factor $(R - R_T)\times (2R_T)^{-1}$ may be omitted in the x approximation, the left-hand side of Eq. (2.99) does not depend on the choice of S. The same holds for the right-hand side. Relationship (2.99) can be written in the form

$$\int P dV = \frac{\pi R \mathcal{J}^2}{c^2} S_1,$$

where S_1 is the corresponding integral appearing on the right-hand side of Eq. (2.99).

Since S_1 is independent of the choice of B_2 and, in particular, $B_2 = 0$ may be chosen, the value of S_1 can be expressed in terms of the total poloidal field measured outside the plasma. The latter equation is a generalization of the integral equilibrium condition (2.9) in a cylindrical plasma column for the case of a toroidal configuration:

$$\beta_{\mathcal{J}} = S_1 + \mu_{\mathcal{J}}. \tag{2.102}$$

The difference between S_1 and 1 is due to the noncircularity of the cross section. One can easily see this if one assumes that $B_2 = B_{\text{ext}}$. Then $B_{\tau 1} \approx 2\mathcal{J}/c\rho$ and $B_{n1} \approx 0$ (in the low toroidicity approximation) and the surface integral in Eq. (2.99) is equal to $\pi R \mathcal{J}^2/c^2$. For S_1 we then obtain

$$S_1 = 1 - \frac{c}{2\pi R \mathcal{J}^2} \int j_s [B_{z,\text{ext}} (r - R) - B_{r,\text{ext}} z] dV. \tag{2.103}$$

The uniform part of the maintaining field $B_\perp e_z$, which serves to compensate the toroidal repelling force, does not contribute to S_1. The difference between S_1 and 1 becomes appreciable if a field deforming the plasma cross section is included in B_{ext}. For example, for a straight plasma column with elliptical cross section, ratio of semiaxes λ, and flat current distribution the value of S_1 is

$$S_1 = 1 - (\lambda - 1)^2/(\lambda^2 + 1). \tag{2.104}$$

As follows from Eq. (2.102) in the absence of poloidal current $\mu_{\mathcal{J}} = 0$, the value of $\beta_{\mathcal{J}}$ differs from unity in general. Sometimes, it is convenient instead of $\beta_{\mathcal{J}}$ to use some parameter β_p, which is equal to 1, when there are no poloidal currents. In the approximation of low toroidicity one may assume that

$$\beta_p = \beta_{\mathcal{J}}/S_1, \quad \mu_p = \mu_{\mathcal{J}}/S_1. \tag{2.105}$$

Then, $\beta_p = 1 + \mu_p$. In order to introduce β_p invariantly for the general case we shall use the equilibrium equation (2.61) averaged over the angles, which can be written in the form

$$cdp/dV = -\frac{dF}{dV}\frac{d\Phi}{dV} + \frac{d\mathcal{J}}{dV}\frac{d\Psi}{dV}. \qquad (2.106)$$

Here the variable V has the meaning of the volume enclosed by a toroidal surface, so that $dV = 4\pi^2 \langle\sqrt{g}\rangle da$. Multiplying Eq. (2.106) by V and integrating by parts over the complete volume occupied by the plasma we obtain

$$c\int p\,dV = -\int V \frac{d\mathcal{J}}{dV}\frac{d\Psi}{dV}\,dV + \int V \frac{dF}{dV}\frac{d\Phi}{dV}\,dV. \qquad (2.107)$$

All the integrals here do not depend on the choice of the volume of integration since outside the plasma $p = 0$, $d\mathcal{J}/dV = 0$, $dF/dV = 0$. Defining β_p and μ_p in the following way:

$$\beta_p = -\frac{c\int p\,dV}{\int V \frac{d\mathcal{J}}{dV}\frac{d\Psi}{dV}\,dV}, \qquad (2.108)$$

$$\mu_p = -\frac{\int V \frac{dF}{dV}\frac{d\Phi}{dV}\,dV}{\int V \frac{d\mathcal{J}}{dV}\frac{d\Psi}{dV}\,dV}, \qquad (2.109)$$

we obtain a universal relation for arbitrary toroidal configurations:

$$\beta_p = 1 + \mu_p. \qquad (2.110)$$

For a plasma with an elliptical cross section and flat current density we have in accordance with Eq. (2.104)

$$\beta_{\mathcal{J}}/\beta_p = \mu_{\mathcal{J}}/\mu_p = 2\lambda/(\lambda^2 + 1).$$

In contrast to $\beta_{\mathcal{J}}$ and $\mu_{\mathcal{J}}$, the parameters β_p and μ_p contain integrals of the magnetic field over the plasma volume. Therefore, they are less convenient. They may be useful for optimizing the equilibrium configurations by numerical methods.

For $m = 1, 2, \ldots$, the integral equilibrium equation (2.90) allows us to relate the moments of the P value, such as $\int P x\,dV$, $\int P(x^2 - z^2)dV$, and so on, to the magnetic fields outside the plasma. This makes it possible to obtain these moments experimentally from probe measurements. In particular, Eq. (2.102) allows one to determine $\beta_{\mathcal{J}}$ from the measurement of a diamagnetic signal which gives the value of $\mu_{\mathcal{J}}$.

Instead of measuring the higher moments of the plasma pressure it is more convenient in practice to compare the measured value of the poloidal fields outside the plasma with the two-dimensional numerical calculations of the plasma equilibrium with a small number of fitting parameters [28, 29].

Together with the integral relation (2.90) relating the quadratic combinations of the magnetic fields one can use the theorem of mutuality of electrodynamics to obtain the relationship between the moments of the current density and the external field.

Multiply the equation

$$\frac{4\pi}{c} \mathbf{j} = \operatorname{curl} \mathbf{B} \tag{2.111}$$

by an auxiliary vector $\mathbf{q}$ which satisfies the following condition:

$$\operatorname{curl} \mathbf{q} = \nabla g. \tag{2.112}$$

The vector $\mathbf{q}$ and the scalar g can be considered as a vector and scalar potential of some vortex-free magnetic field. Using the vector identity

$$\operatorname{div} [\mathbf{B}\mathbf{q}] = \mathbf{q} \operatorname{curl} \mathbf{B} - \mathbf{B} \operatorname{curl} \mathbf{q}, \tag{2.113}$$

we integrate Eq. (2.111) multiplied by $\mathbf{q}$ over the volume V bounded by the surface S. The result is

$$\frac{4\pi}{c} \int \mathbf{j}\mathbf{q}\,dV = \oint \{[\mathbf{B}\mathbf{q}] + g\mathbf{B}\}\,d\mathbf{S}. \tag{2.114}$$

In the axisymmetric case $\partial/\partial\zeta = 0$, assuming $\mathbf{q} = (f/r)\mathbf{e}_s$, $dV = 2\pi r\,dS_\zeta$ we get

$$\frac{4\pi}{c} \int j_\zeta f\,dS_\zeta = \oint (fB_\tau + grB_n)\,dl. \tag{2.115}$$

On the left-hand side of Eq. (2.115) the integration is in fact over the cross section of the plasma; on the right-hand side it is around the contour defined by the intersection of the surface S with the meridional plane, and B_τ and B_n are the tangential and normal components of the magnetic field. The functions f and g satisfy the equations

$$\Delta^* f = \frac{\partial^2 f}{\partial r^2} - \frac{1}{r}\frac{\partial f}{\partial r} + \frac{\partial^2 f}{\partial z^2} = 0, \tag{2.116}$$

$$\frac{\partial g}{\partial z} = \frac{1}{r}\frac{\partial f}{\partial r}, \quad \frac{\partial g}{\partial r} = -\frac{1}{r}\frac{\partial f}{\partial z}. \tag{2.117}$$

For $f = 1$, $g = 0$ relation (2.115) gives the expression for the total current in terms of a contour integral. For $f \neq 1$ one can get information about the moments of the current distribution.

If we transfer to a rectilinear system $r = R - x$, $R \to \infty$, the functions f and g become harmonic and can therefore be taken in the form

$$f_m = (x + iz)^m, \quad g_m = -if_m/R \quad (m = 0, 1, 2, \ldots). \tag{2.118}$$

In the toroidal case we can take the linearly independent solutions of Eqs. (2.116) and (2.117) to be

$$\begin{aligned}
f^{(0)} &= 1, \quad g^{(0)} = 0; \\
f^{(1)} &= r^2, \quad g^{(1)} = 2z; \\
f^{(2)} &= r^4 - 4r^2 z^2, \quad g^{(2)} = 4r^2 z - \frac{8}{3}z^3; \\
f^{(3)} &= r^6 - 12r^4 z^2 + 8r^2 z^4, \quad g^{(3)} = 6r^4 z - 16r^2 z^3 + \frac{16}{5}z^5
\end{aligned} \tag{2.119}$$

and so on (where we have restricted ourselves to the case of configurations that are symmetric about the plane $z = 0$). To make a direct transition to a straight system one should take the following combinations for the toroidal multipole moments:

$$\begin{aligned}
f_1 &= -(f^{(1)} - R^2 f^{(0)})/2R; \\
f_2 &= (f^{(2)} - 2R^2 f^{(1)} + R^4 f^{(0)})/4R^2; \\
f_3 &= -(f^{(3)} - 3R^2 f^{(2)} + 3R^4 f^{(1)} - R^6)/8R^3,
\end{aligned} \tag{2.120}$$

where $R = \text{const}$ is the characteristic major radius of the configuration. Writing $x = r - R$, we obtain

$$\begin{aligned}
f_1 &= x(1 - x/2R), \quad g_1 = -z/R; \\
f_2 &= x^2(1 - x/2R)^2 - z^2(1 - x/R)^2, \quad g_2 = -2xz/R \\
&\quad + \left(x^2 z - \frac{2}{3}z^3\right)\Big/R^2; \\
f_3 &= x^3(1 - x/2R)^3 - 3xz(1 - x/2R)(1 - x/R)^2 - z^4/R(1 - x/R)^2; \\
g_3 &= -\frac{3x^2 z}{R}(1 - x/2R)^2 + (z^3/R)(1 - 4x/R + 2x^2/R^2) - (2/5)(z^5/R^3).
\end{aligned} \tag{2.121}$$

The following values of Y_m will be the moments of the current distribution for a toroidal plasma column:

$$Y_m = \frac{1}{\mathscr{J}} \int j_s f_m \, dS_\zeta = \frac{c}{4\pi \mathscr{J}} \oint (f_m B_\tau + rg_m B_n) \, dl. \qquad (2.122)$$

For the major radius R it is natural to choose the radius of the "center of the current channel," defined by the condition

$$R_{\mathscr{J}}^2 = \frac{1}{\mathscr{J}} \int j_s r^2 \, dS_\zeta. \qquad (2.123)$$

At $R = R_{\mathscr{J}}$ the first moment Y_1 becomes zero, $Y_1 = 0$.

Note in conclusion that the right-hand side of the integral relations for the current density is independent of the external maintaining fields, $\oint (f_m B_{\tau, \text{ext}} + rg_m B_{n, \text{ext}}) dl = 0$, and is determined by the self-field of the plasma current only. Moreover, in deriving these integral relations we have not used the equilibrium equations which limit the allowed current distribution j_s. Their inclusion permits the use of the method of moments to solve equilibrium problems (Section 3.6).

In contrast to the moments of the pressure, those of the current density (2.123) include only linear combinations of the magnetic fields and therefore they can be readily measured experimentally with Rogowski-coil-type probes with specially designed windings. This allows the characteristics of the geometry of the current channel to be followed in machines with a noncircular cross section [30].

CHAPTER 3

EXACT SOLUTIONS OF THE EQUILIBRIUM

EQUATIONS

In this chapter typical configurations described by a complete exact solution of the equilibrium equations both inside and outside the plasma will be considered. The exact solutions allow for a quantitative description of the interaction of the plasma current with the external maintaining fields. The equilibrium of a straight plasma column with an elliptical cross section and a flat current distribution gives an insight into the effect of the maintaining field on the plasma shape. The existence of a critical amplitude of the maintaining field will be established above which equilibrium becomes impossible. The analytical solution for a plasma column with an elliptical cross section and helical symmetry shows the effect of the longitudinal magnetic field on the equilibrium. The exact solution for a toroidal plasma with circular cross section describes the main toroidal effect, which, first of all, consists of the compensation of the hoop force of the closed plasma column by the transverse maintaining field, and secondly, of a $\beta_{\mathscr{Y}}$ limitation. A spherical equilibrium configuration of the Hill-vortex type is the limiting case of a small-aspect-ratio toroidal plasma equilibrium in a uniform maintaining field.

Some typical magnetostatic problems relevant to equilibrium theory are also considered in this chapter, as well as the basic numerical methods developed for equilibrium problems.

3.1. Straight Plasma Column with an Elliptical Cross Section

The equilibrium equation (2.29) reduces to a linear one in the case of a uniform current density $j_s(A_s) = j_{pl} = \text{const}$. Let us use the elliptical cylinder coordinates u, v related to the Cartesian coordinates x, y by

$$x = d \,\text{sh}\, u \cos v, \quad y = d \,\text{ch}\, u \sin v, \tag{3.1}$$

where $u = u_0$ is assumed at the plasma boundary. Denote the minor semiaxis by a and the ratio of semiaxes by λ. Then

$$a = d \,\text{sh}\, u_0, \quad \lambda = \text{cth}\, u_0 \tag{3.2}$$

and

$$d^2 = a^2(\lambda^2 - 1); \quad \text{sh}\, 2u_0 = 2\lambda/(\lambda^2 - 1);$$
$$\text{ch}\, 2u_0 = (\lambda^2 + 1)/(\lambda^2 - 1); \quad \exp(2u_0) = (\lambda + 1)^2/(\lambda^2 - 1). \tag{3.3}$$

Equation (2.29) for the longitudinal component $A = A_s$ of the vector potential assumes the form

$$\Delta A = \frac{2}{d^2 (\operatorname{ch} 2u + \cos 2v)} \left(\frac{\partial^2 A}{\partial u^2} + \frac{\partial^2 A}{\partial v^2} \right) = \begin{cases} -\dfrac{4\pi}{c} j_{\text{pl}}, & u < u_0; \\ 0, & u > u_0. \end{cases} \quad (3.4)$$

The solution that satisfies the boundary conditions (2.22) and (2.25) is given inside the plasma [31] by

$$A_i = -\frac{4\pi}{c} j_{\text{pl}} \frac{d^2}{8} \left[\operatorname{ch} 2u - \operatorname{ch} 2u_0 - \cos 2v + \frac{\operatorname{ch} 2u}{\operatorname{ch} 2u_0} \cos 2v \right]$$
$$= -\frac{4\pi}{c} j_{\text{pl}} \frac{\lambda^2 a^2}{2(\lambda^2 + 1)} \left(\frac{x^2}{a^2} + \frac{y^2}{\lambda^2 a^2} \right) \quad (3.5)$$

and in the outside vacuum region by

$$A_e = -\frac{4\pi}{c} j_{\text{pl}} \frac{d^2}{8} \left[2(u - u_0) \operatorname{sh} 2u_0 + \frac{\operatorname{sh} 2u_0}{\operatorname{ch} 2u_0} \operatorname{sh} 2(u - u_0) \cos 2v \right] \quad (3.6)$$

The magnetic surfaces A = const inside the plasma have elliptical cross sections with constant ratio of the semiaxes, while the external surfaces have a separatrix with X points at $v = \pi/2$, $u = 2u_0$ which separates nested closed magnetic surfaces near the plasma from open ones (Fig. 2). Only the magnetic configuration of a straight plasma column with a circular cross section has no separatrix.

Let us extract the maintaining field from the total vacuum field. It is regular inside the plasma and contains singularities which are located only outside the plasma boundary including infinity. The vector potential of the maintaining field is of the form

$$A_{\text{ext}} = -\frac{4\pi}{c} j_{\text{pl}} \frac{d^2}{8} \frac{\operatorname{sh} 2u_0}{\operatorname{ch} 2u_0} \exp(-2u_0) \operatorname{ch} 2u \cos 2v$$
$$= \frac{2\pi}{c} j_{\text{pl}} (y^2 - x^2) \frac{\lambda(\lambda - 1)}{(\lambda^2 + 1)(\lambda + 1)}. \quad (3.7)$$

The maintaining field is now just a quadrupole. Its interaction with the plasma current produces the force which tends to stretch the plasma cross section. It is the balance between this force and the plasma self-contraction which determines the cross-sectional shape of the plasma. Let us analyze it in detail.

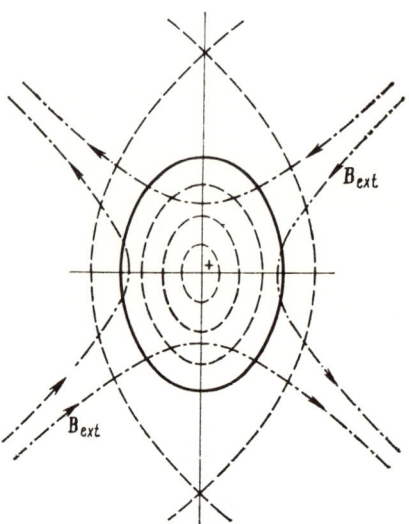

Fig. 2. The equilibrium configuration for a plasma column with an elliptical cross section. The ratio of semiaxes is $\lambda = 1.5$; the solid line shows the cross section of the current channel; the dashed line shows the separatrix; the dashed lines with arrows show the field lines of the maintaining field.

Let A_2 be the amplitude of the vector potential in Eq. (3.7)

$$A_{\text{ext}} = \frac{2\pi}{c} A_2 (y^2 - x^2), \tag{3.8}$$

so that

$$A_2/j_{\text{pl}} = \varphi(\lambda) = \lambda(\lambda - 1)/[(\lambda^2 + 1)(\lambda + 1)]. \tag{3.9}$$

The function $\varphi(\lambda)$ has a maximum at $\lambda = 2.8900$ ($\varphi_{\max} = 0.1501$) as is shown in Fig. 3. It implies two equilibria at $A_2/j_{\text{pl}} < 0.15$ corresponding to $\lambda < 2.89$ and $\lambda > 2.89$ (Fig. 4). One of them can be considered as resulting from the application of the quadrupole maintaining field to a circular plasma cylinder and the other as resulting from the same field but applied to a current-carrying slab. If the amplitude of the qudrupole field is increased above a critical value $A_{2,\text{cr}} = 0.15 j_{\text{pl}}$, the equilibrium becomes impossible because

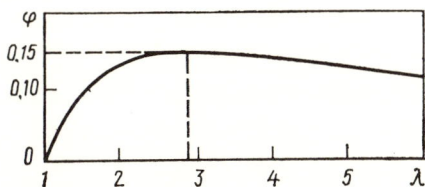

Fig. 3. The function $\varphi(\lambda) = A_2/j_{pl}$ vs the ratio of semiaxes $\lambda = l_y/l_x$.

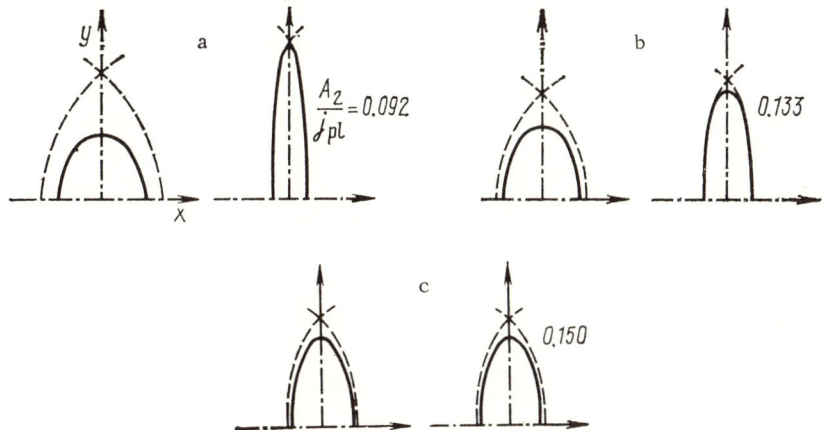

Fig. 4. Possible pairs of equilibria of a plasma column with a uniform current distribution in quadrupole maintaining fields with different A_2/j_{pl}.

the force of interaction with the maintaining field exceeds the self-contraction. In other words, the external field disrupts the plasma column.

The critical amplitude of the quadrupole field depends on the plasma current density only, $A_{2,cr}/j_{pl} = 0.15$. Hence an equilibrium with homogeneous current density distribution becomes impossible everywhere in the cross section.

It is easy to present a picture of the equilibrium in an external quadrupole field also for a nonuniform plasma current distribution which decreases towards the boundary. When the external field is increased, the cross section of the plasma column starts to elongate (we consider the plasma column situated in a strong longitudinal field that preserves the cross section of the magnetic surfaces). When the maintaining field reaches a critical value $A_{2,cr}$ [$A_{2,cr} < 0.15 j_{pl}(0)$, $j_{pl}(0)$ is the current density on the magnetic axis], equilibrium with a given current distribution becomes impossible because the plasma current ceases to confine the periphery of the plasma column. Thus, some inner magnetic surface begins to play the role of the plasma boundary. Finally, when the amplitude A_2 reaches the critical value for the maximum current density $A_2 = 0.15 j_{pl}(0)$ the external field completely disrupts the plasma column.

The relationship between the critical value of the maintaining field and the current density given above can also be derived from the following qualitative arguments. When the amplitude of the quadrupole field increases, the separatrix of the outside magnetic surfaces approaches the plasma column. The coincidence of the separatrix with the plasma boundary gives the critical value of the maintaining field. One can estimate the location of the X point of the separatrix where the poloidal magnetic field is zero, $\mathbf{B} = 0$, by equating the magnetic field of the plasma current $B_{pl} \approx 2\mathcal{I}/c\rho$ with the maintaining field, $B_x = \partial A_{ext}/\partial y = 4\pi y A_2/c$. It follows that the y coordinate of the X point of the separatrix is approximately given by

$$y^2 = (a^2 \lambda/2)(j_{pl}/A_2). \qquad (3.10)$$

Assuming that for a marginal configuration $y = \lambda a$, we obtain the critical amplitude of the maintaining field

$$A_{2,cr} = j_{pl}/2\lambda. \qquad (3.11)$$

Though these qualitative arguments do not determine the critical value of the elongation λ, Eq. (3.11) is in good agreement with the exact formula when $\lambda = 2.89$ is substituted into Eq. (3.11).

It is of importance that similar qualitative arguments give the correct functional dependence of the critical amplitude of the maintaining field on the plasma current density. Hence, they can be used to analyze the effects of other types of multipole external fields on plasma equilibria.

In a multipole field of the form

$$A_s = (2\pi/c)A_m \rho^m \cos m\omega \tag{3.12}$$

the cross section of the plasma column can be described in a linear approximation by the formula

$$\rho = a + \xi \cos m\omega. \tag{3.13}$$

Here a is the minor radius of the plasma column, ξ is the amplitude of the perturbation of the plasma boundary, which, for a uniform current density, j_{pl} = const, is expressed in terms of the amplitude of the maintaining field by

$$A_m = \frac{\xi}{a^{m-1}} \frac{m-2}{m} j_{pl}. \tag{3.14}$$

Now we determine the relationship between the critical value of the maintaining field and the plasma current density by means of the above-mentioned qualitative arguments. Assuming that for m > 2 the deviation of the plasma boundary from a circular cylinder is small even for the critical situation, we have

$$A_{m,\,cr} \approx j_{pl}/ma^{m-2}. \tag{3.15}$$

The formal substitution of $A_{m,\,cr}$ (3.15) into Eq. (3.14) gives $\xi/a = 1/(m-1)$, which justifies the procedure followed. In contrast to the m = 2 case, when m > 2 the radius of the plasma column enters the critical value of the maintaining field.

Equation (3.15) allows us to consider the process of the formation of the plasma column in a previously produced maintaining multipole field, which then plays the role of a magnetic limiter. In the quadrupole field a closed magnetic configuration can be formed only if, from the beginning, the plasma current density exceeds the critical one, $j_{pl} > A_2/0.15$. In multipole fields m ≥ 3 a closed magnetic configuration is possible for an arbitrary plasma current density. Assuming that the current channel is limited by the separatrix during the increase of the total current, due to large transport coefficients along the field lines, it is possible to write at once the relationship between the radius of the plasma column and the total current $\mathcal{J}$

$$a = (\mathcal{J}/\pi m A_m)^{1/m}. \tag{3.16}$$

The resulting current density profile for the situation with a magnetic limiter is of the form

$$j_s(a) = \frac{1}{2\pi a} \frac{d\mathscr{J}}{da} = \frac{m^2}{2} A_m a^{m-2}. \tag{3.17}$$

Thus, in a hexapole external field, m = 3, the current density will be linear in the radius, i.e., $j_s(a) \sim a$, whereas in an octupole field $j_s(a) \sim a^2$, and so on. The simple analysis given here is in good agreement with numerical simulations which take transport processes into account [35].

3.2. Equilibrium of a Plasma Column with Elliptical Cross Section and Helical Symmetry

The equilibrium equation (2.44) for the helical flux function Ψ^* has an explicit exact solution only in the case of vacuum magnetic fields, p' = 0, F' = 0. We do not know of a coordinate system for this equation which would admit the separation of variables and allow one to obtain the solution both inside and outside the plasma current-carrying column by the matching techniques of Section 3.1. However, the main feature of the equilibrium of a plasma column with helical symmetry can be formulated in the long-wavelength approximation. In this limit the helical problem is similar to a straight one [Eq. (2.47)] and it is possible to use the coordinates of an elliptical cylinder to obtain an exact solution for a plasma column with an elliptical cross section and a uniform current density [36].

Let the cross section of the magnetic surfaces be characterized by the same parameters as for a straight column (Section 3.1). For the case of helical symmetry we have

$$\Delta\Psi^* = \frac{2}{d^2(\operatorname{ch} 2u + \cos 2v)} \left(\frac{\partial^2 \Psi^*}{\partial u^2} + \frac{\partial^2 \Psi^*}{\partial v^2} \right) = \begin{cases} -\dfrac{4\pi}{c}(j_{\mathrm{pl}} - j_B), & u < u_0; \\ \dfrac{4\pi}{c} j_B, & u > u_0, \end{cases} \tag{3.18}$$

where j_{pl} = const is the plasma current density and $(4\pi/c)j_B = (2n/mR) \times B_s$ = const. The solution of Eq. (3.18) within the plasma is given by

$$\Psi_i^* = -\frac{4\pi}{c}(j_{\mathrm{pl}} - j_B)\frac{d^2}{8}\left[\operatorname{ch} 2u - \operatorname{ch} 2u_0 - \cos 2v + \frac{\operatorname{ch} 2u}{\operatorname{ch} 2u_0}\cos 2v\right] =$$

$$= -\frac{4\pi}{c}(j_{\text{pl}} - j_B)\frac{\lambda^2 a^2}{2(\lambda^2 + 1)}\left(\frac{x^2}{a^2} + \frac{y^2}{\lambda^2 a^2}\right), \qquad (3.19)$$

and outside it by

$$\Psi_e^* = -\frac{4\pi}{c} j_{\text{pl}} \frac{a^2}{8}\left[2(u - u_0)\operatorname{sh} 2u_0 + \frac{\operatorname{sh} 2u_0}{\operatorname{ch} 2u_0}\operatorname{sh} 2(u - u_0)\cos 2v\right]$$

$$+ \frac{4\pi}{c} j_B \frac{d^2}{8}\left[\operatorname{ch} 2u - \operatorname{ch} 2u_0 - \cos 2v + \frac{\operatorname{ch} 2u}{\operatorname{ch} 2u_0}\cos 2v\right]. \qquad (3.20)$$

Here, to the solution for A_i and A_e, obtained in Section 3.1, the part which takes into account the presence of j_B on the right-hand side of Eq. (3.18) is simply added. Extracting Ψ_{ext}^* one should bear in mind that the total flux Ψ_e^* is the sum of the flux Ψ_{pl}^* of the magnetic field of the plasma current, the flux of the maintaining field Ψ_{ext}^*, and the flux of the longitudinal magnetic field Ψ_{pl}^* through the helical surface:

$$\Psi_e^* = \Psi_{\text{pl}}^* + \Psi_{\text{ext}}^* + \Psi_B^*. \qquad (3.21)$$

By subtracting from Ψ_e^* the flux Ψ_B^* of the longitudinal field

$$\Psi_B^* = \frac{4\pi}{c} j_B \frac{x^2 + y^2}{4} = \frac{4\pi}{c} j_B \frac{d^2}{8}(\operatorname{ch} 2u - \cos 2v) \qquad (3.22)$$

and using the same procedure as in Section 3.1 to separate Ψ_{ext}^* we obtain

$$\Psi_{\text{ext}}^* = -\frac{4\pi}{c}\frac{d^2}{8}\left[j_{\text{pl}}\frac{\operatorname{sh} 2u_0}{\operatorname{ch} 2u_0}\exp(-2u_0) - j_B\frac{1}{\operatorname{ch} 2u_0}\right]\operatorname{ch} 2u \cos 2v$$

$$= \frac{\pi}{c}(y^2 - x^2)\frac{\lambda^2 - 1}{\lambda^2 + 1}\left[\frac{2\lambda}{(\lambda + 1)^2} j_{\text{pl}} - j_B\right]. \qquad (3.23)$$

For $j_B = 0$ ($n = 0$) Eq. (3.23) reduces to Eq. (3.7). The presence of the term $(4\pi/c)j_B = (2n/mR)B_s$ is the main distinctive feature of helical equilibria of a current-carrying plasma column. It indicates that besides the plasma self-contraction force and that of the interaction between the plasma current and the quadrupole external field, in a helical configuration there is also the additional force of interaction between the plasma current flowing along the helical lines and the longitudinal magnetic field B_s (Fig. 5). For these configurations the latter field acts to some extent as the maintaining field.

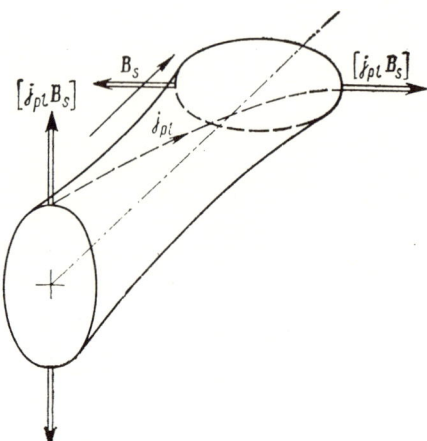

Fig. 5. The force of interaction of the plasma current with a longitudinal magnetic field for a column with helical symmetry.

As a result, the amplitude of the quadrupole field is smaller as compared with the case of a straight plasma column.

For a helical column it is possible to obtain an equilibrium without the maintaining field of helical windings. To realize this possibility it is necessary that

$$j_B = 2\lambda j_{pl}/(\lambda + 1)^2. \qquad (3.24)$$

For $\lambda \to 1$ Eq. (3.24) reduces to $j_B = j_{pl}/2$.

Here, we can see the relationship between the theory of helical equilibria and that of kink instabilities of a circular plasma cylinder. In fact, when the longitudinal field is such that $j_B = j_{pl}/2$ a plasma column which initially has a circular cross section also exhibits an independent equilibrium shape with respect to $m = 2$ helical perturbations. If the longitudinal field is somewhat larger, i.e., $j_B > j_{pl}/2$, then any $m = 2$ deformation of the plasma cross section grows, since the force associated with the longitudinal field exceeds the force of self-contraction. The relation $j_B = j_{pl}/2$ corresponds to the left boundary of the unstable region of the $m = 2$ kink mode for flat plasma current density: $m - 1 < nq < m$ [$q = (c/4\pi)(2B_s/Rj_{pl})$],

since
$$2j_B/j_{pl} = nq = 1. \tag{3.25}$$

Consider the magnetic field configuration outside the plasma. At a large distance from the plasma boundary, where the field of the plasma current is small, the flux function is of the form

$$\Psi^*_{e,\infty} = \Psi^*_B + \Psi^*_{ext} = \frac{\pi}{c} j_B [x^2 + y^2 + \varepsilon_\infty (x^2 - y^2)], \tag{3.26}$$

where

$$\varepsilon_\infty = \frac{\lambda^2 - 1}{\lambda^2 + 1} \left[1 - \frac{2\lambda}{(\lambda+1)^2} \frac{j_{pl}}{j_B} \right] \tag{3.27}$$

The parameter ε_∞ characterizes the geometry of the magnetic surface outside the plasma. For $|\varepsilon_\infty| > 1$ the surfaces are not closed, while for $|\varepsilon_\infty| < 1$ the cross sections of the magnetic surfaces are elliptical. For this case $\varepsilon_\infty = (\lambda_\infty^2 - 1)/(\lambda_\infty^2 + 1)$, where λ_∞ is the ratio of the semiaxes. Depending on the relative magnitudes of j_{pl} and j_B, the following configurations are possible (Fig. 6):

1. Strong plasma current. In this case the current is strong enough to satisfy the condition $|\varepsilon_\infty| > 1$. The longitudinal field plays a minor role for the magnetic configuration and its topology is analogous to that of a straight column with translational symmetry (Fig. 6a).

2. The case of a moderate current. In this case j_{pl} is small enough to satisfy $|\varepsilon_\infty| < 1$ but still so large that $j_{pl} > j_B$. The magnetic surfaces are closed everywhere (Fig. 6b,c). Near the plasma the magnetic field B^* is in the same direction as the field of the plasma current; further away from the plasma there is a closed separatrix, within which two magnetic islands are formed. Beyond this separatrix the magnetic surfaces again have a simple shape like ellipses but the magnetic field B^* changes its direction.

In the case of a moderate current it is useful to distinguish between two possibilities: a) The major semiaxis of the external ellipses is perpendicular to that of the plasma cross section $\varepsilon_\infty < 0$ (Fig. 6b). b) The major semiaxis of the external ellipses and that of the plasma cross section are parallel $\varepsilon_\infty > 0$ (Fig. 6c). The transition from case a) to case b) occurs when the boundary of the kink instability is crossed.

3. The case of a weak plasma current, $j_{pl} < j_B$. In this case the magnetic field B^* is everywhere directed opposite to the field

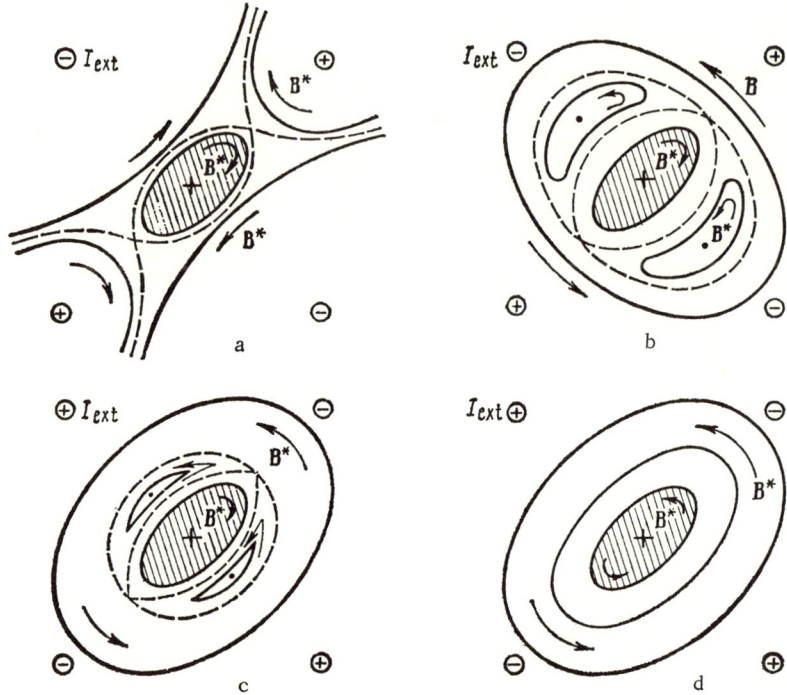

Fig. 6. Possible kinds of equilibria for a straight plasma column with helical symmetry: a) the case of a strong plasma current, $|\varepsilon_\infty| > 1$; b) an equilibrium with a moderate plasma current, $-1 < \varepsilon_\infty < 0$; c) an equilibrium with a moderate plasma current, $0 < \varepsilon_\infty < 1$, $j_{pl} > j_B$; the configuration is unstable to kink modes; d) an equilibrium of the stellarator type with weak plasma current, $j_{pl} < j_B$.

of the plasma $\mathbf{B}_{pl}$. The magnetic configuration is simple, having no separatrices. In configurations of this type separatrices may arise because of deviations from the long-wave approximation.

In the case of weak current the current of the plasma is only of minor importance and the geometry is determined by the longitudinal field Ψ_B^* and the field Ψ_{ext}^* of the helical windings. The possibility of forming closed surfaces in the absence of the longitudinal current is a feature of configurations with a helical symmetry, which is the basis of stellarators with helical windings.

The properties of a helical equilibrium also hold qualitatively for other shapes of the plasma cross section. If the contour of the plasma cross section is described by

$$\rho = a + \xi \cos m\theta, \quad \theta = \omega - \varkappa s, \quad (3.28)$$

then the maintaining field is given in the linear approximation by

$$\Psi^*_{ext} = \frac{\pi}{c} a^2 \frac{\rho^m}{a^m} \cos m\theta \frac{2\xi}{a} \left[\frac{m-1}{m} j_{pl} - j_B \right] \quad (3.29)$$

for a flat current distribution $j_s = j_{pl} = $ const. Outside the plasma the flux function is given by

$$\Psi^*_e = \frac{\pi}{c} j_B \left[\rho^2 - 2a\xi \frac{\rho^m}{a^m} \cos m\theta \right] - \frac{\pi}{c} j_{pl} a^2 \left[\ln \frac{\rho^2}{a^2} - 2 \frac{\xi}{a} \frac{\rho^m}{a^m} \cos m\theta \right]$$
$$- \frac{\pi}{c} j_{pl} \frac{2\xi a}{m} \left(\frac{\rho^m}{a^m} - \frac{a^m}{\rho^m} \right) \cos m\theta. \quad (3.30)$$

As in the case m = 2 the magnetic configuration outside the plasma may be one of three types, corresponding to: 1) the equilibrium of a straight plasma with noncircular cross section and with an open separatrix, 2) an equilibrium with a closed separatrix and magnetic islands, and 3) a stellarator equilibrium with a simple topology.

3.3. Plasma Torus with Circular Cross Section

For the toroidal case an analytical solution of the equilibrium equation (2.21) can be obtained for a quasiuniform current density [38], when

$$2\pi c \frac{dp}{d\Psi} = A = \text{const}, \quad \frac{1}{c} \frac{dF^2}{d\Psi} = B = \text{const}, \quad (3.31)$$

while the plasma column has a circular cross section. In this case we can make use of toroidal coordinates η, ω, ζ,

$$r = R_0 \operatorname{sh} \eta / (\operatorname{ch} \eta - \cos \omega), \quad z = R_0 \sin \omega (\operatorname{ch} \eta - \cos \omega), \quad (3.32)$$

where we consider the surface $\eta = \eta_0$ as the plasma boundary. Within the plasma $\eta > \eta_0$, outside $\eta < \eta_0$. On the axis of the toroidal coordinates, where $r = R_0$, $z = 0$, the coordinate $\eta = \infty$, while in the axis of symmetry $r = 0$ and for $r^2 + z^2 \to \infty$ we have $\eta = 0$. The minor radius a and the major radius R of the torus are given by

$$a = R_0/\operatorname{sh}\eta_0, \quad R = R_0 \operatorname{cth}\eta_0. \tag{3.33}$$

Under condition (3.31) the current density is uniform in the direction of the z axis:

$$j_s = Ar + B/r. \tag{3.34}$$

The toroidal coordinates allow us to separate variables provided Ψ is expressed in terms of an auxiliary function $F(\eta, \omega)$:

$$\Psi(\eta, \omega) = F(\eta, \omega)/\sqrt{2(\operatorname{ch}\eta - \cos\omega)}, \tag{3.35}$$

for which Eq. (2.21) reduces to the form

$$\frac{\partial^2 F}{\partial \eta^2} - \operatorname{cth}\eta \frac{\partial F}{\partial \eta} + \frac{\partial^2 F}{\partial \omega^2} + \frac{1}{4} F$$
$$= -\frac{32\pi^2 R_0^2}{c} \frac{r}{[2(\operatorname{ch}\eta - \cos\omega)]^{3/2}} j_s = S(\eta, \omega). \tag{3.36}$$

When $p'(\Psi)$ and $FF'(\Psi)$ are given by Eq. (3.31), the right-hand side $S(\eta, \omega)$ of Eq. (3.36) is known. If $F(\eta, \omega)$ and $S(\eta, \omega)$ are expanded in a Fourier series

$$F(\eta, \omega) = \sum_{n=0}^{\infty} F_n(\eta) \cos n\omega; \tag{3.37}$$

$$S(\eta, \omega) = \sum_{n=0}^{\infty} S_n(\eta) \cos n\omega, \tag{3.38}$$

we obtain equations for the coefficients $F_n(\eta)$

$$\frac{d^2 F_n}{d\eta^2} - \operatorname{cth}\frac{dF_n}{d\eta} - \left(n^2 - \frac{1}{4}\right) F_n = S_n(\eta), \quad n = 0, 1, 2\ldots \tag{3.39}$$

Linearly independent solutions for the corresponding uniform equation are the Fock functions

$$g_n(\eta) = \frac{1}{2\pi} \int_0^{2\pi} \sqrt{2(\operatorname{ch}\eta - \cos\omega)} \cos n\omega\, d\omega; \tag{3.40}$$

$$j_n(\eta) = \frac{1}{2\pi} \int_{-\eta}^{\eta} \sqrt{2(\operatorname{ch}\eta - \operatorname{ch} t)} \operatorname{ch} nt\, dt, \tag{3.41}$$

which are regular everywhere in space, excluding $\eta = 0$ for $g_n(\eta)$ and $\eta = \infty$ for $f_n(\eta)$. The Fock functions are expressed in terms of associated Legendre functions

$$(n^2 - 1/4) f_n(\eta) = \operatorname{sh} \eta P^1_{n-1/2}(\operatorname{ch} \eta),$$
$$(n^2 - 1/4) g_n(\eta) = \operatorname{sh} \eta Q^1_{n-1/2}(\operatorname{ch} \eta). \qquad (3.42)$$

For a quasiuniform current density (3.34) the functions $S_n(\eta)$ on the right-hand side of Eq. (3.39) can be written in the form

$$S_n(\eta) = \begin{cases} \dfrac{32\pi^2 R_0^2}{c} \left[\dfrac{8AR_0^2}{15} \operatorname{sh} \eta \left(\dfrac{1}{\operatorname{sh} \eta} \left(\dfrac{1}{\operatorname{sh} \eta} \left(\dfrac{g'_n(\eta)}{\operatorname{sh} \eta} \right)' \right)' \right)' \\ + \dfrac{2B}{\operatorname{sh}^2 \eta} \left(n^2 - \dfrac{1}{4} \right) g_n(\eta) \right], \quad \eta > \eta_0; \\ 0, \quad \eta < \eta_0, \end{cases} \qquad (3.43)$$

$$n = 1, 2, 3 \ldots$$

For $n = 0$ the right-hand side of Eq. (3.43) should be multiplied by 2. The Wronskian of the functions f_n, g_n is equal to

$$f_n g'_n - g_n f'_n = \operatorname{sh} \eta / [\pi (n^2 - 1/4)]. \qquad (3.44)$$

Having obtained explicit solutions of the uniform equations (3.39) one can now evaluate the expression for the total flux Ψ and also for the fluxes Ψ_{pl} of the plasma self-field and Ψ_{ext} of the maintaining field

$$\Psi(\eta, \omega) = \dfrac{\pi}{\sqrt{2(\operatorname{ch} \eta - \cos \omega)}} \sum_{n=0}^{\infty} \left(n^2 - \dfrac{1}{4} \right) \Bigg\{ g_n(\eta) \int_{\eta_0}^{\eta} \dfrac{f_n(t) S_n(t)}{\operatorname{sh} t} dt$$
$$+ f_n(\eta) \int_{\eta}^{\infty} \dfrac{g_n(t) S_n(t)}{\operatorname{sh} t} dt - g_n(\eta) \dfrac{f_n(\eta_0)}{g_n(\eta_0)} \int_{\eta_0}^{\infty} \dfrac{g_n(t) S_n(t)}{\operatorname{sh} t} dt \Bigg\} \cos n\omega; \qquad (3.45)$$

$$\Psi_{pl}(\eta, \omega) = \dfrac{\pi}{\sqrt{2(\operatorname{ch} \eta - \cos \omega)}} \sum_{n=0}^{\infty} (n^2 - 1/4) \Bigg\{ g_n(\eta) \int_{\eta_0}^{\eta} \dfrac{f_n(t) S_n(t)}{\operatorname{sh} t} dt$$
$$+ f_n(\eta) \int_{\eta}^{\infty} \dfrac{g_n(t) S_n(t)}{\operatorname{sh} t} dt \Bigg\} \cos n\omega; \qquad (3.46)$$

$$\Psi_{ext}(\eta, \omega) = \dfrac{-\pi}{\sqrt{2(\operatorname{ch} \eta - \cos \omega)}} \sum_{n=0}^{\infty} \left(n^2 - \dfrac{1}{4} \right) \dfrac{f_n(\eta_0)}{g_n(\eta_0)} g_n(\eta) \times$$

$$\times \int_{\eta_0}^{\infty} \frac{g_n(t) S_n(t)}{\operatorname{sh} t} dt \cos n\omega. \qquad (3.47)$$

The function $\Psi(\eta, \omega)$ is regular inside the plasma column and in particular at the origin of the toroidal coordinate system ($\eta = \infty$) and is equal to zero on the plasma boundary ($\eta = \eta_0$): $\Psi(\eta_0, \omega) = 0$. The flux of the plasma field Ψ_{pl} is regular everywhere and is equal to zero at $\eta = 0$, i.e., at $r = 0$ and for $r^2 + z^2 \to \infty$. Outside the plasma $\eta < \eta_0$ the coefficients in front of $g_n(\eta)$ in Eq. (3.46) reduce to zero and Ψ_{pl} is expressed in the form of a series of the functions $f_n(\eta)$:

$$\Psi_{pl}(\eta, \omega)|_{\eta < \eta_0}$$
$$= \frac{\pi}{\sqrt{2(\operatorname{ch}\eta - \cos\omega)}} \sum_{n=0}^{\infty} \left(n^2 - \frac{1}{4}\right) f_n(\eta) \int_{\eta_0}^{\infty} \frac{g_n(t) S_n(t)}{\operatorname{sh} t} dt \cos n\omega. \qquad (3.48)$$

The use of toroidal coordinates not only permits one to obtain the analytical solution for an arbitrary value of the aspect ratio R/a, but also gives the form of the functions $f_n(\eta)$ which decrease for $r^2 + z^2 \to \infty$ and in terms of which the self-field of any other axisymmetric configuration that is finite in space may be expressed. This gives us the possibility to develop a numerical method for solving the equilibrium equation (2.21) for arbitrary current distributions using toroidal coordinates. In the method of expansion with respect to the curvature $[(r - R)^2 + z^2]^{1/2}/R \ll 1$ the known functions f_n and g_n also permit one to separate the parts which decrease and those which increase as $r^2 + z^2 \to \infty$ in the expressions which are valid near the axis of the coordinate system. This cannot be done by exploiting the expansion with respect to curvature only.

Solutions for a toroidal plasma with a circular cross section have been analyzed by means of a computer in [39]. A typical equilibrium configuration is shown in Fig. 7. The external magnetic surfaces of a toroidal column have a separatrix which separates the region of closed magnetic surfaces from that of open ones. The separatrix has an X point where the poloidal field vanishes, $\mathbf{B}_p = 0$, which is located between the plasma and the axis of symmetry. In [39] a comparison has been carried out of analytical formulas obtained in the large-aspect-ratio approximation [$R/a \gg 1$, $\exp(\eta_0) \gg 1$] with exact solutions.

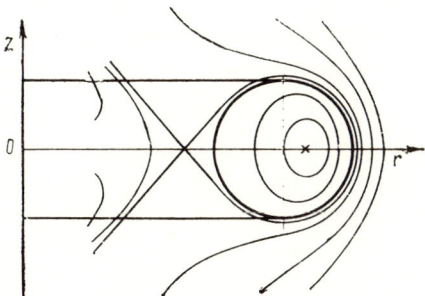

Fig. 7. Magnetic surfaces for a toroidal plasma column with circular cross section, $R/a = 3.7$, $\bar{\beta}_{\mathcal{J}} = 4$ [74].

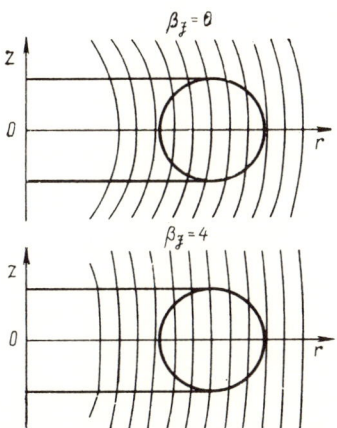

Fig. 8. The field lines of the maintaining magnetic field for a toroidal column with circular cross section, $R/a = 3.7$.

Denote by $\mathcal{J}$ the total plasma current

$$\mathcal{J} = \int j_s dS_\zeta = \int (Ar + B/r) dr dz, \qquad (3.49)$$

while $\bar{\beta}_p$ is the parameter characterizing the plasma pressure

$$\bar{\beta}_p = \frac{1}{\mathcal{J}} \int Ar dS_\zeta = \frac{2\pi c \int \frac{dp}{d\Psi} dS_\zeta}{\mathcal{J}}. \tag{3.50}$$

For a column with a circular cross section and a large aspect ratio $\bar{\beta}_p$ is equal to β_p (Section 2.6). When $\bar{\beta}_p$ increases, the X point of the separatrix approaches the plasma boundary and at

$$\bar{\beta}_p = R/a + 1/2 \tag{3.51}$$

the separatrix coincides with the plasma boundary. Comparison of Eq. (3.51) with the exact solution shows that formula (3.51) provides an accuracy of 10% even for $R/a = 1.5$.

The maintaining field for a column with a circular cross section has a mean value

$$B_\perp = -B_{z,\text{ext}} = \frac{\mathcal{J}}{cR} \left(\ln \frac{8R}{a} + \bar{\beta}_p - \frac{5}{4} \right) \tag{3.52}$$

and a decay index

$$n = -\frac{r}{B_{z,\text{ext}}} \frac{dB_{z,\text{ext}}}{dr}\bigg|_{r=R_0} = \frac{3}{4} \frac{\ln \frac{8R}{a} - \frac{17}{12}}{\ln \frac{8R}{a} + \bar{\beta}_p - \frac{5}{4}}. \tag{3.53}$$

The accuracy of these formulas is the same as that of Eq. (3.51).

3.4. Compact Toroidal Plasma Column

Simple exact solutions for a quasiuniform current can be obtained by making use of spherical coordinates and also of the coordinates associated with the prolate and oblate ellipsoids of revolution. In the first case the boundary of the toroidal column will be a sphere, while in the second and third cases it will be the ellipsoids of revolution with semiaxes denoted by l_r and l_z and having a ratio $l_z/l_r = \lambda$.

Let the current distribution be given by the following form:

$$j_s = 2\pi c \frac{dp}{d\Psi} r = Ar = -\frac{15}{8} \frac{c}{\pi l_r^2} B_0 r. \tag{3.54}$$

The quantity $B_0 = \text{const}$ has the dimension of a magnetic field and will be used in the following instead of the constant A. As will be seen below, B_0 is equal to the amplitude of the maintaining field for a spheroidal configuration: $l_z = l_r$ ($\lambda = 1$).

The internal solution of the equilibrium equation (2.21), which covers the three cases under consideration, may be written in the form

$$\Psi_i = \frac{4\pi^2}{c} A l_r^2 \frac{\lambda^2}{4\lambda^2+1} r^2 \left(1 - \frac{r^2}{l_r^2} - \frac{z^2}{l_z^2}\right)$$

$$= -\pi r^2 B_0 \frac{15}{2} \frac{\lambda^2}{4\lambda^2+1} \left(1 - \frac{r^2}{l_r^2} - \frac{z^2}{l_z^2}\right). \tag{3.55}$$

Now, we apply the method of separation of variables to find the complete solution of Eq. (2.21).

In the spherical coordinates ρ, v, ζ,

$$r = \rho \sin v, \quad z = \rho \cos v \tag{3.56}$$

Eq. (2.21) is of the form

$$\frac{\partial^2 \Psi}{\partial \rho^2} + \frac{\sin v}{\rho^2} \frac{\partial}{\partial v} \frac{1}{\sin v} \frac{\partial \Psi}{\partial v} = -\frac{8\pi^2}{c} j_s r$$

$$= -\frac{8\pi^2}{c} A r^2 = 2\pi r^2 B_0 \frac{15}{2 l_r^2} . \tag{3.57}$$

The general solution for a vacuum field in spherical coordinates is expressed in terms of the derivatives $\dot{P}_n(t)$ and $\dot{Q}_n(t)$ of the Legendre functions $P_n(t)$ and $Q_n(t)$,

$$\dot{P}_n(t) \equiv \frac{dP_n(t)}{dt}, \quad \dot{Q}_n(t) = \frac{dQ_n(t)}{dt}, \quad \dot{Q}_n \frac{d\dot{P}_n}{dt} - \dot{P}_n \frac{d\dot{Q}_n}{dt} = -\frac{n(n+1)}{(t^2-1)^2}, \tag{3.58}$$

as follows:

$$\Psi_{vac} = \rho^2 \sin^2 v \sum_{n=1}^{\infty} \left(A_n \rho^{n-1} + B_n \frac{1}{\rho^{n+2}}\right) \dot{P}_n(\cos v) + A_0 \rho \cos v. \tag{3.59}$$

If the plasma boundary is assumed to be a spherical surface $\rho = l_r$, then the solution will read [13]

$$\left. \begin{array}{l} \Psi_i = \dfrac{3}{2} \pi B_0 \rho^2 \sin^2 v \, (\rho^2/l_r^2 - 1); \\[4pt] \Psi_e = \pi B_0 \rho^2 \sin^2 v \, (1 - l_r^3/\rho^3). \end{array} \right\} \tag{3.60}$$

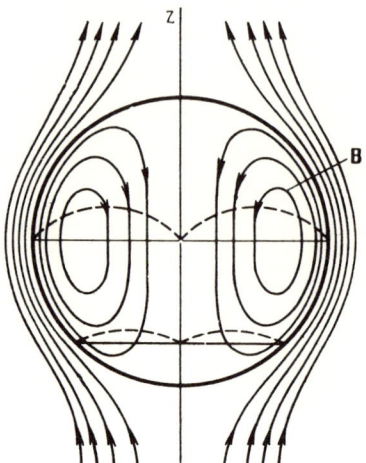

Fig. 9. The equilibrium configuration for a plasma column with a spherical boundary [13]. The dashed line shows the pressure distribution.

Magnetic surfaces of this configuration are shown in Fig. 9.

From the expression for Ψ_e it is seen that the maintaining field for a spherical configuration is uniform,

$$\mathbf{B}_{ext} = B_z \mathbf{e}_z = B_0 \mathbf{e}_z. \qquad (3.61)$$

Note that such an equilibrium magnetic configuration has a hydrodynamical analog called the Hill vortex.

In the coordinates of elongated ellipsoids ($\lambda > 1$) u, v, ζ,

$$r = d \, \text{sh} \, u \sin v, \quad z = d \, \text{ch} \, u \cos v \qquad (3.62)$$

Eq. (2.21) assumes the form

$$\frac{1}{d^2 (\text{sh}^2 u + \sin^2 v)} \left[\text{sh} \, u \frac{\partial}{\partial u} \frac{1}{\text{sh} \, u} \frac{\partial \Psi}{\partial u} + \sin v \frac{\partial}{\partial v} \frac{1}{\sin v} \frac{\partial \Psi}{\partial v} \right]$$
$$= -\frac{8\pi^2}{c} j_s r = 15 \pi \frac{r^2}{l_r^2} B_0. \qquad (3.63)$$

In these coordinates the flux function of the vacuum field is written as

$$\Psi_{vac} = d^2 \operatorname{sh}^2 u \sin^2 v \sum_{n=1}^{\infty} [A_n \dot{P}_n(\operatorname{ch} u) + B_n \dot{Q}_n(\operatorname{ch} u)] \dot{P}_n(\cos v)$$
$$+ A_0 d \operatorname{ch} u \cos v. \qquad (3.64)$$

If we assume that the plasma boundary is situated at the coordinate surface $u = u_0$

$$l_r = d \operatorname{sh} u_0, \quad l_z = d \operatorname{ch} u_0, \quad d^2 = l_z^2 - l_r^2, \qquad (3.65)$$

the internal and external solutions will read

$$\Psi_i = \frac{15}{2} \pi B_0 d^2 \operatorname{sh}^2 u \sin^2 v \lambda^2 \left\{ \frac{\operatorname{ch}^2 u}{\operatorname{ch}^2 u_0} \cos^2 v \right.$$
$$\left. - \frac{\dot{P}_3(\operatorname{ch} u)}{\dot{P}_3(\operatorname{ch} u_0)} \left(\cos^2 v - \frac{1}{5} \right) - \frac{1}{5} \right\};$$

$$\Psi_e = \frac{15}{2} \pi B_0 d^2 \operatorname{sh}^2 u \sin^2 v \lambda^2 \frac{\operatorname{sh}^4 u_0}{\operatorname{ch} u_0} \left\{ \frac{\cos^2 v - 1/5}{6(5 \operatorname{ch}^2 u_0 - 1)} [\dot{Q}_3(\operatorname{ch} u_0) \dot{P}_3(\operatorname{ch} u) \right.$$
$$\left. - \dot{P}_3(\operatorname{ch} u_0) \dot{Q}_3(\operatorname{ch} u)] + \frac{1}{5} [\dot{Q}_1(\operatorname{ch} u) - \dot{Q}_1(\operatorname{ch} u_0)] \right\}. \qquad (3.66)$$

In explicit notation:

$$\left. \begin{array}{l} \dot{P}_3(t) = \dfrac{3}{2}(5t^2 - 1); \\[4pt] \dot{Q}_1(t) = \dfrac{1}{2} \ln \dfrac{t+1}{t-1} - \dfrac{t}{t^2-1}; \\[4pt] \dot{Q}_3(t) = \dfrac{3}{4}(5t^2-1) \ln \dfrac{t+1}{t-1} - \dfrac{3}{2} \dfrac{(5t^2-1)t}{t^2-1} + \dfrac{5t}{(t^2-1)}. \end{array} \right\} \qquad (3.67)$$

The flux Ψ_{ext} of the maintaining field for an elongated ellipsoid is

$$\Psi_{ext} = \frac{15}{2} \pi B_0^2 d^2 \operatorname{sh}^2 u \sin^2 v \lambda^2 \frac{\operatorname{sh}^4 u_0}{\operatorname{ch} u_0} \left\{ \frac{\cos^2 v - 1/5}{6(5 \operatorname{ch}^2 u_0 - 1)} \dot{Q}_3(\operatorname{ch} u_0) \right.$$
$$\left. \times \dot{P}_3(\operatorname{ch} u) - \frac{1}{5} \dot{Q}_1(\operatorname{ch} u_0) \right\}. \qquad (3.68)$$

It may be written as a sum of two terms:

$$\Psi_{ext} = -\pi r^2 B_0 \left[\varphi_1(\lambda) + \varphi_2(\lambda) \frac{r^2 - 4z^2}{2 l_r^2} \right], \qquad (3.69)$$

where the first term corresponds to a uniform field, while the second term is associated with a quadrupole field. The coefficients

$\varphi_1(\lambda)$ and $\varphi_2(\lambda)$ depend on λ. For $\lambda > 1$ they are given by

$$\varphi_1(\lambda) = \frac{15}{2} \frac{\lambda}{\lambda^2-1} \left[\frac{1}{2\sqrt{\lambda^2-1}} \ln(\lambda + \sqrt{\lambda^2-1}) - \frac{1}{2}\lambda + \lambda \frac{\lambda^2-1}{4\lambda^2+1} \right];$$
$$\varphi_2(\lambda) = \frac{15}{4} \frac{\lambda}{(\lambda^2-1)^2} \left[\frac{3}{2\sqrt{\lambda^2-1}} \ln(\lambda + \sqrt{\lambda^2-1}) - \frac{3}{2}\lambda + 5\lambda \frac{\lambda^2-1}{4\lambda^2+1} \right]. \quad (3.70)$$

In this case the maintaining magnetic field is given by

$$B_{z,\text{ext}} = -B_0 \varphi_1(\lambda) - B_0 \varphi_2(\lambda) \frac{r^2 - 2z^2}{l_r^2};$$
$$B_{r,\text{ext}} = -B_0 \varphi_2(\lambda) \frac{2rz}{l_r^2}. \quad (3.71)$$

In the coordinates of a flattened ellipsoid ($\lambda < 1$) w, v, ζ,

$$r = d \operatorname{ch} w \sin v, \quad z = d \operatorname{sh} w \cos v. \quad (3.72)$$

Equation (2.21) assumes the form

$$\frac{1}{d^2(\operatorname{sh}^2 w + \cos^2 v)} \left[\operatorname{ch} w \frac{\partial}{\partial w} \frac{1}{\operatorname{ch} w} \frac{\partial \Psi}{\partial w} + \sin v \frac{\partial}{\partial v} \frac{1}{\sin v} \frac{\partial \Psi}{\partial v} \right]$$
$$= -\frac{8\pi^2}{c} j_s r = 15\pi \frac{r^2}{l_r^2} B_0. \quad (3.73)$$

In the general case the flux function for the vacuum magnetic field is the series

$$\Psi_{\text{vac}} = d^2 \operatorname{ch}^2 w \sin^2 v \sum_{n=1}^{\infty} \left[A_n \dot{P}_n(i \operatorname{sh} w) + B_n \frac{1}{i} \dot{Q}_n(i \operatorname{sh} w) \right] \dot{P}_n(\cos v)$$
$$+ A_0 d \operatorname{sh} w \cos v. \quad (3.74)$$

Assuming that the plasma boundary is situated at $w = w_0$

$$l_r = d \operatorname{ch} w_0, \quad l_z = d \operatorname{sh} w_0, \quad d^2 = l_r^2 - l_z^2, \quad (3.75)$$

we have

$$\Psi_i = \frac{15}{2} \pi B_0 d^2 \operatorname{ch}^2 w \sin^2 v \lambda^2 \left\{ \frac{\operatorname{sh}^2 w}{\operatorname{sh}^2 w_0} \cos^2 v - \frac{\dot{R}_3(i \operatorname{sh} w)}{\dot{P}_3(i \operatorname{sh} w_0)} \right.$$
$$\left. \times \left(\cos^2 v - \frac{1}{5} \right) - \frac{1}{5} \right\};$$

$$\Psi_e = \frac{15}{2} \pi B_0 d^2 \operatorname{ch}^2 w \sin^2 v \lambda^2 \frac{\operatorname{ch}^4 w_0}{\operatorname{sh} w_0} \left\{ \frac{\cos^2 v - 1/5}{6(5 \operatorname{sh}^2 w_0 + 1) i} \right. \quad (3.76)$$
$$\times \left[\dot{P}_3(i \operatorname{sh} w_0) \dot{Q}_3(i \operatorname{sh} w) - \dot{Q}_3(i \operatorname{sh} w_0) \dot{P}_3(i \operatorname{sh} w) \right]$$
$$\left. + \frac{1}{5i} \left[Q_1(i \operatorname{sh} w_0) - \dot{Q}_3(i \operatorname{sh} w) \right] \right\}.$$

The flux Ψ_{ext} of the maintaining field for the flattened ellipsoid is readily extracted from Ψ_e and is a sum of uniform and quadrupole fields (3.69) and (3.71) as in the previous case. The functions $\varphi_1(\lambda)$ and $\varphi_2(\lambda)$ at $\lambda < 1$ are the analytic continuation of formulas (3.70)

$$\begin{aligned}\varphi_1(\lambda) &= \frac{15}{2}\frac{\lambda}{\lambda^2-1}\left[\frac{1}{2\sqrt{1-\lambda}}\arctg\frac{\sqrt{1-\lambda^2}}{\lambda}\right.\\ &\quad\left. -\frac{1}{2}\lambda + \lambda\frac{\lambda^2-1}{4\lambda^2+1}\right];\\ \varphi_2(\lambda) &= \frac{15}{4}\frac{\lambda}{(\lambda^2-1)^2}\left[\frac{3}{2\sqrt{1-\lambda^2}}\arctg\frac{\sqrt{1-\lambda^2}}{\lambda}\right.\\ &\quad\left. -\frac{3}{2}\lambda + 5\lambda\frac{\lambda^2-1}{4\lambda^2+1}\right].\end{aligned} \quad (3.77)$$

The form of the functions φ_1 and φ_2 vs the parameter $\varepsilon = (\lambda^2 - 1)/(\lambda^2 + 1)$ is shown in Fig. 10.

As in the case of a straight plasma column there is also a critical value of the quadrupole field beyond which the equilibrium becomes impossible. For $\lambda \to \infty$ this configuration transforms to a θ pinch with a reversed field. In this case the maintaining field is purely uniform as in the case of the spherical configuration and is equal to

$$\mathbf{B}_{ext} = B_z \mathbf{e}_z = \frac{15}{8} B_0 \mathbf{e}_z. \quad (3.78)$$

In the opposite case, $\lambda \to 0$, the configuration becomes a thin disk with a distributed current and the maintaining field in the $z = 0$ plane becomes

$$B_{z,\,ext} = \frac{2\mathscr{J}}{cl_r}\left[-\frac{3}{8}\pi^2 + \frac{9}{16}\pi^2\frac{r^2}{l_r^2}\right], \quad (3.79)$$

where $\mathscr{J}$ is the total plasma current.

3.5. Magnetostatic Problems Related to Plasma Equilibrium

A number of magnetostatic problems arise in a natural way within the theory of plasma equilibrium of toroidal systems. In the

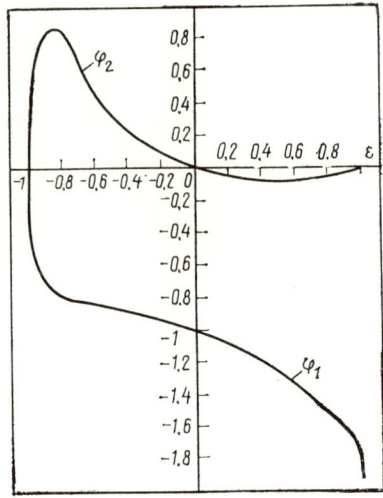

Fig. 10. The amplitudes of the homogeneous $\varphi_1(\varepsilon)$ and the quadrupole $\varphi_2(\varepsilon)$ maintaining fields for compact toroidal configurations with a boundary of ellipsoidal form, $\varepsilon = (\lambda^2 - 1)/(\lambda^2 + 1)$, $\lambda = l_z/l_r$.

present section we will describe the solution of some of these problems as applied to axisymmetric systems.

<u>1. Optimization of the Shape of the Toroidal Field Coils.</u> In systems with a high longitudinal field, such as tokamaks and stellarators (and also toroidal magnetic energy storage devices) the longitudinal field coils are the most highly stressed components of the equipment. Coils of circular cross section currently employed are most convenient for production. However, a change in their shape can be important to optimize some characteristics of toroidal solenoids.

In particular, one can ask how to contour the coil to compensate the forces associated with the pressure of the magnetic field by tension of the conductor only [40]. For a multi-turn solenoid this requirement is equivalent to the condition

$$T/\rho = (B_s^2/8\pi)2\pi r, \tag{3.80}$$

where T is the total force of tension in the conductor, and ρ is the radius of curvature of the contour of the meridional cross section. Taking the radial dependence of the longitudinal magnetic field into account, we get an equation for the radius of curvature:

$$\rho = Cr, \quad C = \text{const.} \tag{3.81}$$

A similar requirement for mechanical equilibrium may be posed for a single-turn coil which permits tension in the toroidal direction. In this case for the condition of equal tension in the toroidal and poloidal directions, we have instead of (3.80)

$$T(k_1 + k_2) = B_s^2 \, r/4, \tag{3.82}$$

where k_1 and k_2 are the main curvatures of a toroidal shell:

$$k_1 = \frac{1}{\rho} = \frac{z''}{(1+z'^2)^{3/2}}, \quad k_2 = \frac{z'}{r\sqrt{1+z'^2}} \tag{3.83}$$

[z = z(r) is the equation for a contour of the cross section].

The lines satisfying Eq. (3.81) (the l contour) and Eq. (3.82) (the s contour) are shown in Fig. 11. Note that the shape of the family of l contours coincides with the trajectory of a charged particle with only a poloidal velocity component $\mathbf{v} = v_r \mathbf{e}_r + v_z \mathbf{e}_z$, in a vacuum toroidal field $\mathbf{B} = B_s \mathbf{e}_s$. In fact, the condition for equality of the centrifugal force and the Lorentz force for the field $B_s \sim 1/r$ is reduced to Eq. (3.81). The curves are not closed, and this reflects the fact that it is impossible to produce a solenoid with a purely toroidal field in which the magnetic field pressure is completely compensated by the tension of the conductor. An additional external structure is needed which takes up the centripetal force of the solenoid towards the symmetry axis. One can take as a coil contour the closed loop formed by the curve shown in Fig. 11. In this case the centripetal force should be compensated by a structure located on the external side of the solenoid (Fig. 12a). The other possibility is to connect the points of the curve nearest to the

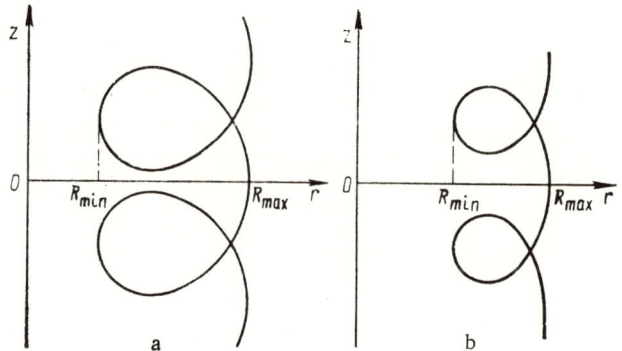

Fig. 11. The curves z(r) for the optimal contours of the toroidal field coils: a) l curve, which is a solution of Eq. (3.81) for $(R_{max} - R_{min})/(R_{max} + R_{min}) = 0.5$; b) s curve, determined by Eq. (3.82) for $(R_{max} - R_{min})/(R_{max} + R_{min}) = 0.3$.

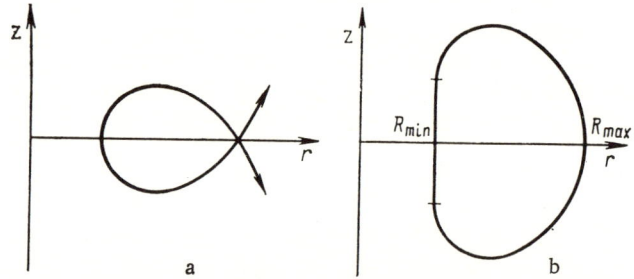

Fig. 12. Possible variants of l curves as contours of the toroidal field coils: a) a loop of the l curve as a coil contour; b) a coil contour with internal straight section.

axis of symmetry by a straight line (Fig. 12b). In this case the centripetal force which acts upon the straight section should be taken up by a central support.

The compensation of the pressure of the magnetic field by the tension of the conductor and, hence, the elimination of bending forces is attractive for the construction of superconducting systems where relative motion of the current-carrying elements is undesirable. In the absence of bending moments the relative shifts of the superconducting layers prove to be quantities of second order compared with the total deformation of the structure which results from the increase of the magnetic field. From what has been said above, it follows that even in optimized toroidal solenoids such a compensation of the pressure can only be realized in the curved part of the cross-sectional contour. Hence, it is impossible to eliminate completely shearing forces in the bulk of the superconductor.

Note, that the mechanical problem considered is equivalent to an electrodynamical one, where the problem is to obtain the maximum magnetic energy in a toroidal solenoid with a given contour length or with a given surface [41]. The solution of the first problem is given by the l contour which is determined by Eq. (3.80), and of the second problem by the s contour which corresponds to Eqs. (3.82) and (3.83).

2. Current Distribution on a Superconducting Torus. In a superconducting torus the current is distributed over the surface in such a way that the magnetic field is zero inside the superconductor, $\mathbf{B} = 0$. In the axisymmetric case the magnetic field produced by a longitudinal surface current on the inner side of the contour L of the cross section can be written in the following form:

$$B_{\tau i}(l) = -\frac{2\pi}{c} i_s(l) + \oint_L b_\tau(l; l') i_s(l') \, dl', \qquad (3.84)$$

where $B_{\tau i}(l)$ is the tangential field component and $b_\tau(l; l')$ is the tangential field component at the point l due to unit current flowing at the point l' of the contour:

$$b_\tau(l; l') = b_r(l; l') n_z(l) - b_z(l; l') n_r(l); \qquad (3.85)$$

n is the outer normal to the contour, and $b_r(l; l')$ and $b_z(l; l')$ may be calculated according to the formulas

$$b_r(r, z; r', z') = \frac{2}{c} \frac{z-z'}{r\sqrt{(r+r')^2 + (z-z')^2}} \left[-K(k) \right.$$
$$\left. + \frac{r'^2 + r^2 + (z-z')^2}{(r-r')^2 + (z-z')^2} E(k) \right]; \qquad (3.86)$$

$$b_z(r, z; r', z')$$

$$= \frac{2}{c} \frac{1}{\sqrt{(r+r')^2 + (z-z')^2}} \left[K(k) + \frac{r'^2 - r^2 - (z-z')^2}{(r-r')^2 + (z-z')^2} E(k) \right], \quad (3.87)$$

$K(k)$ and $E(k)$ are the complete elliptic integrals, and

$$k^2 = \frac{4rr'}{(r+r')^2 + (z-z')^2}. \quad (3.88)$$

Taking into account that

$$\oint_L b_\tau(l'; l) \, dl' = \frac{2\pi}{c}, \quad (3.89)$$

Eq. (3.84) may be written in the form

$$B_{\tau i}(l) = \oint_L [b_\tau(l; l') i_s(l') - b_\tau(l'; l) i_s(l)] \, dl'. \quad (3.90)$$

Equation (3.90) is a convenient expression because the integrand vanishes as $l' \to l$ and, therefore, the integral permits the direct use of standard formulas for numerical integration.

In the absence of field sources inside the torus the current distribution $i_s^0(l)$ in the superconducting torus is defined by Eq. (3.80) with $B_{\tau i} = 0$. For such a current distribution, magnetic flux threads around the outside of the torus and through the center while the magnetic field inside the torus is zero.

Such a configuration is of interest as a version of an inductor which can drive the plasma current and which at the same time does not produce stray fields in the plasma volume. Numerical calculations of the current distribution according to Eq. (3.90) have been performed [42, 43]. Figure 13 shows the distribution $i_s^0(l)$ for a torus with the same shape as the l contour with $(R_{max} - R_{min}) \times (R_{max} + R_{min})^{-1} = 0.7$.

For $B_{\tau i}(l) \neq 0$ Eq. (3.88) defines the distribution of the current $i_s(l)$ which produces a prescribed field inside the torus. The solution $i_s(l)$ is not unique since the addition of the superconductor current $i_s^0(l)$ does not change the field inside the torus. For $B_{\tau i} = -B_{\tau,ext}$, where $B_{\tau,ext}$ is the tangential component of the external field, Eq. (3.88) gives the current distribution in the superconductor located in the field B_{ext} of the external sources.

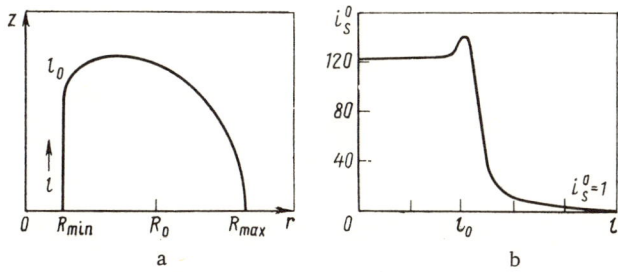

Fig. 13. a) A contour of the torus cross section in the form of an l curve with a straight section, $(R_{max} - R_{min})/(R_{max} + R_{min}) = 0.7$; b) the distribution of the superconductor current $i_s^0(l)$. The upper half of the torus is shown. On the outer side of the torus $z = 0$, $r = R_{max}$; the current density $i_s^0(l)$ is normalized at 1.

The problem of the current distribution for a superconductor i_s^0 may also be solved from the condition for a constant flux function on the surface of the torus

$$\oint_L \Psi_0(l; l') i_s^0(l') dl' = \text{const}, \qquad (3.91)$$

where $\Psi_0(l; l')$ is the value of the flux function at the point l due to a unit ring current through the point l' of the contour

$$\Psi_0(r, z; r', z') = \frac{8\pi}{c} \sqrt{\frac{rr'}{k^2}} \left[\left(1 - \frac{k^2}{2}\right) K(k) - E(k) \right]. \qquad (3.92)$$

Equation (3.91) relates to a class of Fredholm equations of the first kind and is poorly conditioned for numerical integration. However, the logarithmic singularity as $l \to l'$ in the kernel $\Psi_0(l; l')$ simplifies the regularization of this equation.

Here we also give a method to solve the problem of screening the field of sources located inside the toroidal shell (casing), by surface currents i_s in the shell. The equation defining the distribution i_s may be derived by setting the tangential field component on the outer surface of the torus equal to zero:

$$B_{\tau,\text{int}}(l) + \frac{2\pi}{c} i_s(l) + \oint_L b_\tau(l; l') i_s(l') \, dl' = 0, \qquad (3.93)$$

where $B_{\tau,\text{int}}$ is the field of the current located inside the torus. Taking Eq. (3.89) into account this equation may be rewritten so that the integrand possesses no indefiniteness:

$$B_{\tau,\text{int}}(l) + \frac{4\pi}{c} i_s(l) + \oint_L [b_\tau(l; l') i_s(l') - b_\tau(l'; l) i_s(l)] \, dl' = 0; \qquad (3.94)$$

thus, it permits a direct numerical solution.

3. *Determination of the Equilibrium Field outside the Plasma.* The complete analytical solution of the equilibrium equations is possible for a quasiuniform current only for particular cases, when the plasma boundary coincides with a coordinate surface of the coordinate system, which allows the separation of variables (Sections 3.1-3.4). However, if we are not interested in the external region, the solution of the equilibrium equation (2.21) for a quasiuniform current density $2\pi c \, dp/d\Psi = A$, $(1/c)(dF^2/d\Psi) = B$ is readily found by adding to the solution of the inhomogeneous equation

$$\Psi = (-(4\pi^2/c)) (Ar^2 + B) z^2 \qquad (3.95)$$

a solution of the homogeneous equation, e.g.,

$$r^2, \ r^4 - 4r^2 z^2, \ z^6 - 12 r^4 z^2 + 8 r^2 z^4 \ldots \qquad (3.96)$$

This allows one to obtain, in particular, the solution which models a plasma column with a D-shaped cross section:

$$\Psi_i = (1 - r^2/R_1^2 - z^2/Z_1^2)(r^2/R_2^2 - 1). \qquad (3.97)$$

To obtain a complete solution it is necessary to find the field outside the plasma.

Using the given Ψ_i inside the plasma it is possible to find a maintaining field $\mathbf{B}_{\text{ext}}$ with the help of the principle of virtual casing (Section 2.5).

In order to continue $\mathbf{B}_{\text{ext}}$ beyond the plasma boundary one can imagine it to be enclosed by a toroidal contour L with a surface current $i_{s,\text{ext}}$ adjusted to produce the same magnetic field inside the plasma. For this it is sufficient to equate the tangential component of the maintaining field, calculated by means of the principle of virtual casing, and the field of the current $i_{s,\text{ext}}$ to be found [42]:

$$B_{\tau,\text{ext}}(l) = \oint_L b_\tau(l; L) i_{s,\text{ext}}(L) dL. \tag{3.98}$$

Here, the point l is situated on the plasma surface, while the point L is located on the surface of the auxiliary shell. Equation (3.98) is a Fredholm equation of the first kind. Such equations are ill-posed, since the addition of a quickly oscillating term to $i_{s,\text{ext}}$ does not change the value of the integral. In order to eliminate the occurrence of such an additional term in the process of numerical solution it is necessary to apply regularization techniques [44].

An algorithm for the solution is based on the following. Denote the integral operator on the right-hand side of Eq. (3.98) by $\hat{C}i_{s,\text{ext}}$. The problem of solving Eq. (3.98) is equivalent to a minimization of the discrepancy functional

$$d = \oint_l [\hat{C}i_{s,\text{ext}} - B_{\tau,\text{ext}}]^2 dl. \tag{3.99}$$

According to the regularization techniques, a smoothing functional is added to the functional (3.99) and the minimum of the following functional is then investigated:

$$\alpha \oint_L \left[k_1 \left(\frac{di_{s,\text{ext}}}{dL} \right)^2 + k_2 i_{s,\text{ext}}^2 \right] dL + \oint_l [\hat{C}i_{s,\text{ext}} - B_{\tau,\text{ext}}]^2 dl. \tag{3.100}$$

Here, k_1 and k_2 are positive coefficients of the order of unity and α is the regularization parameter. The search for the minimum of (3.100) is a stable problem even for a small α. For the choice of α one can use the method described in [45]. The accuracy of the solution of Eq. (3.98) is determined by the discrepancy d (3.99). If one wants to obtain a given accuracy, for example, $d = d_0 = 10^{-6} \times \oint B_{\tau,\text{ext}}^2 dl$, the regularization parameter may be chosen by means of the following iterative process:

$$\alpha_{n+1} = \alpha_n \sqrt{d_0/d}. \tag{3.101}$$

The procedure for choosing α is finished when the given accuracy is achieved. The regularization method chooses from all the solutions $i_{s,\text{ext}}$ of Eq. (3.98) those which correspond to the minimum of

$$\oint_L \left[k_1 \left(\frac{di_{s,\text{ext}}}{dL} \right)^2 + k_2 i_{s,\text{ext}}^2 \right] dL.$$

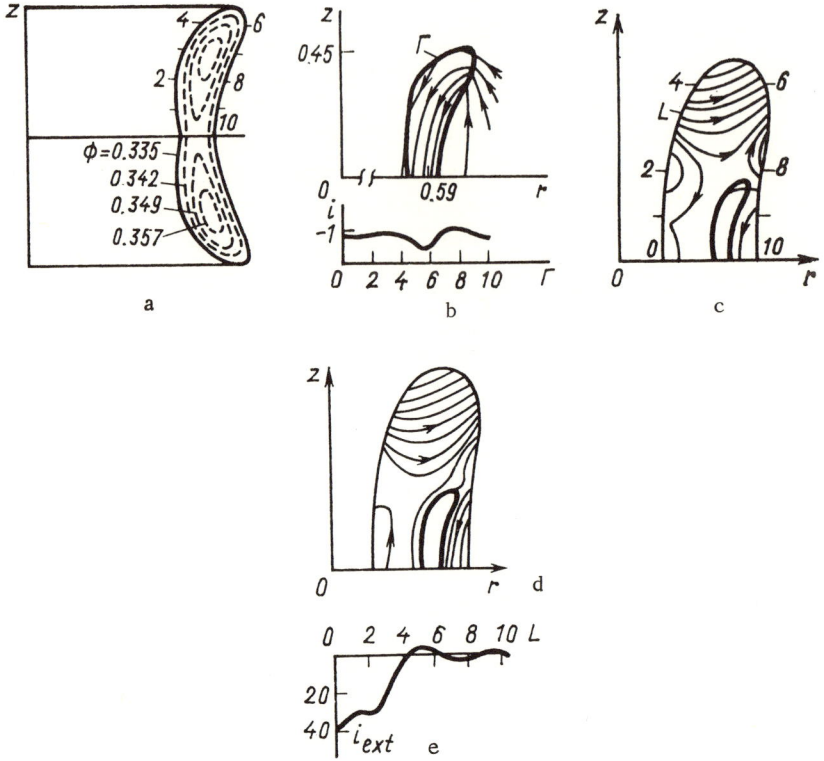

Fig. 14. An equilibrium configuration of the "doublet" type with three magnetic axes inside the plasma [42]: a) internal magnetic surfaces; b) the field of the current in the virtual casing; c) the continuation of the maintaining field outside the plasma boundary; d) external magnetic surfaces of the equilibrium configuration; e) the distribution of $i_{ext}(L)$.

Figure 14 shows the maintaining field for a configuration with three magnetic axes obtained by means of the principle of virtual casing together with a continuation beyond the plasma boundary. Since the self-field B_{pl} of the plasma current outside the column is known and equal to that of the current in the virtual casing, the total field of the equilibrium configuration outside the plasma is presented as a field of two surface currents.

4. Determination of the Equilibrium Field outside the Plasma for a Straight Plasma Column. The method of continuation of the equilibrium field beyond the boundary of the column considered in the previous case applies for both toroidal and straight geometry. In the case of straight geometry this problem can also be solved by conformal mapping [46].

Assume that

$$w \equiv u + iv = w(x + iy) \tag{3.102}$$

is the function which maps the region outside the plasma on a semi-plane $v > v_0$ and that the plasma boundary Γ is given by

$$v(x, y) = v_0. \tag{3.103}$$

We shall look for a vector potential A_s in the form $A_s(u, v)$.

Since outside the plasma A_s satisfies the Laplace equation, it can be written as the imaginary part of some analytic function $\varphi(w)$:

$$A_s = \text{Im } \varphi(w). \tag{3.104}$$

Denote

$$g(u, v) = \left| \frac{dw}{d(x + iy)} \right|^{-1}. \tag{3.105}$$

Mutually orthogonal components of the magnetic field B_u and B_v are of the form

$$B_u = \frac{1}{g} \frac{\partial A_s}{\partial v} = \frac{1}{g} \frac{\partial \text{Re } \varphi}{\partial u}, \quad B_v = -\frac{1}{g} \frac{\partial A_s}{\partial u} = \frac{1}{g} \frac{\partial \text{Re } \varphi}{\partial v}. \tag{3.106}$$

Hence,

$$d\varphi/dw = g(u, v)(B_u - iB_v). \tag{3.107}$$

On the plasma boundary $B_v = 0$, while the tangential field distribution B_u is given by the internal solution, $B_u = B_u^0(u)$. With the help of the known value at the plasma boundary $v = v_0$ the analytic function $d\varphi/dw$ is uniquely continued into the range $v > v_0$:

$$d\varphi/dw = g(u + i(v - v_0), v_0)B_u^0(u + i(v - v_0)). \tag{3.108}$$

Consequently, for the vector potential we have

$$A_s(u, v) = A_{s0} + \text{Im} \int_0^{u+i(v-v_0)} g(w', v_0) B_u^0(w') dw'$$

$$= A_{s0} + \frac{1}{2i} \int_{u-i(v-v_0)}^{u+i(v-v_0)} g(w', v_0) B_u^0(w') dw'. \quad (3.109)$$

Here, A_{s0} is the value of A_s on the plasma boundary. Using Eqs. (3.107) and (3.108) we obtain for the magnetic field

$$B_u - iB_v = \frac{g(u+i(v-v_0), v_0)}{g(u,v)} B_u^0(u + i(v-v_0)). \quad (3.110)$$

Equations (3.109) and (3.110) give the solution of the problem of the continuation of the internal solution beyond the plasma boundary. It follows from Eq. (3.109) that the occurrence of singularities in the external field may have two causes. Some singularities of A_s are associated with singularities of the geometric factor $g(w', v_0)$. Other singularities of Eq. (3.109) may be present in the function $B_u^0(w') = B_u^0[u + i(v - v_0)]$. Note that some singularities associated with the cross-sectional geometry of the column may disappear due to corresponding zeros of the function $B_u^0(w')$.

5. **Penetration of a Magnetic Field through the Gaps of a Shell.**
Consider the screening of magnetic fields when the shells have gaps. In well-conducting shells, which serve to maintain equilibrium, such gaps permit the penetration of an electric field (meridional gaps) and of a longitudinal magnetic field B_s (longitudinal gaps). These may also essentially affect the penetration of the magnetic fields which provide plasma equilibrium.

We assume that the shell is superconducting and consider the effect of meridional gaps. Suppose that a current-carrying column is present inside the shell and denote its field on the surface of the shell by $B_{\tau, \text{int}}$. In the absence of a gap the surface current $i_s(l)$ in the shell is determined by Eq. (3.93) and is related to the total field on the surface of the shell by

$$\frac{4\pi}{c} i_s(l) = -B_\tau(l). \quad (3.111)$$

When there is a gap, the "symmetric" component of this current $i_s^0(l)$ which corresponds to the current distribution in a supercon-

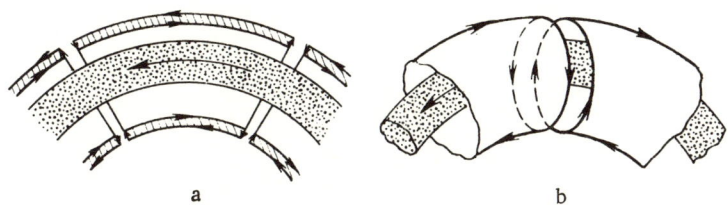

Fig. 15. The circuit of (a) the symmetrical component of the surface current in the casing with gaps and (b) the asymmetric component [66].

ducting torus (see Section 2) is closed along the outer surface of the shell, as shown in Fig. 15a. The integral of the current i_s^0 is equal to the plasma current. The circulation of the remaining asymmetric part $\tilde{i}_s(l)$ of the surface current, which is given by

$$\frac{4\pi}{c} \tilde{i}_s(l) = -B_\tau(l) + \frac{4\pi}{c} i_s^0(l), \qquad (3.112)$$

is completed around the gap (Fig. 15b).

A similar asymmetric component $\tilde{i}_s(l)$ circulating around the gap is also produced when the flux of an external poloidal field B_{ext} is applied to the shell. In this case (neglecting the width of the gap) the distribution $\tilde{i}_s(l)$ is determined by Eq. (3.84) with the additional condition $\oint \tilde{i}_s(l) dl = 0$.

Denote by $i_p(l, s)$ the distribution of the poloidal component of a surface current, where $s = r\zeta$ is the longitudinal coordinate, and s is measured from the middle of the gap. This is determined by the condition for the component of the magnetic field normal to the surface of the shell to be zero and by that for the closure of the longitudinal current $\tilde{i}_s(l)$. Near the gap, the main contribution to the field is given by the current $i_p(l, s)$. If the width h(l) of the gap is considered to be small compared to the dimensions of the torus, the basic variation of $i_p(l, s)$ is associated with s, while l can be treated as a parameter. Therefore, the dependence on s can be determined from the solution of a model problem for the current distribution in the vicinity of the edges of two superconducting semiplanes situated a distance h* from each other (Fig. 16), where currents I flow

---
*In the figure the distance is given as 2h — Editor.

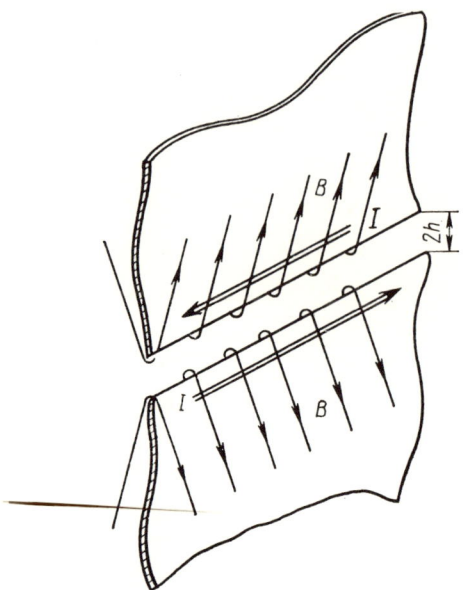

Fig. 16. The model problem for the magnetic field distribution between two ideally conducting semiplanes.

in opposite directions along these edges. The vector potential A of such a field is readily found in the elliptical cylinder coordinates (Section 3.1), where d = h and the coordinate lines v = $\pi/2$ and v = $-\pi/2$ are taken along the edges of the semiplanes:

$$A = \frac{2\pi}{c} v \frac{I}{\ln 2L/h}. \qquad (3.113)$$

The normalization is chosen such that I becomes equal to the total current in each of the semiplanes and L represents the reconnection length of the surface current. This expression corresponds to the following current distribution over the plates [47]:

$$i_p(l, s) = \frac{I(l)}{\ln 2L/h} \frac{1}{\sqrt{s^2 - h^2}} \qquad (3.114)$$

and also gives the distribution of the field component in the gap normal to the surface of the plates:

$$B_{n,g} = \frac{2\pi}{c} \frac{I}{\ln 2L/h} \frac{1}{\sqrt{h^2 - s^2}}. \quad (3.115)$$

The reconnection current is distributed over the l contour according to the expression

$$I(l) = \int_0^l \tilde{i}_s(l)\, dl. \quad (3.116)$$

Here, the lower limit of the integral $l = 0$ is determined either by the longitudinal gap in the shell or by symmetry.

The flux $\Psi_g(l)$ of the magnetic field (3.115) through the gap per unit length of the contour is

$$\frac{d\Psi_g(l)}{dl} = \frac{2\pi}{c} \pi \frac{I(l)}{\ln 2L/h}, \quad (3.117)$$

or

$$\Psi_g = \frac{2\pi}{c} \pi \frac{\int_0^l I(l)\, dl}{\ln 2L/h}. \quad (3.118)$$

When there are a number of gaps, it is necessary to sum the fluxes through each of them.

The magnitude of the flux through the gaps is comparable with the incident flux even when h is small. For example, consider a cylindrical shell of radius b located in a uniform field $\mathbf{B}_{ext} = B_\perp \mathbf{e}_y$. Then, the current distribution $\tilde{i}_s(\theta)$ over the surface of the shell will be

$$\frac{2\pi}{c} \tilde{i}_s(\theta) = B_\perp \cos\theta. \quad (3.119)$$

Substituting this into Eq. (3.116) and Eq. (3.118) we get the following value for the flux passing through the gap:

$$\Psi_g = 2\pi b^2 B_\perp / \ln \frac{2L}{h}. \quad (3.120)$$

This value corresponds to the passage of a uniform unperturbed magnetic field $B_\perp \mathbf{e}_y$ through a gap of width

$$h_{\text{eff}} = \pi b / \ln \frac{2L}{h}. \quad (3.121)$$

Thus, in a thin shell each gap has an effective width comparable with the diameter of the shell.

For a shell with a finite thickness d which is comparable with the width of the gap the result is somewhat changed [48]. In this case in Eqs. (3.117), (3.118), (3.120), and (3.121) the logarithmic term should be supplemented by the term $\sqrt{2}d/h$. In particular, the effective width of the gap for a shell with a finite thickness becomes smaller:

$$h_{\text{eff}} = \frac{\pi b}{\ln 2L/h + \sqrt{2d/h}}. \quad (3.122)$$

The results obtained above for meridional gaps can be easily extended to longitudinal gaps. In particular, if the longitudinal gap interrupts circulating current in the shell, its effective width for field penetration will be appreciably higher than the geometrical width.

6. <u>Force-Free Toroidal Solenoid.</u> In the absence of plasma pressure, $\nabla p = 0$, the current density **j** in the plasma column is everywhere parallel to the magnetic field [**jB**] = 0 (force-free magnetic field). Consider a force-free configuration in the case where the current flows only on the surface of the torus. In this case, the equilibrium condition is reduced to pressure balance on the surface of the torus:

$$B_{si}^2 = B_{se}^2 + B_p^2, \quad (3.123)$$

where B_{si} is the longitudinal field on the inner side of the torus, and B_{se} and B_p are the longitudinal and poloidal fields on the outer side of the torus.

Equality (3.123) may be satisfied also for $B_{se} = 0$. In this case the distribution of the longitudinal current i_s and that of the poloidal field is

$$(4\pi/c)i_s = B_p = 2F/cr, \quad (3.124)$$

where F is the total current.

Such a force-free configuration can be considered as a version of a toroidal solenoid in which the forces acting upon current-carrying elements are compensated. This property may be used for designing systems with extremely high fields [49]. The analogy with the equilibrium of a plasma column shows that the equilibrium of such a solenoid requires an external maintaining field B_{ext}. For a torus with a circular cross section in the large aspect ratio approximation $R/a \gg 1$ the maintaining field is given by

$$\mathbf{B}_{ext} = B_z \mathbf{e}_z = -\frac{\mathcal{J}}{cR}\left(\ln\frac{8R}{a} - \frac{3}{2}\right)\mathbf{e}_z, \quad \mathcal{J} = aF/R. \quad (3.125)$$

The forces acting on the maintaining field coils should be taken up by a mechanical structure. Since only the toroidal effects appear to be uncompensated, such a system reduces the maximum forces by about a factor of R/a.

3.6. Numerical Methods for Solving Equilibrium Problems

Analytic solutions which can be rather simply obtained for typical cases (Sections 3.1-3.4) help to reveal the basic effects associated with the interaction of the plasma currents with external fields. The analytical solutions apply to a uniform current distribution. This is a good model, allowing one to extend the main results to a nonuniform current distribution at least qualitatively. However, in addition to analytical methods, providing an understanding of the basic effects, numerical methods are required for exact calculations of equilibrium configurations for generalized distributions of $p(a)$ and $\mathcal{J}(a)$ and for an arbitrary cross section. A number of numerical methods have been developed for axisymmetric systems. These cover all reasonable formulations of the equilibrium problem. For systems without symmetry progress is much more modest and it is achieved mainly by using high-power computers of the CRAY type. A review of numerical methods for calculation of equilibria is given in [50].

There are two main approaches to the numerical solution of equilibrium problems. One is based on using the Euler mesh for solving the two-dimensional equilibrium equation (2.21). Here there are algorithms using Green's functions, eigenfunction series, and the finite-difference method. In most cases the functions $p'(\Psi)$ and $FF'(\Psi)$ are considered as data in the problem. The other approach which has been developed lately formulates the equilibrium problem directly in natural coordinates (Section 2.4).

All the methods associated with the Euler mesh are based on the fact that Eq. (2.21) with a given right-hand side is simply a magnetostatics equation with a nonlinear current density. The iteration procedure is also common for all cases. It should be noted once more that the theory of equilibrium does not ascribe any special meaning to the functions $p'(\Psi)$ and $FF'(\Psi)$ entering in Eq. (2.20) for the current density j_s, but only points to the requirements for the functional dependence of j_s. In the process of numerical solution it seems reasonable to choose the longitudinal current density in the form

$$j_s(r, \Psi) = j_0 \frac{r}{R} \left[f_1\left(\frac{\Psi - \Psi_b}{\Psi_0 - \Psi_b}\right) + \frac{R^2}{r^2} f_2\left(\frac{\Psi - \Psi_b}{\Psi_0 - \Psi_b}\right) \right], \quad (3.126)$$

because the range of variation of the Ψ function is not known beforehand. In Eq. (3.126) R is a constant and the functions $f_1(\overline{\Psi})$ and $f_2(\overline{\Psi})$ coincide within a numerical factor with $p'(\Psi)$ and $FF'(\Psi)$. The introduction of a dimensionless argument $\overline{\Psi} \equiv (\Psi - \Psi_b)/(\Psi_0 - \Psi_b)$ (Ψ_b and Ψ_0 are the values of Ψ on the plasma boundary and on the magnetic axis) defines the range of its variation: $\overline{\Psi} = 0\text{-}1$, and helps one to represent the current distribution in the column before the equation is solved.

When the current density is chosen in the form (3.126), the nonlinear equation (2.21) can be readily solved by the simplest iterations, when the current density in each iteration is distributed in accordance with the function Ψ taken from the previous iteration. In this case the values of j_0 and Ψ_b which determine the normalization of the current density and the cross section of the column may be chosen in each iteration from additional considerations (the conservation of the total current, the conservation of the current density on the magnetic axis, the adjustment of the boundary of the current channel to the limiter aperture).

The Euler methods differ only in the form of the representation of the solution of Eq. (2.21) with the known right-hand side. The main methods are the method of integral equations (the method of Green's function), the use of special coordinate systems and series in eigenfunctions, and the use of finite-difference representation of Eq. (2.21).

1. The Method of Integral Equations (the method of Green's functions).

Using Eq. (3.92) for the flux function $\Psi_0(r, z; r', z')$ of a unit ring current the self magnetic flux Ψ_{pl} of the plasma current $j_s(r, \Psi)$ can be represented in the form of Biot–Savart's integral [51, 52]

$$\Psi_{pl}(r, z) = \iint \Psi_0(r, z; r', z') j_s(r', \Psi_{pl} + \Psi_{ext}) \, dr \, dz'. \quad (3.127)$$

The integral is taken over the cross section of the plasma column. This expression can be used directly to find the equilibrium of a plasma column with a free boundary in the given magnetic field of external currents I_i ($i = 1, 2, \ldots$). The flux Ψ_{ext} of the maintaining field is determined by the following relation:

$$\Psi_{ext}(r, z) = \sum_i \Psi_0(r, z; r', z') I_i. \quad (3.128)$$

For the calculation of $\Psi_{pl}(r, z)$ it is necessary to replace the integral (3.127) by a finite sum where the plasma current is represented by a set of equivalent ring currents with values which are proportional to the area of the element of the cross section and to the current density. The diagonal elements of the sum at $r = r'$, $z = z'$ are the inductances of the corresponding elementary currents and they can be calculated with the help of the formulas for the self-inductance of ring currents with a finite cross section (and with a uniform current distribution)

$$L = 4\pi R \left[\ln 8R/a - 3/2 + O(a^2/R^2) \right], \quad (3.129)$$

where a is the typical transverse dimension of the area of the element $a = (\Delta S/\pi)^{1/2}$; $R = r'$ and corrections $O(a^2/R^2)$ associated with the shape of the element of area can be refined by means of either numerical calculations of the self-inductance or reference books.

For the free boundary problem it is necessary to impose additional requirements on the boundary of the current channel. It is natural, for example, to require that the current channel be contained within the aperture of the limiter or of the chamber of the machine. In a number of cases it is natural to require that the current channel be limited by a separatrix. These requirements impose appropriate conditions on the choice of Ψ_b in Eq. (3.126), the value of Ψ on the plasma boundary, and on the external currents

I_i, which should provide, in particular, the required major radius of the column, the required elongation of the cross section, and so on. In most cases the requirements on the external currents can be reduced to the conditions for obtaining some characteristic points of an external magnetic surface. This is sufficient to fix the position and the form of the cross section of the plasma column [53]. This principle is also valid for the maintenance of plasma equilibria in real installations by feedback. In numerical calculations these requirements are imposed at every iteration. This maintains the current channel inside the chosen numerical mesh.

Examples of calculations of equilibria with a free boundary are given in Fig. 17.

The equilibrium equation can also be written in the integral form for the problem of plasma equilibrium in a perfectly conducting shell. In this case the maintaining field Ψ_{ext} will be produced by image currents $i_s(l)$ in the shell. They can be found from the condition for screening; namely, the field outside the shell is zero. Denote by $B_\tau(l)$ the tangential component of the plasma field on the surface of the shell:

$$B_\tau(l) = \int\int b_\tau(l; r', z') j_s(r', \Psi) dr' dz'. \tag{3.130}$$

Then the current distribution i_s would be determined by Eq. (3.94). Having calculated the inverse integral operator for Eq. (3.94), we can write its solution in the form

$$i_s(l) = \oint_{l'} K(l; l') B_\tau(l') dl'$$

$$= \oint_{l'} K(l; l') \int\int b_\tau(l'; r', z') j_s(r', \Psi) dr' dz'. \tag{3.131}$$

The flux Ψ_{ext} of the maintaining field produced by the surface current $i_s(l)$ is

$$\Psi_{ext}(r, z) = \oint_l \Psi_0(r, z; l) i_s(l) dl. \tag{3.132}$$

Adding it to the flux of the self-field of the plasma (3.125) we get an integral equation

$$\Psi(r, z) = \int\int \Psi_b(r, z; r', z') j_s(r', \Psi) dr' dz' \tag{3.133}$$

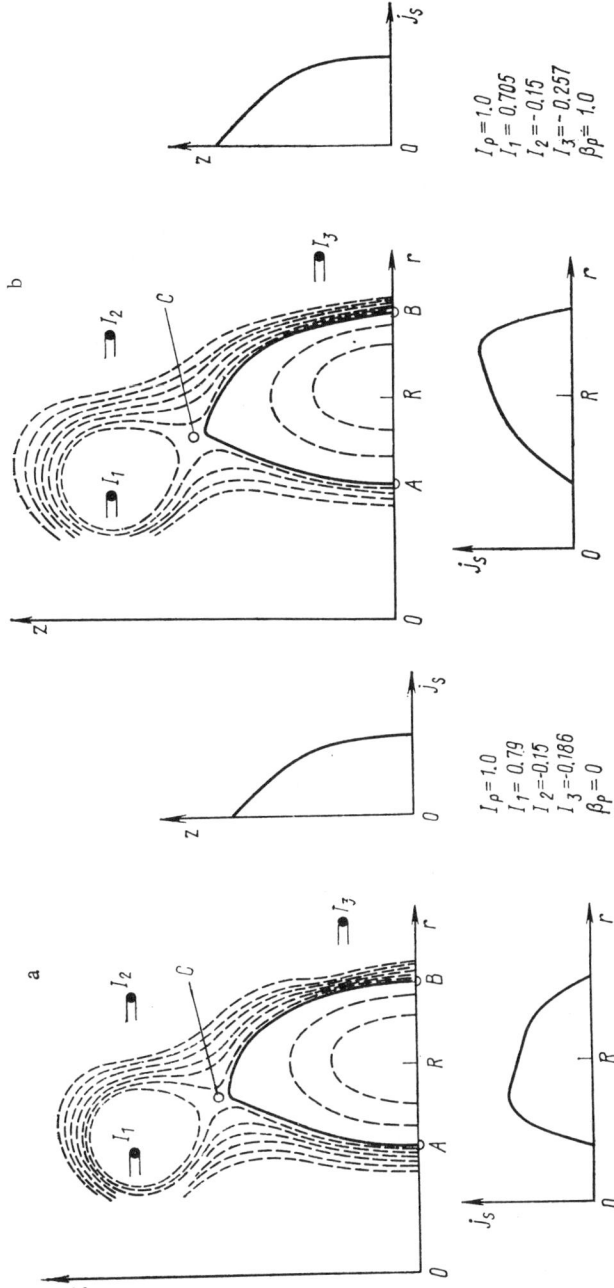

Fig. 17. Plasma equilibria with a free boundary for two values of β_p. The upper half of the configuration is shown. The solid line shows the plasma surface. The unfilled circles indicate the fixed characteristic points A, B, C: the external currents are determined from the condition that the boundary magnetic surface passes through the fixed points A, B, C.

for the flux function of the total field. In Eq. (3.133) the kernel is equal to

$$\Psi_b(r, z; r', z') = \Psi_0(r, z; r', z')$$
$$+ \oint_l \oint_{l'} \Psi_0(r, z; l) K(l, l') b_\tau(l'; r', z') dl dl'. \qquad (3.134)$$

The integration in Eq. (3.133) and in Eq. (3.127) is performed over the cross section of the current channel. In the case of a fixed boundary the equality $\Psi = 0$ on the surface of the shell is automatic.

As an example of the solution of an equilibrium problem for a plasma inside a given shell we shall consider the question of the influence of the elongation of the shell on the shape of the magnetic surfaces. If the current in the column is uniform, then all magnetic surfaces inside the column will reflect the shell elongation. However, if the current density is concentrated in the vicinity of the magnetic axis, the internal magnetic surfaces will have a limited elongation even for high values of the elongation of the shell. Figure 18 shows the results of calculations of equilibria for a current density of the type

$$j_{pl} = j_0 (\Psi/\Psi_0)^n/r, \quad n = 1, 2. \qquad (3.135)$$

For $n = 1$ the elongation of the shell cross section results in elongation of the magnetic surfaces in the vicinity of the magnetic axis. For $n = 2$ the magnetic surfaces near the axis have a limited elongation, $l_z/l_r = 1.5$, independent of the shell elongation.

The method of integral equations can be readily extended to the combined problem when the maintaining field of external currents and magnetic screens such as a well-conducting shell are both present.

2. <u>The Method of Eigenfunctions.</u> To solve the equilibrium problem for a plasma with a free boundary it is necessary to have an expression for the plasma self-field. This can be obtained in the form of a series if one uses a coordinate system which allows the separation of variables in the homogeneous equation (2.21). For a toroidal column one can take either spherical ρ, v, ζ or toroidal η, ω, ζ coordinates (Section 3.3).

In spherical coordinates the general solution of the homogeneous equation (2.21) is written in the form of a series (3.59). The so-

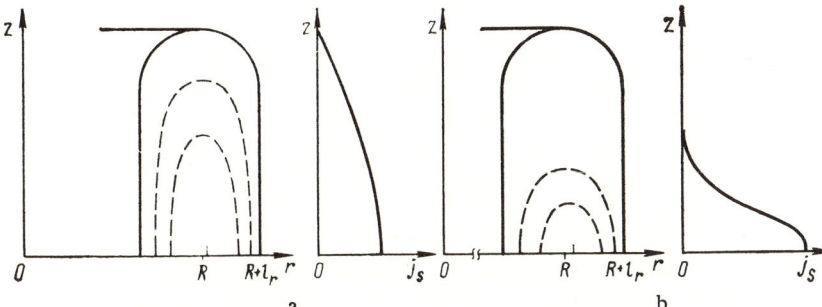

Fig. 18. Plasma equilibria in a casing of racetrack form with inhomogeneous current distribution, $R/l_r = 3$, $l_z/l_r = 4$: a) $j_s \sim \Psi/r$. The internal magnetic surfaces increase their elongation when l_z/l_r increases. b) $j_s \sim \Psi^2/r$. The internal magnetic surfaces have a limited elongation independent of the elongation of the casing.

lution of the inhomogeneous equation (2.21) with the condition $\Psi_{pl} \to 0$ as $\rho \to \infty$ can be written in the form

$$\Psi_{pl}(\rho, v) = \frac{4\pi^2}{c} \sum_{n=1}^{\infty} \frac{\sin^2 v}{n(n+1)} \left[\frac{1}{\rho^n} \int_0^\rho \rho^{n+2} S_n \, d\rho + \rho^{n+1} \int_\rho^\infty \frac{S_n \, d\rho}{\rho^{n-1}} \right] \dot{P}_n(\cos v)$$

$$+ \frac{4\pi^2}{c} \left[\rho \int_\rho^\infty S_0 \rho \, d\rho - \int_\rho^\infty S_0 \rho^2 \, d\rho \right] \cos v, \qquad (3.136)$$

where

$$S_n(\rho) = \int_0^\pi j_s(\rho, v) \sin^2 v \dot{P}_n(\cos v) \, dv;$$

$$S_0(\rho) = \frac{2}{\pi} \int_0^\pi j_s(\rho, v) \sin 2v \, dv. \qquad (3.137)$$

If ρ exceeds its maximum in the cross section of the current channel, the function $S_n(\rho)$ reduces to zero and only terms explicitly decreasing as $\rho \to \infty$ remain in Eq. (3.136). Having the expression for the plasma self-field, one can obtain the solution of the nonlinear equilibrium equation by means of iterations, as has been considered for the integral equations method.

Spherical coordinates and the series (3.136) have been used by Feneberg and Lackner [54], who for the first time have solved the equilibrium problem for a plasma column in given external fields and, in particular, have shown the existence of a critical value in a quadrupole external field (Section 3.1).

Spherical coordinates are natural for configurations like compact toroids. For a column with a large aspect ratio it is reasonable to use the toroidal coordinates described in Section 3.3. Appropriate expressions for the self-field of the plasma current have already been written in Section 3.3 [Eq. (3.46)]. To calculate the eigenfunctions $f_n(\eta)$, one can use a hypergeometric series

$$f_n(\eta) = \frac{1}{2} \operatorname{sh}^2 \eta \exp\left[-\left(n + \frac{3}{2}\right)\eta\right] F\left(n + \frac{3}{2}, \frac{3}{2}, 1 - \exp(-2\eta)\right), \quad (3.138)$$

and for the functions $g_n(\eta)$,

$$g_n(\eta) = -\frac{2}{\sqrt{\pi}} \operatorname{sh}^2 \eta \exp\left[-\left(n + \frac{3}{2}\right)\eta\right] \frac{\Gamma(n-1/2)}{n!} F\left(n + \frac{3}{2}, \frac{3}{2}, n+1, \exp(-2\eta)\right). \quad (3.139)$$

3. The Method of Finite Differences. If it is required to solve the equilibrium equation (2.21) in a limited region as, for example, in the fixed boundary problem, one can use the finite-difference method. The boundary condition which is required in this method is expressed explicitly: $\Psi|_\Gamma = \text{const}$. The iteration procedure can be standard when the current density is distributed according to $\Psi^{(n-1)}$ taken from the previous iteration.

It is of interest to extend this method to the problem of equilibrium with a free boundary [57]. To do this it is necessary to overcome a difficulty associated with the finiteness of the computation mesh and to find an appropriate boundary condition. The disadvantage of the finite-difference method as compared to those described above is that it does not allow us to separate the self-field of the plasma current and the maintaining field. However, one can use the virtual casing principle to find the self-field and so it becomes possible even with a computational mesh limited in space to obtain a boundary condition for which the solution will correspond to the equilibrium in a given maintaining field Ψ_{ext}.

Let Γ be the contour delimiting the computational mesh. The value of the flux function Ψ on Γ is composed of the given flux Ψ_{ext} and the flux Ψ_{pl} due to the plasma current. The plasma field outside Γ may be found in accordance with the virtual casing principle as the field of a surface current $i_s(l)$ determined by

$$\frac{4\pi}{c} i_s(l) = B_\tau^1(l) = -\frac{1}{2\pi r}\frac{\partial \Psi^1}{\partial n}. \tag{3.140}$$

In order that the virtual casing principle can be applied, it is necessary that the contour Γ should be the cross section of a magnetic surface. Therefore, to determine Ψ^1 and i_s it is sufficient to solve an auxiliary boundary problem [58], namely, to solve Eq. (2.21) inside Γ with a given right-hand side and a boundary condition $\Psi^1|_\Gamma$ = const. Having calculated $\Psi_{pl}|_\Gamma$ which is the flux due to the surface current $i_s(l)$ (3.140), we get the boundary condition for the flux Ψ of the equilibrium field:

$$\Psi|_\Gamma = \Psi_{pl}|_\Gamma + \Psi_{ext}|_\Gamma.$$

All the methods considered above have been widely used for the solution of equilibrium problems. In the integral equations method all the sources of the magnetic field enter explicitly, since the plasma current is replaced by a set of ring currents. This makes it possible to adjust the method for various formulations of the problem and makes it reliable in use. The advantage of the method is its insensitivity to the choice of a computational mesh, which can be defined by any curvilinear coordinates. It is only important to provide a sufficiently accurate integration. For a given mesh the kernel of the integral equations is calculated only once and it can be used for further solution of problems with various current distributions in the plasma and in the external circuits. The disadvantage of the method of integral equations is the necessity to store the kernel of the integral equation in the operative memory of the computer in the form of a two-dimensional matrix. This limits the potential for increasing the accuracy of the calculations. Typically 100 to 150 grid points inside the plasma are used. This is sufficient for the solution of a great number of equilibrium problems and, in particular, for calculations of magnetic systems serving to provide the equilibrium of the plasma. The replacement of the plasma current by a set of ring currents complicates the description of the magnetic surfaces inside the plasma.

The solutions obtained using series of eigenfunctions typically with about 50 terms, ~100 points in the radial direction, and ~50 points in the angular direction are smoother. This allows quite accurate interpolation and a detailed description of the structure of the configuration both outside and inside the plasma.

The finite-difference method is convenient because it allows the use of a simple rectangular computational mesh for which effective algorithms for solution of the Poisson equation have been developed. The choice of such a computational mesh facilitates the matching of equilibrium calculations with other problems of plasma theory, e.g., with studies of MHD instabilities, when the calculations of the equilibrium yield the initial conditions for the time-dependent problem. This is the reason why the finite-difference method is currently widely used. The method is somewhat inconvenient when an equilibrium with a fixed boundary is studied, due to the necessity of approximating the boundary conditions on a curvilinear contour. Therefore, equilibrium problems with a fixed boundary are solved by this method in the same way as problems with a free boundary, while the currents producing the maintaining field are chosen so that the boundary of the column passes through given points in the mesh [57].

4. The Method of Inverse Variables. A common disadvantage of Euler methods is the fact that they only give the values of the flux function at the fixed points of the computational mesh. However, a quasicylindrical description of the plasma equilibrium seems to be natural from a physical point of view (Section 2.4). In order to make a transition to this description it is necessary to determine magnetic surfaces and to calculate integral characteristics by means of integration over magnetic surfaces. In the Euler methods such a procedure requires an additional interpolation of the flux function and construction of the contours Ψ = const (a cubic spline technique is customarily applied to perform this). We shall now show the relationship between the Eulerian solution and functions entering the quasicylindrical description.

The label "a" of the magnetic surfaces can be chosen arbitrarily. Therefore, we shall consider the dependences $\Psi(a)$ to be chosen. In addition, there is an arbitrariness in the choice of the angle θ on each of the magnetic surfaces. Therefore, the function $\theta(r, z)$ may be chosen from the additional arguments. The integral characteristics of magnetic surfaces such as $\mathcal{J}(a)$ and $\Phi(a)$ can be found by direct integration. The metric coefficients g_{ik} can be found using the relations

$$g_{11} = \hat{g} g^{22}, \quad g_{22} = \hat{g} g^{11}, \quad g_{12} = -\hat{g} g^{12}, \tag{3.141}$$

where

$$\hat{g} = \frac{1}{\Psi'^2} \left(\frac{\partial \Psi}{\partial r} \frac{\partial \theta}{\partial z} - \frac{\partial \Psi}{\partial z} \frac{\partial \theta}{\partial r} \right)^{-2}, \quad g^{11} = \frac{1}{\Psi'^2} |\nabla \Psi|^2,$$
$$g^{22} = |\nabla \theta|^2, \quad g^{12} = \frac{1}{\Psi'} \nabla \Psi \nabla \theta. \tag{3.142}$$

It is possible to write the following explicit formulas for some averaged quantities, important for evolution problems and for MHD stability:

$$\langle \sqrt{g} \rangle = -\frac{\Psi'}{4\pi^2} \oint \frac{dl}{rB_p}, \quad \left\langle \frac{g_{22}}{\sqrt{g}} \right\rangle = -\frac{1}{\Psi'} \oint B_p \, dl;$$
$$\left\langle \frac{\sqrt{g}}{g_{33}} \right\rangle = -\frac{\Psi'}{4\pi^2} \oint \frac{dl}{r^2 B_p}, \quad q = -\frac{\Phi'}{\Psi'} = \frac{4\pi}{c} \frac{F}{4\pi^2} \oint \frac{dl}{r^2 B_p}. \tag{3.143}$$

Equations (3.141)–(3.143) allow one to transfer from the solution on the Eulerian mesh to the quasicylindrical description.

If the solutions of the equilibrium equations are considered as the initial data for studies of MHD stability or for the evolution of equilibria, it is natural to formulate the problem from the beginning in natural coordinates, while the Eulerian coordinates $r(a, \theta)$, $z(a, \theta)$ of magnetic surfaces are the result of the computation [59]. This can most easily be done in the orthogonal coordinates a, θ. The condition for orthogonality $g_{12} = 0$ is written in the following way:

$$\frac{\partial r}{\partial a} \frac{\partial r}{\partial \theta} + \frac{\partial z}{\partial a} \frac{\partial z}{\partial \theta} = 0, \tag{3.144}$$

or

$$\frac{\partial z}{\partial \theta} \bigg/ \frac{\partial r}{\partial a} = -\frac{\partial r}{\partial \theta} \bigg/ \frac{\partial z}{\partial a} = \alpha(a, \theta). \tag{3.145}$$

In this case

$$g_{11} = \left(\frac{\partial r}{\partial a}\right)^2 + \left(\frac{\partial z}{\partial a}\right)^2, \quad g_{22} = \alpha^2 g_{11}, \quad \sqrt{g} = \alpha r g_{11}. \quad (3.146)$$

The equilibrium equation (2.71), with the inclusion of orthogonality, can be written in the form

$$\frac{\partial}{\partial a}\frac{\alpha}{r}\Psi' = \frac{d}{da}\left\langle\frac{\alpha}{r}\right\rangle\Psi' + 16\pi^3 \frac{p'}{\Psi'}\left[\langle\sqrt{g}\rangle\frac{(\widetilde{\sqrt{g}}/g_{33})}{\langle\sqrt{g}/g_{33}\rangle} - (\widetilde{\sqrt{g}})\right]$$

$$+ \frac{(\widetilde{\sqrt{g}}/g_{33})}{\langle\sqrt{g}/g_{33}\rangle}\frac{d}{da}\left\langle\frac{\alpha}{r}\right\rangle\Psi', \quad (3.147)$$

while the condition for orthogonality is

$$\frac{\partial}{\partial a}\alpha\frac{\partial z}{\partial a} + \frac{\partial}{\partial \theta}\frac{r}{\alpha}\frac{\partial z}{\partial \theta} = 0; \quad (3.148)$$

$$\frac{\partial}{\partial a}\alpha\frac{\partial r}{\partial a} + \frac{\partial}{\partial \theta}\frac{1}{\alpha}\frac{\partial r}{\partial \theta} = 0. \quad (3.149)$$

These three equations serve to determine the three unknown functions $\alpha(a, \theta)$, $z(a, \theta)$, and $r(a, \theta)$.

Let us consider the additional conditions for the system of equations (3.147)-(3.149). Firstly, the solutions of Eqs. (3.147)-(3.149) should be periodic in θ. Secondly, the regularity condition

$$\alpha \partial r/\partial a = 0, \quad \alpha \partial z/\partial a = 0 \quad (3.150)$$

should be satisfied at $a = 0$, since $\alpha = 0$ on the magnetic axis. At the plasma boundary $a = a_0$ the functions $r(a_0, \theta)$ and $z(a_0, \theta)$ are related by the equation for the contour of the cross section

$$z = z(r). \quad (3.151)$$

The dependence $r(a_0, \theta)$ [or $z(a_0, \theta)$] for a given function $\alpha(a, \theta)$ cannot be arbitrary and should satisfy, for example, the following relation:

$$\left(\frac{\partial r}{\partial \theta}\right)^2 + \left(\frac{\partial z}{\partial \theta}\right)^2 = \alpha^2 \left[\left(\frac{\partial r}{\partial a}\right)^2 + \left(\frac{\partial z}{\partial a}\right)^2\right]. \quad (3.152)$$

For a given function $\alpha(a, \theta)$ these boundary conditions determine the solutions $r(a, \theta)$ and $z(a, \theta)$ of Eqs. (3.148) and (3.149) uniquely.

In turn, the function $\alpha(a, \theta)$ is found from Eq. (3.147). This system of equations with nonlinear boundary conditions can be solved by the method of successive relaxations. Figure 19 shows a calculation for a compact torus with a given dependence $p(a)$ using the method of inverse variables.

5. The Method of Moments (the Method of Prescribed Magnetic Surfaces). The numerical methods described above are universal and are applicable to the calculation of equilibria with arbitrary cross sections. However, in most cases it is sufficient to consider only basic effects associated with the deformation of the cross sections of magnetic surfaces, allowing simplifications in the description of the equilibrium. In many cases the shape of the magnetic surfaces can be approximated by simple formulas. For example, we can take a parametric dependence in the form of a series

$$r = R - \Delta(a) + \sum_{n=1}^{\infty} r_n(a) \cos n\theta; \quad z = \sum_{n=1}^{\infty} z_n(a) \sin n\theta, \quad (3.153)$$

and restrict the number of terms to a minimum admissible number without losing the main effects. Thus, for a tokamak with a circular cross section, the shift of the magnetic surfaces is the main effect of toroidicity, while the deformation of their shape is not very important. Therefore, to describe such an equilibrium it is sufficient to consider a model with circular magnetic surfaces shifted with respect to the magnetic axis (the Clark–Sigmar model [60]). For a column with an elongated cross section one can take as a basis a model consisting of nested ellipses with a variable ratio of semiaxes. A model in which the displacement, elongation, and triangularity of the magnetic surfaces are taken into account is widely applicable:

$$r = R - \Delta(a) - \rho(a, \theta) \cos\theta; \quad z = \lambda(a)\rho(a, \theta) \sin\theta;$$
$$\rho = a + \delta(a) \cos 3\theta. \quad (3.154)$$

The triangularity parameter δ is important for the MHD stability of ballooning modes, and it should be taken into account when the optimization of toroidal configurations is carried out.

There are many ways to obtain the equations for the parameters of the magnetic surfaces. One of them is based on the relationships for the moments of the equilibrium equations (Section 2.6). To ob-

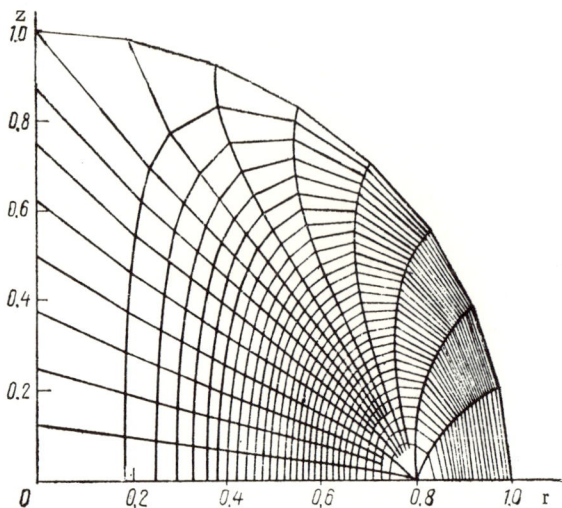

Fig. 19. The calculation of equilibria using the inverse variables method. The upper half of the compact toroidal plasma column cross section is shown. The pressure distribution is $p = C \left(\oint \frac{dl}{rB_p} \right)^{-\gamma}$, $\gamma = 5/3$.

tain them we use Eq. (2.115) in which the region of integration is assumed to be the cross section of a magnetic surface. Then we obtain

$$\frac{4\pi}{c} \int j_s f_k \frac{\sqrt{g}}{r} \, da d\theta = \oint B_\tau f_k \sqrt{g_{22}} \, d\theta, \tag{3.155}$$

where f_k are, for example, the functions of the series (2.121).

After differentiating with respect to a the system of equations (3.155) reduces in the natural coordinate system to the following:

$$-4\pi^2 c \frac{4\pi}{c} p' \langle f_k \sqrt{g} \rangle - \left(\frac{4\pi}{c} \right)^2 FF' \left\langle f_k \frac{\sqrt{g}}{g_{33}} \right\rangle$$
$$= \Psi' \frac{d}{da} \left\langle f_k \frac{g_{22}}{\sqrt{g}} \right\rangle \Psi'. \tag{3.156}$$

Here the brackets $\langle \ \rangle$ denote averaging with respect to θ. As was mentioned above, the function FF' is adjusted to the pressure $p(a)$ and the current distribution in the plasma, while the latter are determined in real systems by heating and transport processes. For convenience the function $\Psi'(a)$ may be used instead of $j_s(a)$ as data. Recall that the relationship between them is

$$\frac{4\pi}{c} j_s(a) = -\frac{1}{a}\frac{d}{da}\left\langle \frac{g_{22}}{\sqrt{g}}\frac{R}{a}\right\rangle a\frac{\Psi'}{2\pi R}. \qquad (3.157)$$

The function FF' may be eliminated from Eq. (3.156) if one expresses it in terms of p' and Ψ' using the averaged equilibrium equation ($f_0 = 1$) and substitutes this value into the remaining equations. Then

$$\frac{4\pi}{c} c 4\pi^2 p' \left\langle f_k \left[\frac{\sqrt{g}/g_{33}}{\langle \sqrt{g}/g_{33}\rangle} \langle \sqrt{g}\rangle - \sqrt{g}\right]\right\rangle =$$

$$= \Psi' \frac{d}{da}\left\langle f_k \frac{g_{22}}{\sqrt{g}}\right\rangle \Psi' - \Psi' \frac{\langle f_k \sqrt{g}/g_{33}\rangle}{\langle \sqrt{g}/g_{33}\rangle} \frac{d}{da}\left\langle \frac{g_{22}}{\sqrt{g}}\right\rangle \Psi'. \qquad (3.158)$$

For the simplest models the averaging can be done analytically. In this case a system of ordinary differential equations for the parameters of the magnetic surfaces can be obtained. Let us consider a model consisting of nested ellipses as an application to the equilibrium of a straight column. Suppose the equations for the cross sections of magnetic surfaces are the following:

$$x = a\cos\theta, \quad y = \lambda(a)a\sin\theta, \qquad (3.159)$$

where $\lambda(a)$ is the ratio of semiaxes. Let us write the main relationships which are contained in the equations for the moments:

$$\left.\begin{array}{l} g_{22} = \left(\dfrac{\partial x}{\partial \theta}\right)^2 + \left(\dfrac{\partial y}{\partial \theta}\right)^2 = \dfrac{a^2}{2}[\lambda^2 + 1 + (\lambda^2 - 1)\cos 2\theta]; \\[6pt] \sqrt{g} = \dfrac{\partial x}{\partial a}\dfrac{\partial y}{\partial \theta} - \dfrac{\partial y}{\partial a}\dfrac{\partial x}{\partial \theta} = a\lambda\left[1 + \dfrac{1}{2}\dfrac{a\lambda'}{\lambda} - \dfrac{1}{2}\dfrac{a\lambda'}{\lambda}\cos 2\theta\right]; \\[6pt] f_0 = 1, \quad f_2 = y^2 - x^2 = \dfrac{a^2}{2}[\lambda^2 - 1 - (\lambda^2 + 1)\cos 2\theta]. \end{array}\right\} \qquad (3.160)$$

The corresponding averaged values are

$$\langle \sqrt{g} \rangle = \frac{1}{2}(a^2\lambda)', \quad \langle f_2\sqrt{g}\rangle = \frac{1}{8}[a^4(\lambda^3-\lambda)]';$$

$$\left\langle \frac{g_{22}}{\sqrt{g}} \right\rangle = \frac{a}{2\lambda} \cdot \frac{(\lambda^2+1)\left(1+\frac{1}{2}\frac{a\lambda'}{\lambda}+\sqrt{1+\frac{a\lambda'}{\lambda}}\right)+(\lambda^2-1)\frac{1}{2}\frac{a\lambda'}{\lambda}}{\sqrt{1+\frac{a\lambda'}{\lambda}}\left(1+\frac{1}{2}\frac{a\lambda'}{\lambda}+\sqrt{1+\frac{a\lambda'}{\lambda}}\right)};$$

$$\left\langle f_2 \frac{g_{22}}{\sqrt{g}} \right\rangle = \frac{a^3}{4\lambda} \cdot \frac{(\lambda^4-1)\sqrt{1+\frac{a\lambda'}{\lambda}}-2\lambda^2\frac{a\lambda'}{\lambda}}{\sqrt{1+\frac{a\lambda'}{\lambda}}\left(1+\frac{1}{2}\frac{a\lambda'}{\lambda}+\sqrt{1+\frac{a\lambda'}{\lambda}}\right)}$$

(3.161)

The relationship between $\Psi'(a)$ and the current density $j(a)$ is of the form

$$-\Psi'\left\langle \frac{g_{22}}{\sqrt{g}} \right\rangle = \frac{4\pi}{c}\int_0^a j(a)\frac{1}{2}(a^2\lambda)'\,da, \quad \Psi = A_s. \quad (3.162)$$

The equation for the ellipticity $\lambda(a)$ can be written in the following way:

$$\frac{a\lambda'}{\lambda}\left\{\left[\lambda^4+6\lambda+1+\frac{a\lambda'}{\lambda}\cdot\frac{\lambda^4-1}{\left(1+\sqrt{1+\frac{a\lambda'}{\lambda}}\right)^2}\right]\lambda j(a)\right.$$

$$-\left[(\lambda^2+1)\left(1+\frac{2}{1+\sqrt{1+\frac{a\lambda'}{\lambda}}}\right)+\lambda^2-1\right]\frac{1}{a^4}\int_0^a j'(a)a^4\lambda(\lambda^2-1)\,da$$

$$\left.+\left[4(\lambda^4-1)\frac{1}{1+\sqrt{1+\frac{a\lambda'}{\lambda}}}-8\lambda^2\right]\frac{1}{a^2}\int_0^a j'(a)a^2\lambda\,da\right\}$$

$$= 4(\lambda^2+1)\frac{1}{a^4}\int_0^a j'(a)a^4\lambda(\lambda^2-1)\,da$$

$$-4(\lambda^4-1)\frac{1}{a^2}\int_0^a j'(a)a^2\lambda\,da. \quad (3.163)$$

When the current density profile j(a) is given, then Eq. (3.163) can be solved by iteration if we substitute the parameters taken from the previous iteration into the braces and into the right-hand side of Eq. (3.163).

If the functions $p(a)$, $\Psi'(a)$, and $\lambda(a)$ are known, then all variables which characterize the configuration may be evaluated using general relations (Section 2.4). Let us write the expressions for the rotational transform $\mu = -\Psi'/\Phi'$ and the quadrupole component of the maintaining field for the model under consideration:

$$\mu = \frac{4\pi}{c} \frac{R}{B_s} \frac{j(a) a^2 \lambda - \int_0^a j'(a) a^2 \lambda\, da}{(a^2 \lambda)' \langle g_{22}/\sqrt{g} \rangle} ; \qquad (3.164)$$

$$A_{\text{ext}} = \frac{4\pi}{c} (y^2 - x^2) \frac{\lambda j_s(a) - \frac{1}{a_0^2} \int_0^{a_0} j_s' a^2 \lambda\, da}{(\lambda+1)^2}$$

$$\times \frac{\lambda^2 - 1 + \lambda^2 a\lambda'/\lambda}{(\lambda^2+1)(1+\sqrt{1+a\lambda'/\lambda}) + \lambda^2 a\lambda'/\lambda}. \qquad (3.165)$$

All quantities in Eq. (3.165) are evaluated at the plasma boundary $a = a_0$. For a plasma column with a uniform current distribution expression (3.165) reduces to Eq. (3.7), which was obtained from an exact solution.

Another method to obtain the equations for the parameters $u(a)$ of the magnetic surfaces has been proposed by Khait [61]. The equilibrium equation (2.21) can be considered as an Euler–Ostrogradskii equation for the variation of the following functional:

$$Q = \iint \left\{ \frac{1}{r} (\nabla \Psi)^2 - \frac{16\pi^2}{c} \left[2\pi crp(\Psi) + \frac{F^2(\Psi)}{cr} \right] \right\} dS_\zeta \qquad (3.166)$$

with respect to the function $\Psi(r, z)$. The integration in Eq. (3.166) is performed over the cross section S_ζ of the plasma column.

Let us make a transform to the natural coordinate system a, θ, ζ, where the function $\Psi(r, z)$ is one-dimensional: $\Psi = \Psi(a)$. Then the functional (3.166) takes the form

$$Q = \iint \left\{ \frac{g_{22}}{\sqrt{g}} \Psi'^2 - \frac{16\pi^2}{c} \left[2\pi c p(a) + \frac{F^2(a)}{cr^2} \right] \sqrt{g} \right\} da\, d\theta. \quad (3.167)$$

Now its extremum should be found with respect to the parameters $u(a)$ of the cross sections of magnetic surfaces $[a(r, z) = \text{const}]$. Let us introduce the notations

$$K \equiv \left\langle \frac{g_{22}}{\sqrt{g}} \right\rangle = \frac{1}{2\pi} \oint \frac{|\nabla a|}{r} dl; \quad (3.168)$$

$$V \equiv \int_0^a \langle \sqrt{g} \rangle\, da = \frac{1}{2\pi} \iint_{S_\zeta(a)} r\, dS_\zeta; \quad (3.169)$$

$$L \equiv \int_0^a \left\langle \frac{\sqrt{g}}{g_{33}} \right\rangle da = \frac{1}{2\pi} \iint_{S_\zeta(a)} \frac{dS_\zeta}{r}. \quad (3.170)$$

From the expressions for K, V, and L it follows that K depends both on the parameters u_i and on their derivatives $u_i'(a)$: $K = K(a, u_i, u_i')$, while the functions V and L do not depend on derivatives: $V = V(a, u_i)$, $L = L(a, u_i)$. Functional (3.167) can be rewritten in the form of a one-dimensional integral

$$Q = 2\pi \int_0^{a_0} \left[K \Psi'^2 - \frac{16\pi^2}{c} \left(2\pi c p V' + \frac{F^2}{c} L' \right) \right] da. \quad (3.171)$$

The last two terms can be integrated by parts

$$Q = 2\pi \int_0^{a_0} \left[K \Psi'^2 + \frac{16\pi^2}{c} \left(2\pi c p' V + \frac{2FF'}{c} L \right) \right] da$$
$$- \frac{32\pi^3}{c^2} F^2 L \bigg|_{a=a_0}. \quad (3.172)$$

For a fixed plasma boundary the term outside the integral (3.172) is constant and does not contribute to variation of the functional. The Euler–Lagrange equations for the one-dimensional functions $u_i(a)$ reduce to

$$\frac{d}{da}\left(\Psi'^2 \frac{\partial K}{\partial u_i'} \right) - \Psi'^2 \frac{\partial K}{\partial u_i} - \frac{16\pi^2}{c}\left(2\pi c p' \frac{\partial V}{\partial u_i} + \frac{2FF'}{c} \frac{\partial L}{\partial v_i} \right) = 0. \quad (3.173)$$

Together with the averaged equilibrium equation

$$\Psi' \frac{d}{da} K\Psi' + \frac{8\pi^2}{c}\left(2\pi c p' V' + \frac{2FF'}{c} L'\right) = 0 \tag{3.174}$$

Eqs. (3.173) provide a full set of equations for the parameters $u_i(a)$ of the magnetic surfaces.

3.7. Equilibrium with Anisotropic Pressure

In rarefied high-temperature plasmas anisotropy of the plasma pressure, which may be connected, for example, with neutral beam injection or high-frequency heating, can be maintained. In the drift approximation the equilibrium of plasma with an anisotropic pressure is described by formulas (1.24) and (1.28). As was shown by Grad [62] these equations can be expressed by one vector equation if one introduces the vector function

$$\frac{4\pi}{c} \mathbf{K} = \operatorname{curl} \sigma \mathbf{B}, \tag{3.175}$$

instead of the current density. Here

$$\sigma = 1 - 4\pi (p_\| - p_\perp)/B^2. \tag{3.176}$$

The equilibrium equation takes the form

$$\nabla p_\| = \frac{p_\| - p_\perp}{2B^2} \nabla B^2 + \frac{1}{c} [\mathbf{K}\mathbf{B}]. \tag{3.177}$$

In fact, the scalar product of $\mathbf{B}$ and Eq. (3.177) gives the equation of equilibrium along the magnetic field (1.28). On the other hand, taking the vector product of $\mathbf{B}$ and Eq. (3.177) and taking into account that

$$\mathbf{K}_\perp = \sigma \mathbf{j}_\perp + \frac{c}{B^2}[\mathbf{B}, \nabla(p_\| - p_\perp)] - \frac{c(p_\| - p_\perp)}{B^4}[\mathbf{B}, \nabla B^2], \tag{3.178}$$

we obtain Eq. (1.24). The solutions of Eq. (3.177) were examined in paper [63].

The transverse and longitudinal pressures are not now constant on the magnetic surfaces $\Psi = \text{const}$, $\mathbf{B}\nabla\Psi = 0$. In the case of axial symmetry which we shall consider below, the components of the pressure tensor are functions of two variables Ψ and u. The variable u may have a meaning, for example, the strength $|\mathbf{B}|$ of magnetic field, the angle θ, the radial distance r from the axis of symmetry, and so on.

As before, let us write the magnetic field as a sum of the poloidal and toroidal components:

$$\mathbf{B} = \frac{1}{2\pi r}[\nabla\Psi \mathbf{e}_s] + \frac{2F}{cr}\mathbf{e}_s \qquad (3.179)$$

and substitute this expression into the equilibrium equation (3.177)

$$\nabla p_\| = \frac{p_\| - p_\perp}{2B^2}\nabla B^2 - \frac{1}{2\pi rc}\left\{\nabla\Psi(\mathbf{K}\mathbf{e}_s) - (\mathbf{K}\nabla\Psi)\mathbf{e}_s + \frac{4\pi}{c}F[\mathbf{K}\mathbf{e}_s]\right\}. \qquad (3.180)$$

Since due to axial symmetry $p_\|$, B^2, and Ψ do not depend on ζ the scalar product of Eq. (3.180) and $\mathbf{e}_s$ gives

$$\mathbf{K}\nabla\Psi = 0, \qquad (3.181)$$

i.e., the lines of the vector $\mathbf{K}$ lie on the magnetic surface. In accordance with the definition (3.175) the vector $\mathbf{K}$ is expressed in terms of the poloidal flux Ψ and the poloidal current F:

$$\mathbf{K} = \frac{1}{2\pi r}[\nabla\sigma F, \mathbf{e}_s] - \frac{c}{8\pi^2 r}\left(r^2 \operatorname{div}\frac{\sigma\nabla\Psi}{r^2}\right)\mathbf{e}_s. \qquad (3.182)$$

The previous condition $\mathbf{K}\nabla\Psi = 0$ leads to the requirement

$$[\nabla\sigma F, \nabla\Psi] = 0, \qquad (3.183)$$

i.e., σF is a surface function which depends only on Ψ.

Now writing the gradients of $p_\|$ and B^2 appearing in Eq. (3.180) in the form

$$\nabla p_\| = \frac{\partial p_\|}{\partial u}\nabla u + \frac{\partial p_\|}{\partial \Psi}\nabla\Psi \qquad (3.184)$$

and taking into account that

$$[\mathbf{K}\mathbf{e}_s] = -\frac{\nabla(\sigma F)}{2\pi r} = -\frac{\nabla\Psi}{2\pi r}\frac{d\sigma F}{d\Psi}, \qquad (3.185)$$

we obtain for the components of the vector equation (3.180) along ∇u and $\nabla \Psi$

$$p_\perp = p_\| - B \frac{\partial p_\|}{\partial u} \Big/ \frac{\partial B}{\partial u}; \qquad (3.186)$$

$$rK_{s'} = r(Ke_s) = 2\pi c \frac{\partial p_\|}{\partial \Psi} r^2 + \frac{F}{\pi c} \frac{d\sigma F}{d\Psi} - \pi c \frac{p_\| - p_\perp}{B^2} \frac{\partial B^2}{\partial \Psi} r^2. \qquad (3.187)$$

The former permits us to calculate the oscillating part of $p_\|$ if $p_\perp(\Psi, u)$ and the average value of $p_\|$ are known while the latter connects the toroidal current density

$$j_s = K_s/\sigma + \frac{c}{4\pi}[\nabla \ln \sigma, \mathbf{B}]_s \qquad (3.188)$$

with the parameters of the equilibrium configuration.

We can obtain the equation for Ψ from the s component of Eq. (3.182)

$$r^2 \operatorname{div} \frac{\sigma \nabla \Psi}{r^2} = -\frac{8\pi^2}{c} rK_s, \qquad (3.189)$$

or [62]

$$\Delta^* \Psi = r^2 \operatorname{div} \frac{\nabla \Psi}{r^2} = -16\pi^3 r^2 \frac{\partial p_\|}{\partial \Psi} - \frac{16\pi^2}{c^2} F \frac{d\sigma F}{d\Psi}$$

$$+ 8\pi^3 r^2 \frac{p_\| - p_\perp}{B^2} \frac{\partial B^2}{\partial \Psi} - 4\pi \nabla \Psi \nabla \frac{p_\| - p_\perp}{B^2}. \qquad (3.190)$$

Examples of the solution of Eq. (3.190) are presented in [64, 65].

In the case of anisotropic pressure the integral relationships preserve their form if P and T are assumed to be

$$P = p_\perp + \frac{B_s^2 - B_{se}^2}{8\pi} + \frac{B_p^2}{B^2}(p_\| - p_\perp); \qquad (3.191)$$

$$T = p_\| + \frac{B_p^2}{8\pi} - \frac{B_s^2 - B_{se}^2}{8\pi} - \frac{B_p^2}{B^2}(p_\| - p_\perp). \qquad (3.192)$$

The main distinction between Eqs. (3.191), (3.192), and Eq. (2.90) is that now P contains $p_\perp$ and T contains $p_\parallel$ in the principal terms. The last terms in P and T are small, at least for tokamak conditions $B_p^2/B^2 \ll 1$. As a result, instead of β_p the perpendicular component $\beta_{\mathcal{J}\perp}$ enters the equation of equilibrium along the minor radius (2.102), while the longitudinal component $\beta_{\mathcal{J}\parallel}$ enters instead of $\beta_{\mathcal{J}}$ in the equation of equilibrium along the major radius (2.98), where

$$\beta_{\mathcal{J}\perp} = \frac{2c^2 \int p_\perp \, dS\, \zeta}{\mathcal{J}^2} \,, \quad \beta_{\mathcal{J}\parallel} = \frac{2c^2 \int p_\parallel \, dS\, \zeta}{\mathcal{J}^2} \,. \tag{3.193}$$

Having eliminated $\mu_{\mathcal{J}}$ from Eq. (2.98) using Eq. (2.102) we get for the maintaining field

$$B_\perp = \frac{\mathcal{J}}{2cR}\left(\beta_{\mathcal{J}\perp} + \beta_{\mathcal{J}\parallel} + s_1 + \frac{\partial L}{2\pi \partial R}\right). \tag{3.194}$$

Correspondingly, $(\beta_{\mathcal{J}\perp} + \beta_{\mathcal{J}\parallel})/2$ will enter instead of $\beta_{\mathcal{J}}$ into the expressions for the relative displacement of the external magnetic surfaces (see Section 4.3).

CHAPTER 4

EQUILIBRIUM OF THE PLASMA COLUMN WITH CIRCULAR CROSS SECTION

The equilibrium of a toroidal plasma column with a circular cross section can be described using an expansion in powers of the curvature. The basic factors affecting the equilibrium of such a plasma have been systematically considered in review [66] with emphasis on the interaction between the plasma current and the tokamak structural units. This material has been used in writing the present chapter. The equilibrium of plasma with a circular cross section and a nonplanar magnetic axis is also considered here.

4.1. Approximation of Low Toroidicity for an Axisymmetric Plasma Column

The deviation of the equilibrium of a toroidal plasma column with a circular cross section from the corresponding cylindrical one depends on the parameters a/R and $\beta_\mathcal{J} a/R$, which can be used as expansion parameters in solving the problem. To describe the equilibrium of a circular torus we shall use both a laboratory, quasicylindrical coordinate system ρ, ω, ζ

$$r = R - \rho \cos \omega; \quad z = \rho \sin \omega, \qquad (4.1)$$

and a natural coordinate system a, θ, ζ associated with the magnetic surfaces. At low plasma pressure, $\beta_\mathcal{J} a/R \ll 1$, the toroidicity effects mainly cause a shift $\Delta(a)$ of the magnetic surfaces along the major radius. The distortion of the magnetic surfaces proves to be small and can be characterized by the ellipticity parameter α. The relation between cylindrical and natural coordinates in this case can be expressed in the form

$$r = R + \Delta(a) - (a - \alpha \cos 2\theta) \cos \theta;$$
$$z = (a - \alpha \cos 2\theta) \sin \theta. \qquad (4.2)$$

The parameter α is small, $\alpha \ll a$, and is related to λ [Eq. (3.154)] by $\alpha = a(\lambda - 1)/2$. Here $\Delta(0)$ is the shift of the magnetic axis. Note that the coordinate system used here differs from that used in [66] in the definition of the angle ω. The transformation from the corresponding formulas of [66] is performed by substituting the angle $\omega \to \pi - \omega$ ($\cos \omega \to -\cos \omega$, $\sin \omega \to -\sin \omega$) and the current $\mathcal{J} \to -\mathcal{J}$.

The equilibrium equation (2.21) expressed in quasicylindrical coordinates will be of the following form, if one restricts oneself to the quadratic approximation in the curvature:

$$\rho \frac{\partial}{\partial \rho} \rho \frac{\partial \Psi}{\partial \rho} + \frac{\partial^2 \Psi}{\partial \omega^2} = -\frac{8\pi^2}{c} \left(2\pi c \frac{dp}{d\Psi} r^2 + \frac{1}{c} \frac{dF^2}{d\Psi} \right) \rho^2$$
$$- \left(\frac{\rho}{R} \cos \omega + \frac{1}{2} \frac{\rho^2}{R^2} + \frac{1}{2} \frac{\rho^2}{R^2} \cos 2\omega \right) \rho \frac{\partial \Psi}{\partial \rho}$$
$$+ \left(\frac{\rho}{R} \sin \omega + \frac{1}{2} \frac{\rho^2}{R^2} \sin 2\omega \right) \frac{\partial \Psi}{\partial \omega}. \qquad (4.3)$$

The poloidal components of the magnetic field are related to the flux function by

$$B_\rho = \frac{1}{2\pi r} \frac{1}{\rho} \frac{\partial \Psi}{\partial \omega}, \quad B_\omega = -\frac{1}{2\pi r} \frac{1}{\rho} \frac{\partial \Psi}{\partial \rho}. \quad (4.4)$$

Further we shall need the following partial solutions of the homogeneous magnetostatic equation (4.3) ($dp/d\Psi = 0$, $dF/d\Psi = 0$):

$$\begin{aligned}
\Psi^0 &= \frac{4\pi R}{c} \mathcal{J} \left[\ln \frac{8R}{\rho} - 2 - \frac{1}{2}\left(\ln \frac{8R}{\rho} - 1\right) k\rho \cos\omega \right. \\
&\quad \left. + \frac{1}{16} k^2 \rho^2 + \frac{1}{8} k^2 \rho^2 \ln \frac{8R}{\rho} - \frac{1}{16}\left(\ln \frac{8R}{\rho} - 2\right) k^2 \rho^2 \cos 2\omega \right]; \\
B_\omega^0 &= \frac{2\mathcal{J}}{c\rho} \left[1 + \frac{1}{2} \ln \frac{8R}{\rho} k\rho \cos\omega + \left(\frac{3}{8} \ln \frac{8R}{\rho} - \frac{5}{16}\right) \right. \\
&\quad \left. \times k^2 \rho^2 \sin 2\omega \right]; \\
B_\rho^0 &= \frac{2\mathcal{J}}{c\rho} \left[\frac{1}{2}\left(\ln \frac{8R}{\rho} - 1\right) k\rho \sin\omega \right. \\
&\quad \left. + \left(\frac{3}{8} \ln \frac{8R}{\rho} - \frac{1}{2}\right) k^2 \rho^2 \sin 2\omega \right]; \\
\Psi_{ext}^1 &= \frac{4\pi R}{c} \mathcal{J} \left(k\rho \cos\omega - \frac{1}{4} k^2 \rho^2 - \frac{1}{4} k^2 \rho^2 \cos 2\omega \right); \\
B_{\omega,ext}^1 &= -\frac{2\mathcal{J}}{c\rho} k\rho \cos\omega, \\
B_{\rho,ext}^1 &= -\frac{2\mathcal{J}}{c\rho} k\rho \sin\omega; \\
\Psi_{int}^1 &= \frac{4\pi R}{c} \mathcal{J} \left[k \frac{a^2}{\rho} \cos\omega - \frac{1}{4} k^2 a^2 \cos 2\omega - \frac{1}{2} k^2 a^2 \ln\frac{a}{\rho} \right]; \\
B_{\omega,int}^1 &= \frac{2\mathcal{J}}{c\rho} \left[k \frac{a^2}{\rho} \cos\omega + \frac{1}{2} k^2 a^2 \cos 2\omega \right]; \\
B_{\rho,int}^1 &= -\frac{2\mathcal{J}}{c\rho} k \frac{a^2}{\rho} \sin\omega; \\
\Psi_{ext}^2 &= \frac{4\pi R}{c} \mathcal{J} k^2 \rho^2 \cos 2\omega; \\
B_{\omega,ext}^2 &= -\frac{2\mathcal{J}}{c\rho} 2k^2 \rho^2 \cos 2\omega; \quad B_{\rho,ext}^2 = -\frac{2\mathcal{J}}{c\rho} 2k^2 \rho^2 \sin 2\omega; \\
\Psi_{int}^2 &= \frac{4\pi R}{c} \mathcal{J} k^2 \frac{a^4}{\rho^2} \cos 2\omega; \\
B_{\omega,int}^2 &= \frac{2\mathcal{J}}{c\rho} 2k^2 \frac{a^4}{\rho^2} \cos 2\omega; \quad B_{\rho,int}^2 = -\frac{2\mathcal{J}}{c\rho} 2k^2 \frac{a^4}{\rho^2} \sin 2\omega.
\end{aligned} \quad (4.5)$$

Here and hereinafter $k = 1/R$. If not prescribed otherwise, the normalized current $\mathcal{J}$ means the plasma current.

The function Ψ^0 represents an expansion, near the axis of the quasicylindrical coordinate system, of the flux function of the current ring. It coincides within a constant with the first term $f_0(\eta) \times [2(\cosh \eta - \cos \omega)]^{-1/2}$ of the series (3.48). Away from the coordinate axis, Ψ^0 transforms into a function which tends to zero as $r^2 + z^2 \to \infty$.

The functions Ψ^1_{ext} and Ψ^1_{int} are the magnetic fluxes inside and outside the toroidal shell with radii R and a which contains a surface current of the following form:

$$i_s^1(\omega) = \frac{\mathcal{J}}{\pi a}\left(ka \cos \omega + \frac{1}{4} k^2 a^2 \cos 2\omega\right). \tag{4.6}$$

The flux Ψ^1_{ext} corresponds to the uniform magnetic field

$$\mathbf{B}^1_{ext} = B_z \mathbf{e}_z = -\frac{2\mathcal{J}}{ca} ka \mathbf{e}_z. \tag{4.7}$$

The functions Ψ^2_{ext} and Ψ^2_{int} are the magnetic fluxes inside and outside a toroidal shell with a surface current of the form

$$i_s^2(\omega) = \frac{2\mathcal{J}}{\pi a} k^2 a^2 \cos 2\omega. \tag{4.8}$$

The flux function Ψ^2_{ext} corresponds to a quadrupole field

$$B_r = \frac{4\mathcal{J}}{ca} k^2 a\rho \sin \omega, \quad B_z = -\frac{4\mathcal{J}}{ca} k^2 a\rho \cos \omega. \tag{4.9}$$

In what follows we shall normalize dipole surface currents and uniform field fluxes to $i_s^1(\omega)$: $i_s(\omega) = i i_s^1(\omega)$ and Ψ^1_{ext}: $\Psi_{ext} = C\Psi^1_{ext}$ respectively.

For a quasiuniform current $dp/d\Psi = \text{const}$, $dF^2/d\Psi = \text{const}$ an explicit solution to the inhomogeneous equation (4.3) can be readily obtained for a circular torus:

$$\Psi_i(\rho, \omega) = \frac{2\pi R}{c} \mathcal{J} \left[1 - \frac{\rho^2}{\rho_0^2} - \frac{1}{4}\left(1 - \frac{\rho^2}{\rho_0^2}\right) k\rho \cos \omega \right.$$
$$\left. + \frac{1}{32} k^2 \frac{\rho^4}{\rho_0^2} + \frac{5}{16} k^2 \rho^2 - \frac{1}{16}\left(1 - \frac{\rho^2}{\rho_0^2}\right) k^2 \rho^2 \cos 2\omega \right] -$$

$$-\frac{2\pi R}{c}\beta_{\mathcal{J}}\mathcal{J}\left[\left(1-\frac{\rho^2}{\rho_0^2}\right)k\rho\cos\omega+\frac{1}{4}k^2\frac{\rho^4}{\rho_0^2}\right.$$
$$\left.-\frac{1}{4}\left(1-\frac{\rho^2}{\rho_0^2}\right)k^2\rho^2\cos 2\omega\right]. \tag{4.10}$$

Here ρ_0 is the minor radius of the plasma column, and $\mathcal{J}$ is the total plasma current. The center of the plasma column is assumed to coincide with the origin of the coordinate system. On the plasma boundary ($\rho = \rho_0$) Ψ = const. The displacement of the magnetic surfaces is

$$\Delta(\rho) = \frac{1}{2}\left(\beta_{\mathcal{J}}+\frac{1}{4}\right)\frac{\rho_0^2-\rho^2}{R}. \tag{4.11}$$

and the ellipticity parameter is

$$\alpha = \rho\left(\frac{1}{4}\beta_{\mathcal{J}}^2+\frac{3}{64}\right)\frac{\rho_0^2-\rho^2}{R^2}. \tag{4.12}$$

The internal solution (4.10) allows us to determine the critical value of $\beta_{\mathcal{J}}$ at which the poloidal magnetic field vanishes on the inner side of the toroidal surface. For a quasiuniform current the critical value $\beta_{\mathcal{J},\mathrm{cr}}$ is

$$\beta_{\mathcal{J},\mathrm{cr}} = R/\rho_0 + 1/2. \tag{4.13}$$

In this case the displacement of the magnetic axis $\Delta(0) = \rho_0/2$ and the ellipticity of the internal surfaces $\alpha = \rho_0/4$ (the ratio of semi-axes is $l_z/l_r = 1.6$).

In the case of a quasiuniform current, Eq. (4.3) is also readily solved for an elliptic cross section. In the linear approximation in the curvature, the internal solution is of the form

$$\Psi_i = \frac{2\pi R}{c}\mathcal{J}\frac{2\lambda}{\lambda^2+1}\left(1-\frac{x^2}{l_r^2}-\frac{z^2}{l_z^2}\right)\left[1-\frac{\lambda^2}{3\lambda^2+1}\frac{x}{R}\right.$$
$$\left.-\beta_{\mathcal{J}}\frac{(\lambda^2+1)^2}{\lambda(3\lambda^2+1)}\left(\frac{x}{R}-\frac{1}{2}\frac{x^2}{R^2}-\frac{1}{8}\frac{l_r^2}{R^2}\frac{3\lambda^2+1}{\lambda^2+1}\right)\right],$$
$$x = \rho\cos\omega. \tag{4.14}$$

Here l_r is the radial semiaxis of the ellipse and λ is the ratio of semiaxes. From this expression one can obtain the critical equilibrium value $\beta_{\mathcal{J},\mathrm{cr}}$ for a plasma with an elliptic cross section and a quasiuniform current density:

$$\beta_{\mathcal{J},\text{cr}} = \frac{R}{l_r} \frac{\lambda(3\lambda^2+1)}{(\lambda^2+1)^2} \left[1 + \frac{l_r}{2R}\left(\frac{\lambda^2+1}{3\lambda^2+1} + \frac{1}{4}\frac{3\lambda^2+1}{\lambda^2+1}\right)\right]. \quad (4.15)$$

The coefficients in Eq. (4.15) associated with the elongation of the cross section are weak functions of λ and can be assumed equal to 1 with an accuracy of 10% (for $\lambda \leq 2$).

For an arbitrary current distribution, Eq. (4.3) becomes inconvenient to use because of the nonlinear dependences $p'(\Psi)$ and $FF'(\Psi)$, and it is expedient to transfer to natural coordinates. Metric coefficients entering the expressions relating physical quantities (Section 2.4), in the quadratic approximation in the curvature, are of the following form:

$$\left.\begin{aligned}
g_{11} &= 1 - 2\Delta'\cos\theta + \Delta'^2 - 2\alpha'\cos 2\theta; \\
g_{12} &= \Delta' a \sin\theta + 2\alpha \sin 2\theta; \\
g_{22} &= a^2\left(1 - 2\frac{\alpha}{a}\cos 2\theta\right); \\
g_{33} &= r^2; \\
\sqrt{g} &= ra\left[1 - \Delta'\cos\theta - \left(\frac{\alpha}{a} + \alpha'\right)\cos 2\theta\right].
\end{aligned}\right\} \quad (4.16)$$

Here we assume $\Delta' \approx a/R$, $\alpha \approx a^2/R^2$.

The equations for the displacement and the ellipticity can be obtained from Eq. (2.71) reduced to the form

$$\frac{\mathcal{J}}{\langle g_{22}/\sqrt{g}\rangle}\left[\frac{\partial}{\partial a}\frac{\widetilde{(g_{22}/\sqrt{g})}}{\langle g_{22}/\sqrt{g}\rangle}\mathcal{J} - \mathcal{J}'\frac{\widetilde{(\sqrt{g}/g_{33})}}{\langle\sqrt{g}/g_{33}\rangle}\right]$$
$$-\frac{\mathcal{J}^2}{\langle g_{22}/\sqrt{g}\rangle^2}\frac{\partial}{\partial\theta}\frac{g_{12}}{\sqrt{g}} = \pi c^2 p'\langle\sqrt{g}\rangle\left[\frac{\widetilde{(\sqrt{g}/g_{33})}}{\langle\sqrt{g}/g_{33}\rangle} - \frac{\widetilde{(\sqrt{g})}}{\langle\sqrt{g}\rangle}\right]. \quad (4.17)$$

For the combinations of metric coefficients entering in this equation we can write

$$\frac{g_{22}}{\sqrt{g}} = \frac{a}{R}\left[1 + \frac{1}{2}\Delta'^2 + \frac{1}{2}\frac{a}{R}\Delta' + \frac{1}{2}\frac{a^2}{R^2} + \left(\frac{a}{R} + \Delta'\right)\right.$$
$$\left.\times\cos\theta + \left(\alpha' - \frac{\alpha}{a} + \frac{1}{2}\frac{a^2}{R^2} + \frac{1}{2}\Delta'^2 + \frac{1}{2}\frac{a}{R}\Delta'\right)\cos 2\theta\right];$$

$$\frac{\sqrt{g}}{g_{33}} = \frac{a}{R}\left[1 + \frac{1}{2}\frac{a^2}{R^2} - \frac{1}{2}\Delta'\frac{a}{R} + \left(-\Delta' + \frac{a}{R}\right)\cos\theta\right.$$
$$\left. + \left(\frac{1}{2}\frac{a^2}{R^2} - \frac{1}{2}\Delta'\frac{a}{R} - \alpha' - \frac{\alpha}{a}\right)\cos 2\theta\right];$$

$$\sqrt{g} = aR\left[1 + \frac{1}{2}\Delta'\frac{a}{R} - \left(\Delta' + \frac{a}{R}\right)\cos\theta\right. \qquad (4.18)$$
$$\left. + \left(-\alpha' - \frac{\alpha}{a} + \frac{1}{2}\Delta'\frac{a}{R}\right)\cos 2\theta\right];$$

$$\frac{g_{12}}{\sqrt{g}} = \frac{1}{R}\left[\Delta'\sin\theta + \left(2\frac{\alpha}{a} + \frac{1}{2}\Delta'^2 + \frac{1}{2}\frac{a}{R}\Delta'\right)\sin 2\theta\right].$$

When deriving equations for Δ and α from Eq. (4.17) one can use a basic (cylindrical) approximation for the averaged values. Equating the coefficients of both $\cos\theta$ and $\cos 2\theta$ yields

$$a\Delta'' + \left(\frac{2\mathcal{F}'a}{\mathcal{F}} - 1\right)\Delta' = -\frac{a}{R} + 2\frac{\pi c^2 p' a^4}{R\mathcal{F}^2}; \qquad (4.19)$$

$$a\alpha'' + \left(\frac{2a\mathcal{F}'}{\mathcal{F}} - 1\right)\alpha' - 3\frac{\alpha}{a} = -\frac{\pi c^2 p' a^4}{R\mathcal{F}^2}\left(3\Delta' + \frac{1}{2}\frac{a}{R}\right)$$
$$-\frac{1}{2}\frac{a^2}{R^2} + \frac{a}{R}\Delta' - \frac{a\mathcal{F}'}{\mathcal{F}}\frac{3}{2}\Delta'. \qquad (4.20)$$

Having multiplied Eq. (4.19) by $\mathcal{F}^2/a^2$, we can readily integrate it with the result

$$\Delta'(a) = -\frac{a}{R}\left[\frac{1}{2}l_i(a) + \beta_{\mathcal{F}}(a)\right], \qquad (4.21)$$

where

$$l_i(a) = \frac{2}{\mathcal{F}^2(a)}\int_0^a \frac{\mathcal{F}^2}{a}da = \frac{\langle B_\theta^2\rangle_a}{B_\theta^2(a)}; \qquad (4.22)$$

$$\beta_{\mathcal{F}}(a) = \frac{2c^2}{\mathcal{F}^2}2\pi\int_0^a [p(\rho) - p(a)]\rho d\rho. \qquad (4.23)$$

The solution of Eq. (4.20) for ellipticity cannot be obtained in an explicit form for an arbitrary current distribution. In the particular case of a uniform current and a parabolic pressure its solution is expressed by formula (4.12).

Let us now consider the basic relations between the integral equilibrium characteristics. The relation between the longitudinal current $\mathcal{J}(a)$, enclosed by the magnetic surface with radius a and the poloidal flux $\Psi'(a)$ is of the following form:

$$\frac{4\pi}{c} \mathcal{J}(a) = -\left\langle \frac{g_{22}}{\sqrt{g}} \right\rangle \Psi' = -\frac{a}{R_a}\left(1 + \frac{1}{2}\frac{a^2}{R^2}\right.$$
$$\left. + \frac{1}{2}\Delta'^2 + \frac{1}{2}\frac{a}{R}\Delta'\right)\Psi'(a). \tag{4.24}$$

Here $R_a = R + \Delta(a)$ is the major radius of the magnetic surface under consideration. The longitudinal flux $\Phi'(a)$ is connected with the poloidal current F:

$$\Phi'(a) = \frac{4\pi}{c} F \left\langle \frac{\sqrt{g}}{g_{33}} \right\rangle = \frac{4\pi}{c} F \frac{a}{R}\left(1 + \frac{1}{2}\frac{a^2}{R^2} - \frac{1}{2}\frac{a}{R}\Delta'\right). \tag{4.25}$$

Taking into account the relation between the longitudinal magnetic field B_s and the poloidal current F,

$$B_s = \frac{2F}{cr} = \frac{2F}{cR_a\left(1 - \frac{a}{R}\cos\theta\right)}, \tag{4.26}$$

one can obtain an expression for the safety factor $q(a) = -\Phi'/\Psi'$ in terms of the total current and the longitudinal field:

$$q(a) = \frac{F}{\mathcal{J}}\left\langle \frac{\sqrt{g}}{g_{33}} \right\rangle \left\langle \frac{g_{22}}{\sqrt{g}} \right\rangle \approx \frac{a}{R_a}\frac{B_{s0}}{2\mathcal{J}/ca}\left(1 + \frac{a^2}{R^2} + \frac{1}{2}\Delta'^2\right). \tag{4.27}$$

Here B_{s0} is the value of B_s at $\theta = \pi/2$. The factor before the bracket is equal to the value of q for a corresponding straight plasma column. The toroidal corrections between the brackets should be taken into account for a high-pressure plasma ($\beta_{\mathcal{J}} \sim R/a$).

The distribution of the poloidal magnetic field over a magnetic surface is determined by the following expression:

$$B_\theta(a, \theta) = \frac{2\mathcal{J}}{c\sqrt{g_{22}}} \frac{\frac{g_{22}}{\sqrt{g}}}{\left\langle \frac{g_{22}}{\sqrt{g}} \right\rangle} = \frac{2\mathcal{J}}{ca}\left[1 - \frac{a}{R}\Lambda\cos\theta + \right.$$

$$+ \left(\alpha' + \frac{1}{2} \frac{a^2}{R^2} + \frac{1}{2} \Delta'^2 + \frac{1}{2} \frac{a}{R} \Delta' \right) \cos 2\theta \bigg]. \quad (4.28)$$

Here

$$\Lambda(a) = \beta_{\mathscr{J}}(a) + \frac{1}{2} l_i(a) - 1 \quad (4.29)$$

is the factor of asymmetry of the poloidal field. At low pressure ($\Lambda < 0$) the poloidal magnetic field is larger on the inner side of the torus, while at high pressure ($\Lambda > 0$) the field is larger on the outer side.

The longitudinal current density distribution over a magnetic surface can be found using (2.64)

$$j_s = j_{s0}(a) \left(1 + \frac{a}{R} \cos \theta \right) + \frac{\mathscr{J}}{\pi a^2} \frac{\pi c^2 p' a^3}{\mathscr{J}^2} \frac{a}{R} \cos \theta, \quad (4.30)$$

where $j_{s0}(a) = \mathscr{J}'(a)/2\pi a$. As follows from Eq. (4.30) the distribution of the longitudinal current density is asymmetric. When the pressure profile is parabolic, $p = p_0(1 - a^2/\rho_0^2)$, the third term in Eq. (4.30) is equal to $2\beta_{\mathscr{J}} \frac{\mathscr{J}}{\pi a^2} \frac{a}{R} \cos \theta$. At low plasma pressure, $\beta_{\mathscr{J}} \ll 1$, the current density is higher on the inner side of the magnetic surface ($\theta = 0$). If the pressure rises, the current density increases on the outer side of the torus ($\theta = \pi$) and decreases correspondingly on the inner side. For current and pressure profiles falling towards the plasma periphery and having a finite gradient at the plasma boundary, the current density on the inner side of the torus can even change its direction.

Now we describe the magnetic configuration outside the plasma column. The flux function Ψ_e of the total poloidal field can be found solving the homogeneous equation (4.3) under the condition that $\Psi_e = $ const on the plasma boundary:

$$\rho \cos \omega = (a - \alpha \cos 2\theta) \cos \theta - \Delta(\rho);$$
$$\rho \sin \omega = (a - \alpha \cos 2\theta) \sin \theta \quad (4.31)$$

(a is the plasma minor radius), and matching the vacuum poloidal field and the field of Eq. (4.28) on the plasma boundary.

However, it is more easily done using Eq. (4.24) relating Ψ' to the total current $\mathcal{J}$ if we insert into Eq. (4.24) the solution of Eq. (4.19) for Δ' in the vacuum region:

$$\Delta'(\rho) = -\frac{\rho}{R}\left(\ln\frac{\rho}{a} + \frac{l_i}{2} + \beta_{\mathcal{J}}\right);$$
$$\Delta(\rho) = -\frac{\rho^2}{2R}\ln\frac{\rho}{a} - \frac{\rho^2}{2R}\left(1 - \frac{a^2}{\rho^2}\right)\left(\frac{l_i}{2} + \beta_{\mathcal{J}} - \frac{1}{2}\right) + \Delta. \quad (4.32)$$

Here $l_i = l_i(a)$ and $\beta_{\mathcal{J}} = \beta_{\mathcal{J}}(a)$ are calculated for the whole cross section of the plasma column; $\Delta = \Delta(a)$ on the right-hand side denotes the displacement of the plasma boundary relative to the laboratory coordinate system. For $\Psi'(a)$ we have

$$\Psi'(a) = -\frac{4\pi R}{ca}\mathcal{J}\left(1 + \frac{\Delta}{R} - \frac{1}{2}\frac{a}{R}\Delta' - \frac{1}{2}\frac{a^2}{R^2} - \frac{1}{2}\Delta'^2\right). \quad (4.33)$$

Further, we restrict ourselves mainly to the linear approximation in the curvature. Then for the flux Ψ_e we have

$$\Psi_e(a) = \frac{4\pi R}{c}\mathcal{J}\left(\ln\frac{8R}{a} - 2\right) + \text{const}, \quad (4.34)$$

or in the laboratory coordinate system

$$\Psi_e(\rho, \omega) = \frac{4\pi R}{c}\mathcal{J}\left[\ln\frac{8R}{\rho} - 2 - \frac{\Delta(\rho)}{\rho}\cos\omega\right]$$
$$+ \text{const} = \frac{4\pi R}{c}\mathcal{J}\left\{\ln\frac{8R}{\rho} - 2 - \frac{\Delta}{\rho}\cos\omega\right.$$
$$\left. + \frac{1}{2}\left[\ln\frac{\rho}{a} + \left(1 - \frac{a^2}{\rho^2}\right)\left(\frac{l_i}{2} + \beta_{\mathcal{J}} - \frac{1}{2}\right)\right]\frac{\rho}{R}\cos\omega\right\} + \text{const.} \quad (4.35)$$

The nonessential additive constant in Ψ_e, which does not affect the description of the equilibrium, will be dropped in further calculations.

The flux function Ψ_e of the equilibrium field can also be written as

$$\Psi_e = \Psi_{pl} + \Psi_{ext}, \quad (4.36)$$

where Ψ_{pl} is the self-field of the plasma current:

$$\Psi_{pl} = \Psi^0(\rho, \omega) + \left(C_{pl}^1 - \frac{R\Delta}{a^2}\right)\Psi_{int}^1(\rho, \omega);$$

$$C^1_{pl} \equiv -\frac{1}{2}\left(\frac{l_i}{2} + \beta_{\mathscr{J}} - \frac{1}{2}\right); \qquad (4.37)$$

and Ψ_{ext} is the flux of the maintaining field:

$$\left.\begin{aligned}\Psi_{ext} &= C^1_{ext}\,\Psi^1_{ext}\,(\rho,\omega);\\ C^1_{ext} &\equiv \frac{1}{2}\left(\ln\frac{2R}{a} + \beta_{\mathscr{J}} + \frac{l_i}{2} - \frac{3}{2}\right);\\ B_{z,\,ext} &= -\frac{2\mathscr{J}}{cP}\,C^1_{ext} = -\frac{\mathscr{J}}{cR}\left(\ln\frac{8R}{a} + \beta_{\mathscr{J}} + \frac{l_i}{2} - \frac{3}{2}\right).\end{aligned}\right\} \quad (4.38)$$

Thus, the first approximation in the curvature shows that a uniform maintaining field is necessary to sustain the equilibrium of a toroidal plasma column with a circular cross section.

The formulas presented above allow us to calculate the maintaining field more accurately and in particular to find its decay index. To do this it is sufficient to write the expression for Ψ_e in the quadratic approximation in the curvature and then to represent it in a similar way to Eq. (4.36). However, one can use the virtual casing principle, since a known poloidal field distribution (4.28) over the plasma boundary gives an explicit expression for the surface current generating the maintaining field:

$$\frac{4\pi}{c}\,i_{s,\,ext} = -B_\theta(a,\theta). \qquad (4.39)$$

Before getting down to finding the field of this surface current we will solve two auxiliary problems.

1. <u>Current Distribution $i^0_s(\omega)$ on a Superconducting Toroidal Surface.</u> Let $\rho = a - \alpha\cos 2\omega$ be the equation for a superconducting toroidal surface and $\mathscr{J}$ the total current flowing around the torus. The flux function Ψ_e outside the torus will read

$$\Psi_e(\rho,\omega) = \Psi^0 + \frac{1}{2}\left(\ln\frac{8R}{a} - 1\right)\Psi^1_{int} + \left(\frac{3}{16}\ln\frac{8R}{a} - \frac{1}{4} - \frac{\alpha}{a}\right)\Psi^2_{int}. \quad (4.40)$$

This function satisfies the condition $\Psi_e(a - \alpha\cos 2\omega, \omega) = \text{const}$. It follows from this that

$$i^0_s(\omega) = \frac{\mathscr{J}}{2\pi a}\left[1 + \frac{a}{R}\left(\ln\frac{8R}{a} - \frac{1}{2}\right)\cos\omega + \right.$$

$$+ \frac{a^2}{R^2}\left(\ln\frac{8R}{a} - \frac{17}{16}\right)\cos 2\omega - \frac{\alpha}{a}\cos 2\omega\bigg]. \qquad (4.41)$$

Within an accuracy of 10% this formula is valid up to $R/a = 3$. Such a current distribution does not generate any fields inside the toroidal surface.

2. **Magnetic Field of a Distributed Surface Current.** Let the surface current $i_s(\omega)$ be distributed over a toroidal shell with radii R and a according to the following dependence:

$$i_s(\omega) = \frac{\mathscr{I}}{2\pi a}(1 + kai_1\cos\omega + k^2 a^2 i_2 \cos 2\omega). \qquad (4.42)$$

Let us find the field of this current inside and outside the shell. To do this it is sufficient to rewrite Eq. (4.42) in the form of a sum of $i_s^0(\omega)$ (4.41), $i_s^1(\omega)$ (4.6), and $i_s^2(\omega)$ (4.8):

$$i_s(\omega) = i_s^0(\omega) + \frac{1}{2}\left(i_1 - \ln\frac{8R}{a} + \frac{1}{2}\right)i_s^1(\omega)$$

$$+ \frac{1}{4}\left(i_2 - \frac{1}{4}i_1 - \frac{3}{4}\ln\frac{8R}{a} + \frac{15}{16} + \frac{\alpha}{k^2 a^3}\right)i_s^2(\omega). \qquad (4.43)$$

The fields of each of the terms on the right-hand side of Eq. (4.43) have been found earlier. In particular, $i_s^1(\omega)$ generates a uniform magnetic field (4.7) inside the shell, while $i_s^2(\omega)$ generates a quadrupole field (4.9). The current $i_s^0(\omega)$ does not generate a field inside the shell.

By comparing Eq. (4.39) with Eqs. (4.42) and (4.43) one can write the virtual casing current $i_{s,\text{ext}}$ in the following form:

$$i_{s,\text{ext}} = -i_s^0(\omega) + \frac{1}{2}\left(\ln\frac{8R}{a} + \beta_{\mathscr{I}} + \frac{l_i}{2} - \frac{3}{2}\right)i_s^1(\omega)$$

$$+ \frac{1}{4}\left[\frac{3}{4}\ln\frac{8R}{a} - \frac{19}{16} + \frac{1}{4}\left(\beta_{\mathscr{I}} + \frac{l_i}{2}\right) - \frac{1}{2}\left(\beta_{\mathscr{I}} + \frac{l_i}{2}\right)^2 \right.$$

$$\left. - \frac{\alpha' + \alpha/a}{k^2 a^2}\right]i_s^2(\omega), \qquad (4.44)$$

From this equation it follows that the flux function of the maintaining field is

$$\Psi_{ext} = C_{ext}^1 \Psi_{ext}^1 + C_{ext}^2 \Psi_{ext}^2;$$

$$C_{ext}^2 = \frac{1}{4}\left[\frac{3}{4}\ln\frac{8R}{a} - \frac{19}{16} + \frac{1}{4}\left(\beta_{\mathscr{Y}} + \frac{l_i}{2}\right)\right.$$
$$\left. - \frac{1}{2}\left(\beta_{\mathscr{Y}} + \frac{l_i}{2}\right)^2 - \frac{R^2}{a^2}\alpha' - \frac{R^2}{a^3}\alpha\right]. \quad (4.45)$$

The magnetic field $\mathbf{B}_{ext}$ is characterized by its central value (4.38) found earlier and by the decay index

$$n \equiv -\frac{r}{B_{z,\,ext}}\frac{dB_{z,\,ext}}{dr}\bigg|_{r=R} = \frac{2C_{ext}^2}{C_{ext}^1}. \quad (4.46)$$

For a column with a uniform current and a parabolic pressure Eq. (4.46) reduces to the explicit expression

$$n = \frac{\frac{3}{4}\ln\frac{8R}{a} - \frac{17}{16} - \frac{R^2}{a^2}\left(\frac{l_z}{l_r} - 1\right)}{\ln\frac{8R}{a} + \beta_{\mathscr{Y}} - \frac{5}{4}} \quad (4.47)$$

where l_z and l_r are the semiaxes of the cross section.

A negative decay index gives rise to plasma instability for vertical displacements. The circular toroidal column ($l_z = l_r$) is characterized by n > 0 and is stable to such perturbations.

4.2. Effects of Structural Units on Equilibrium

To provide equilibrium along the major radius it is necessary to produce an approximately uniform maintaining field determined by Eq. (4.38). To produce such a field in a tokamak there are special equilibrium windings. In addition, the eddy Foucault currents induced in the casing or in the liner as well as the magnetization of the iron core are also sources of the maintaining field. The plasma equilibrium in the direction of the major radius is solely affected by the transverse component of this field ($B_{z,\,ext}$) averaged over the azimuth ζ. The quadrupole and higher multipole components only affect the shape of the plasma cross section. Below, we will consider the main factors affecting the equilibrium according mainly to [66].

1. <u>Equilibrium in Complete Ideally Conducting Casing.</u> This case is an idealization of a well-known means of maintaining the

equilibrium using a high-conductivity casing. The position of the plasma column relative to the casing can be determined by imposing the condition $\Psi(b, \omega) = $ const on the flux function (4.35) of the equilibrium field (b is the minor radius of the casing). Then the shift of the plasma is

$$\frac{\Delta}{b} = \frac{b}{2R}\left[\ln\frac{b}{a} + \left(1 - \frac{a^2}{b^2}\right)\left(\frac{l_i}{2} + \beta_{\mathcal{J}} - \frac{1}{2}\right)\right]. \qquad (4.48)$$

The same equation determines the displacement of any vacuum magnetic surface of radius a. Hereinafter, we will denote the shift (4.48) corresponding to the ideally conducting casing by Δ_0.

2. **Equilibrium in an Ideally Conducting Casing with Gaps.**
When there are gaps in the casing, one should take into account the penetration of a portion of the magnetic flux through the gaps [boundary condition (3.118)]. In a toroidal casing of circular cross section the surface current appearing in Eq. (3.118) is distributed according to the law $i_s(\omega) \sim \cos \omega$. It is convenient to normalize the surface current to $i_s^1(\omega)$ (4.6):

$$i_s(\omega) = i_b\, i_s^1(\omega). \qquad (4.49)$$

For generality we consider the following situation. Let a winding be located between the plasma and the casing at a radius b_1. The current in this winding can be approximated by the surface current

$$i_{s,\,\text{int}} = i_{\text{int}}\, i_s^1(\omega). \qquad (4.50)$$

In addition, there is an external winding whose field we divide into a steady part which penetrates through the shell without distortion

$$\Psi_{\text{ext}}^{\text{st}}(\rho, \omega) = C^{\text{st}}\, \Psi_{\text{ext}}^1(\rho, \omega),$$
$$C^{\text{st}} = -B_{z,\,\text{ext}}^{\text{st}} \frac{cR}{2\mathcal{J}} \qquad (4.51)$$

and a "variable" part

$$\Psi_{\text{ext}}^{\text{al}}(\rho, \omega) = C^{\text{al}}\, \Psi_{\text{ext}}^1(\rho, \omega),\quad C^{\text{al}} = -B_{z,\,\text{ext}}^{\text{al}} \frac{cR}{2\mathcal{J}}, \qquad (4.52)$$

affecting the equilibrium only through the gaps; here $B_{z,\,\text{ext}}^{\text{al}}$ is the amplitude of this field component in the absence of the casing.

The flux function for such a system is of the following form:

1) between the plasma and the inner winding: $a \leq \rho < b_1$

$$\Psi^{\mathrm{I}} = \Psi^0 + \left(C_{\mathrm{pl}}^1 - \frac{\Delta R}{a^2}\right) \Psi_{\mathrm{int}}^1 + C_{\mathrm{exi}}^1 \Psi_{\mathrm{ext}}^1 ; \qquad (4.53)$$

2) between the inner winding and the shell: $b_1 \leq \rho < b$

$$\Psi^{\mathrm{II}} = \Psi^{\mathrm{I}} + i_{\mathrm{int}} \left(\Psi_{\mathrm{int}}^1 \frac{b_1^2}{a^2} - \Psi_{\mathrm{ext}}^1\right); \qquad (4.54)$$

3) outside the casing: $b \leq \rho$

$$\Psi^{\mathrm{III}} = \Psi_e^{\mathrm{II}} + i_b \left(\Psi_{\mathrm{int}} \frac{b^2}{a^2} - \Psi_{\mathrm{ext}}^1\right)\left(1 - \frac{\Sigma h_{\mathrm{eff}}}{2\pi R}\right). \qquad (4.55)$$

In the expression for Ψ^{III} the coefficient at Ψ_{ext}^1 should be equal to the given field of the external winding. Using this we get the current value in the shell:

$$i_b = C_{\mathrm{ext}}^1 - i_{\mathrm{int}} - C^{\mathrm{st}} - C^{\mathrm{al}} \bigg/ \left(1 - \frac{\Sigma h_{\mathrm{eff}}}{2\pi R}\right). \qquad (4.56)$$

The boundary condition (3.118) on the shell with gaps, in the presence of a steady-state field, reads

$$\Psi^{\mathrm{III}}(b, \omega) = C^{\mathrm{st}} \Psi_{\mathrm{int}}^1(b, \omega) - \frac{2\pi}{c} b \sum h_{\mathrm{eff}} i_b i_s^1(\omega) + \mathrm{const}, \qquad (4.57)$$

where Σh_{eff} is the total effective width of all the gaps. Substituting Eq. (4.55) into Eq. (4.57) we get the magnitude of the shift

$$\frac{\Delta}{b} = \frac{\Delta_0}{b} - \frac{b}{R} i_{\mathrm{int}} \left(1 - \frac{b_1^2}{b^2}\right) + B_{z,\mathrm{ext}}^{\mathrm{st}} \frac{cb}{2\mathcal{J}} + \frac{\Sigma h_{\mathrm{eff}}}{2\pi R - h_{\mathrm{eff}}} \left[-\frac{b}{R} i_{\mathrm{int}}\right.$$
$$\left. + \frac{cb}{2\mathcal{J}} (B_{z,\mathrm{ext}}^{\mathrm{st}} + B_{z,\mathrm{ext}}^{\mathrm{al}} - B_{z,\mathrm{ext}})\right]. \qquad (4.58)$$

Here $B_{z,\mathrm{ext}}$ is the required value (4.38) of the maintaining field and $h_{\mathrm{eff}} = \pi b/(\ln L/h + \sqrt{2} d/h)$ is the effective width of a gap (3.122).

Note that here we consider the total field of the inner winding to be "variable." The steady part of the field must obviously enter the formula in the same form as the field $B_{z, \text{ext}}^{\text{st}}$ of the outer winding.

When the current circulates around the gap, a force F_z arises due to the presence of the longitudinal magnetic field B_s. This force is applied to the edges of the sections near the gaps and is equal to

$$F_z = \frac{\mathcal{J} B_s}{c} \frac{b^2}{R} i_b = \frac{b^2}{2} B_s \left[-B_{z, \text{ext}} - \frac{2\mathcal{J}}{cR} i_{\text{int}} + B_{z, \text{ext}}^{\text{al}} + B_{z, \text{ext}}^{\text{st}} \right]. \quad (4.59)$$

The direction of this force is opposite on each side of the gap.

3. **Equilibrium in the Presence of an Iron Core.** The current drive system in many tokamak devices uses a transformer with an iron core. The presence of an iron core complicates the calculations of the magnetic system due to, first, its complicated (three-dimensional) geometry, and second, the nonlinearity of the magnetic permeability of the iron. Let us consider in general the technique for taking into account the ferromagnetic units in equilibrium problems.

The magnetic field in a ferromagnetic material is described by the following equation:

$$\text{curl } \frac{1}{\mu} \mathbf{B} = 0, \quad (4.60)$$

where $\mu = \mu(\mathbf{r}, B)$ is the magnetic permeability. The boundary conditions on the surface of the ferromagnetic material are of the form

$$B_{ne} = B_{ni}, \quad \frac{1}{\mu} B_{\tau i} = B_{\tau e}. \quad (4.61)$$

Symbols i and e refer to the region containing ferromagnetic material ($\mu \neq 1$) and to the outer region ($\mu = 1$), respectively.

For axisymmetric systems Eq. (4.60) reduces to the equation for the flux function Ψ:

$$r \frac{\partial}{\partial r} \frac{1}{r\mu} \frac{\partial \Psi}{\partial r} + \frac{\partial}{\partial z} \frac{1}{\mu} \frac{\partial \Psi}{\partial z} = 0. \quad (4.62)$$

In the model case $\mu = \text{const}$ problems with a ferromagnet can be solved in a similar way to vacuum magnetostatic problems (Section 3.5). In this case the currents due to magnetization can be substituted by a surface current i_s^f flowing along the boundary of the ferromagnetic material Γ. The second boundary condition helps us to find the distribution $i_s^f(l)$:

$$\frac{1}{\mu}\left[B_{\tau,\,\text{ext}}(l) - \frac{2\pi}{c} i_s^f(l) + \oint b_\tau(l;l')\, i_s^f(l')\, dl'\right]$$
$$= B_{\tau,\,\text{ext}} + \frac{2\pi}{c} i_s^f(l) + \oint b_\tau(l;l')\, i_s^f(l')\, dl'. \tag{4.63}$$

Here $B_{\tau,\,\text{ext}}$ is the tangential component of the field induced by the currents located outside the ferromagnet and $b_\tau(l;l')$ is defined by Eqs. (3.85)-(3.87). As $\mu \to \infty$ this boundary condition (4.61) reduces to setting the tangential component of the field on the outer surface of the iron to zero, $B_{\tau e} = 0$.

For $\mu \neq \text{const}$ it is necessary to solve Eq. (4.62), which becomes nonlinear when one takes into account the dependence $\mu(B^2)$. In this case the computation of the field inside a ferromagnet becomes as time-consuming as the solution of two-dimensional equilibrium problems. However, in a number of cases one can obtain characteristics of the magnetic system with an iron core, which can be used independently in equilibrium problems without two-dimensional calculations of the field inside the ferromagnet [67].

Let us assume that the magnetic system consists of primary and equilibrium windings with currents I_{ind} and I_{eq}, respectively. The vertical field of these windings in the plasma region can be represented in the following form:

$$B_{z,\,\text{ext}} = b_{\text{ind}}(\Psi_f, \Delta)\, I_{\text{ind}} + b_{\text{eq}}(\Psi_f, \Delta)\, I_{\text{eq}} + b_{\text{pl}}\left(\Psi_f, C_{\text{pl}}^1 - \frac{\Delta R}{a^2}\right)\mathcal{J}. \tag{4.64}$$

Here $\mathcal{J}$ is the total plasma current. The coefficients b_{ind} and b_{eq} may be assumed to be functions of the total flux in the iron Ψ_f and of the plasma position along the major radius. The coefficient $b_{\text{pl}}\left(\Psi_f, C_{\text{pl}}^1 - \frac{\Delta R}{a^2}\right) = b_{\text{pl}}\left(\Psi_f, \beta_\mathcal{J} + \frac{l_i}{2} + \frac{2\Delta R}{a^2}\right)$ characterizes the interaction of the plasma current with the iron core.

Each of the coefficients in Eq. (4.64) can be linearized with respect to the second argument:

$$b_{\text{ind, eq}}(\Psi_f, \Delta) = b_{\text{ind, eq}}(\Psi_f, 0) + \left.\frac{\partial b_{\text{ind, eq}}}{\partial \Delta}\right|_{\Delta=0} \Delta,$$

$$b_{\text{pl}}\left(\Psi_f, \beta_{\mathcal{J}} + \frac{l_i}{2} - \frac{2R\Delta}{a^2}\right) = b_{\text{pl}}(\Psi_f, 0) + \frac{\partial b_{\text{pl}}}{\partial \beta_{\mathcal{J}}}\bigg|_{\substack{\Delta=0 \\ l_i=0,5 \\ \beta_{\mathcal{J}}=0}}$$

$$\times \left(\beta_{\mathcal{J}} + \frac{l_i}{2} + \frac{2R\Delta}{a^2} - \frac{1}{4}\right). \tag{4.65}$$

As a result, it proves to be sufficient to have six characteristics to describe the vertical field, (4.65) which are functions of the magnetic flux Ψ_f in the iron core only. As has been noted in [67], the flux in the iron can be considered as a function of the parameter $I_{\text{ind}} + I_{\text{eq}} + (k_1 + k_2\Delta)\mathcal{J}$. All of these one-dimensional functions can be calculated beforehand for each particular magnetic system by means of two- or three-dimensional codes and can then be used independently in equilibrium problems.

4. **Equilibrium in a Casing with a Finite Conductivity.** In a casing with a finite conductivity an emf is required to maintain the current. If one does not consider the process of penetration of the current into the bulk material of the casing, which is characterized by a small time constant $\tau_{\text{sk}} = \pi\sigma d^2/c^2$ (d is the thickness of the wall), then the relation between the surface current in the shell and the flux variation in time is of the following form:

$$i_s = \sigma dE_s = -\frac{\sigma d}{2\pi Rc} \frac{\partial \Psi}{\partial t}. \tag{4.66}$$

Based on Eq. (4.66) we shall consider some magnetostatic problems.

a) Penetration of an external uniform field through the complete shell. Let the external windings generate a time-varying field with flux $C_e(t)\Psi_{\text{ext}}^1$. Inside the shell the flux is

$$\Psi_i = (C_e(t) + i_b)\Psi_{\text{ext}}^1 = C_i(t)\Psi_{\text{ext}}^1, \tag{4.67}$$

and outside

$$\Psi_e = C_e(t)\Psi_{\text{ext}}^1 + i_b\Psi_{\text{int}}^1, \tag{4.68}$$

where i_b is the amplitude of the current in the shell $i_s(\omega) = i_b i_s^1(\omega)$. Using Eq. (4.66) we get for the shell current

$$i_b = -\tau_b \frac{d}{dt}[C_e(t) + i_b], \quad \tau_b = 2\pi\sigma db/c^2. \tag{4.69}$$

The amplitude of the field inside the shell is

$$C_i(t) = \int_{-\infty}^{t} C_e(t) \exp[(t'-t)/\tau_b] dt'/\tau_b. \quad (4.70)$$

Thus, the penetration of a uniform magnetic field through the shell is characterized by the time constant τ_b.

b) *The motion of plasma in a shell with a finite conductivity.* Let a plasma column be positioned in a steady-state magnetic field $C_e \Psi_{ext}^1$ and be enclosed by a shell with a finite conductivity. The current in the shell will be entirely due to the motion of the column:

$$i_b \, i_s^1 = \frac{\sigma d}{2\pi Rc} \frac{R}{a^2} \frac{d\Delta}{dt} \Psi_{int}(b, \omega). \quad (4.71)$$

Since the value of the current in the shell is determined by the equilibrium condition, we get for the velocity of the plasma column:

$$\frac{d\Delta}{dt} = \frac{b}{\tau_b} \frac{b}{R} (C_{ext}^1 - C_e) = \frac{b}{\tau_b} \frac{b}{2R} \left(\ln \frac{8R}{a} + \beta_{\mathcal{J}} + \frac{l_i}{2} - \frac{3}{2} + \frac{B_z cR}{\mathcal{J}} \right). \quad (4.72)$$

According to Eq. (4.72), the motion of the plasma column in the major radial direction is characterized by the time constant $\tau_R = \tau_b 2R/b$.

c) *The motion of plasma in a shell with a finite conductivity in the presence of gaps.* When there are gaps in the shell with a finite conductivity, first, the magnetic flux penetrates directly through the gaps, and second, the damping of the currents in the shell is changed because the asymmetric part of the currents reconnects around the edges of the gaps. Now there is no well-defined time constant (4.69).

First we shall consider the penetration of an external magnetic field through a shell with gaps. The time dependence of the internal field can generally be described by

$$C_i(t) = \frac{h}{1+h} C_e(t) + \frac{1}{1+h} \int_{-\infty}^{t} C_e(t') g(t-t') dt'. \quad (4.73)$$

Here $h = \Sigma h_{eff}/2\pi R$. The analytical form of the function $g(t - t')$ is not known, and therefore Eq. (4.73) should be used to determine this parameter of the shell. If we suppose that the shell is characterized by only one time constant τ, the function $g(t) = (1/\tau) \times \exp(-t/\tau)$. In this case, Eq. (4.73) is equivalent to the following expression for the amplitude of the current in the shell:

$$i_b = -\frac{\tau}{1+h} \frac{d}{dt}(C(t) + i_b), \qquad (4.74)$$

where $C(t)$ is the amplitude of the flux on the surface of the shell. In comparison with Eq. (4.69) the flux of a magnetic field associated with the gaps is introduced and the time constant is changed ($\tau \neq \tau_b$).

Equation (4.69) makes it possible to find the amplitude of the current in the shell for an arbitrary function $g(t)$:

$$i_b = -\frac{1}{1+h} C(t) + \frac{1}{1+h} \int_{-\infty}^{t} C(t') g(t-t') dt'. \qquad (4.75)$$

The index e of the constant C is omitted since Eq. (4.75) is also valid for the case when the sources of the field are located inside the shell. In this case C is defined as the amplitude of the incident flux.

Now we pass to a description of the motion of the plasma column. For the sake of generality we consider the magnetic system described in Section 2. In order to take into account possible changes in the plasma current $\mathscr{J}(t)$, we include it explicitly in the amplitude of the corresponding fields. The shell current i_b is determined by the equilibrium condition (4.56)

$$i_b = \frac{\mathscr{J}(t)}{\mathscr{J}} C^1_{\text{ext}} - i_{\text{int}} - C^{\text{st}}_{\text{ext}} - C^{\text{al}}_{\text{ext}}, \quad \mathscr{J} = \text{const}. \qquad (4.76)$$

The amplitude of the incident flux is

$$C = -\frac{1}{2} \frac{\mathscr{J}(t)}{\mathscr{J}} \left(\ln \frac{8R}{a} - 1 \right) + \frac{\mathscr{J}(t)}{\mathscr{J}} \left(\frac{a^2}{b^2} C_{\text{pl}}^1 - \frac{\Delta R}{b^2} \right) + C^{\text{al}}_{\text{ext}}$$
$$+ i_{\text{int}} \frac{b_1^2}{b^2} = \frac{\mathscr{J}(t)}{\mathscr{J}} \frac{\Delta_0 - \Delta}{b^2} R + C^{\text{al}}_{\text{ext}} + i_{\text{int}} \frac{b_1^2}{b^2} - \frac{\mathscr{J}(t)}{\mathscr{J}} C^1_{\text{ext}}. \qquad (4.77)$$

Substituting Eqs. (4.77) and (4.76) into (4.75) yields the following equation for the displacement:

$$\frac{2\mathcal{F}(t)}{cb}\frac{\Delta-\Delta_g}{b} = \int_{-\infty}^{t}\left[\frac{2\mathcal{F}(t)}{cb}\frac{\Delta-\Delta_0}{b} - \frac{2\mathcal{F}_0}{cb}i_{\text{int}}\frac{b_1^2}{b^2}\right.$$
$$\left. + B_{z,\text{ext}}^{\text{al}} - B_{z,\text{ext}}(t)\right] g(t-t')\,dt'. \qquad (4.78)$$

Here Δ_g is the displacement in an ideally conducting shell with the same gaps (4.58), and Δ_0 is the displacement (4.48) in a complete, perfectly conducting shell. For the model $g(t) = (1/\tau)\exp(-t/\tau)$ the integral equation is equivalent to the differential equation

$$\frac{d}{dt}\frac{2\mathcal{F}(t)}{cb}\frac{\Delta-\Delta_g}{b} = \frac{1+h}{\tau}(B_z - B_{z,\text{ext}}), \qquad (4.79)$$

where B_z is the total field of all the windings:

$$B_z = B_z^{\text{all}} + B_z^{\text{st}} - \frac{2\mathcal{F}}{cb}i_{\text{int}}. \qquad (4.80)$$

The dependence on the plasma parameters $\beta_\mathcal{F}(t)$ and $l_i(t)$ enters Eqs. (4.78) and (4.79) through Δ_g, Δ_0, and $B_{z,\text{ext}}$.

5. Influence of Currents Reconnecting through the Limiter on the Plasma Equilibrium.

If magnetic surfaces intersect a limiter, then a poloidal current can flow from the upper to the lower edge of the limiter along unclosed magnetic surfaces. The interaction of this current with the longitudinal magnetic field results in a force directed along the major radius:

$$F_r = -\frac{1}{c} I_z B_s 2d, \qquad (4.81)$$

where d is the radius of the limiter opening. This force is equivalent to the introduction of an additional transverse field

$$B_z = -\frac{I_z}{\mathcal{F}} B_s \frac{d}{\pi R}. \qquad (4.82)$$

A detailed consideration of equilibrium in the presence of currents flowing through the limiter is described in papers [68, 69].

Such a mechanism for the maintainance of equilibrium is essential in the initial stage of a discharge, when a closed magnetic configuration has not yet been formed [70].

4.3. Equilibrium of a Plasma Column with a Circular Cross Section and with an Anisotropic Pressure

Consider the equilibrium equations of a plasma with anisotropic pressure in natural coordinates for axisymmetric systems. From the conditions $K\nabla\Psi = 0$ (3.181) and div $K = 0$ we have for contravariant components K^i

$$K^i = \left\{0, \; -\frac{F'_k}{2\pi \sqrt{g}}, \; \frac{\mathscr{J}'_k + \frac{\partial v_k}{\partial \theta}}{2\pi \sqrt{g}}\right\}, \qquad (4.83)$$

where $F_k(a)$ and $\mathscr{J}_k(a)$ are the poloidal and longitudinal fluxes of the vector K. From Eq. (3.175) we get an expression for the covariant components of the vector σB:

$$\sigma B_i = \frac{4\pi}{c} \frac{1}{2\pi} \left\{\frac{\partial \varphi_k}{\partial a} - v_k, \; \mathscr{J}_k + \frac{\partial \varphi_k}{\partial \theta}, \; F_k\right\}. \qquad (4.84)$$

The system of equations for a plasma with an anisotropic pressure similar to that of Eqs. (2.61)-(2.63) is of the following form:

$$4\pi^2 c \sqrt{g} \left(\frac{\partial p_\parallel}{\partial a} - \frac{p_\parallel - p_\perp}{2B^2} \frac{\partial B^2}{\partial a}\right) = -F'_k\left(\Phi' + \frac{\partial \eta}{\partial \theta}\right)$$
$$+ \Psi'\left(\mathscr{J}'_k + \frac{\partial v_k}{\partial \theta}\right); \qquad (4.85)$$

$$\frac{4\pi}{c}\left(\mathscr{J}'_k + \frac{\partial v_k}{\partial \theta}\right) = -\frac{\partial}{\partial a}\frac{g_{22}}{\sqrt{g}}\sigma\Psi' + \frac{\partial}{\partial \theta}\frac{g_{12}}{\sqrt{g}}\sigma\Psi'; \qquad (4.86)$$

$$\frac{4\pi}{c} F_k = \frac{g_{33}}{\sqrt{g}} \sigma \left(\Phi' + \frac{\partial \eta}{\partial \theta}\right); \qquad (4.87)$$

$$\frac{\partial p_\parallel}{\partial \theta} = \frac{p_\parallel - p_\perp}{2B^2} \frac{\partial B^2}{\partial \theta}; \qquad (4.88)$$

$$\sigma = 1 - 4\pi(p_\parallel - p_\perp)/B^2. \qquad (4.89)$$

For plasma with an anisotropic pressure the equilibrium configuration is determined by the external field and by the distributions of the longitudinal pressure $p_\parallel(a, \theta)$ and the current $\mathscr{J}_k(a)$.

Consider the equilibrium of a plasma column with a circular cross section. Suppose that the plasma pressure is of the form

$$p_{\|}(a, \theta) = p_{\|0}(a) + p_{\|1}(a) \cos \theta; \quad (4.90)$$

$$p_{\perp}(a, \theta) = p_{\perp 0}(a) + p_{\perp 1}(a) \cos \theta. \quad (4.91)$$

In a strong longitudinal field

$$|\mathbf{B}| \approx B_s = B_s(1 + ka \cos \theta). \quad (4.92)$$

Taking into account Eq. (4.88) one can write Eq. (4.90) in the form

$$p_{\|}(a, \theta) = p_{\|0}(a) + ka(p_{\|0} - p_{\perp 0}) \cos \theta, \quad p_{\|1} = ka(p_{\|0} - p_{\perp 0}). \quad (4.93)$$

We can take σ to be a magnetic surface function $\sigma = \sigma(a)$ to a good approximation.

Suppose again that the magnetic surfaces are tori with circular cross section. In this case the metric of the natural coordinate system is described by Eqs. (4.16) and (4.18). In the zeroth and first approximations in the parameter ka Eq. (4.85) is of the following form:

$$4\pi^2 caR p'_{\|0} = \Psi' \mathcal{J}'_k - \frac{4\pi}{c} F'_k F_h \frac{a}{R}; \quad (4.94)$$

$$4\pi^2 c \sqrt{\tilde{g}}\, p'_{\|0} + 4\pi^2 caR p'_{\|1} \cos \theta = (p_{\|0} - p_{\perp 0}) 4\pi^2 ca \cos \theta$$
$$+ \Psi' \frac{\partial v_k}{\partial \theta} - \frac{4\pi}{c} F'_k F_h \left(\frac{\sqrt{\tilde{g}}}{g_{33}} \right). \quad (4.95)$$

Equation (4.94) can be rewritten so that it includes the real currents $\mathcal{J}(a)$ and $F(a)$. Since $\mathcal{J}_h(a) \approx \mathcal{J}(a)$ and $F_k(a) = \sigma F(a)$, we get from Eq. (4.94)

$$4\pi^2 caR p'_{\perp 0} = \Psi' \mathcal{J}' - \frac{4\pi}{c} FF' a/R. \quad (4.96)$$

Thus, we see that only the transverse pressure enters the equilibrium equation in the direction of the minor radius (Section 3.7). Correspondingly, the integral condition for the pressure balance (2.9) for a plasma with an anisotropic pressure takes the form

$$\beta_{\mathcal{J}_\perp} = 1 + \mu_{\mathcal{J}}, \quad \beta_{\mathcal{J}_\perp} = \frac{2c^2 \int p_\perp a\, da\, d\theta}{\mathcal{J}^2}. \quad (4.97)$$

Upon eliminating $F_k F'_k$ from Eq. (4.95) using Eq. (4.94) and $p_{\parallel 1}$ using Eq. (4.93), we get the equation for the displacement of magnetic surfaces

$$a\Delta'' + \left(\frac{2a\mathscr{F}'}{\mathscr{F}} - 1\right)\Delta' = -\frac{a}{R} + \frac{\pi c^2 a^4}{R\mathscr{F}^2}(p'_{\parallel 0} + p'_{\perp 0}). \qquad (4.98)$$

The only difference between Eq. (4.98) and the corresponding Eq. (4.19) for an isotropic pressure consists in the substitution of 2p by $(p_{\parallel 0} + p_{\perp 0})$. Thus, the formulas for the plasma displacement and for the maintaining field obtained above remain valid if one makes the following substitution:

$$\beta\mathscr{F} \to (\beta\mathscr{F}_\perp + \beta\mathscr{F}_\parallel)/2, \quad \beta\mathscr{F}_\parallel = \frac{2c^2 \int p_\parallel \, a\,da\,d\theta}{\mathscr{F}^2}. \qquad (4.99)$$

In the approximation considered, the value of $p_{\perp 1}(a)$ (4.91) does not enter. This indicates that the equilibrium equation does not define the displacement of the surfaces $p_\perp$ = const with respect to the magnetic surfaces. The distribution of the longitudinal pressure $p_\parallel(a, \theta)$ is determined by Eq. (4.93). For $p_{\parallel 0} > p_{\perp 0}$ the surfaces $p_\parallel$ = const are shifted inwards with respect to the magnetic surfaces, while for $p_{\parallel 0} < p_{\perp 0}$ they are shifted outwards.

4.4. Stability of the Equilibrium of the Plasma Column

The hydromagnetic stability of the equilibrium with respect to axisymmetric perturbations is now considered. For a plasma column with an arbitrary cross section, study of the stability of the equilibrium is an independent problem, associated with the solution of the two-dimensional equations of motion [71]. For a plasma column with a circular cross section and a large aspect ratio $(R/a \gg 1)$ the problem is simplified since rigid vertical and horizontal displacements of the column are the eigenfunctions of axisymmetric modes.

One can deduce the condition for the stability against rigid displacements by considering the force arising when the plasma is displaced in the external maintaining field $\mathbf{B}_{ext}$. For vertical displacement we have

$$F_z = -\frac{2\pi R}{c} \int j_s \left(\frac{\partial B_{r,\text{ext}}}{\partial z} \xi_z + \delta B_r \right) dS_s. \qquad (4.100)$$

The first term in the parentheses is the force of interaction with the maintaining field, while the second term is the interaction with the fields of image currents excited in the structure elements when the plasma is displaced. In particular, δB_r can result from an active feedback system. The expression in the parentheses in Eq. (4.100) can be assumed constant over the plasma column, and thus, we can write the stability criterion in the following form:

$$\frac{\delta B_r}{\delta \xi_z} + \frac{\partial B_{r,\text{ext}}}{\partial z} > 0. \qquad (4.101)$$

Since $\partial B_{r,\text{ext}}/\partial z = \partial B_{z,\text{ext}}/\partial r$, the second term in Eq. (4.101) can be expressed in terms of the decay index. The result is as follows:

$$\frac{\delta B_r}{\delta \xi_z} + \frac{n}{R} \left| B_{z,\text{ext}} \right| > 0. \qquad (4.102)$$

In the absence of a feedback system or magnetic screens Eq. (4.102) reduces to the condition [72]

$$n > 0. \qquad (4.103)$$

The decay index turns out to be negative for a plasma column with a cross section elongated in the direction of the axis of symmetry. The stability of the equilibrium of such a plasma can only be ensured by an appropriate active or passive feedback system giving a sufficiently large value of $\delta B_r/\delta \xi_z$. Closed vertical frames as shown in Fig. 20 can serve as a passive system. Such frames do not affect the penetration of a quadrupole external field into the plasma and conserve the horizontal magnetic flux.

The condition for stability against horizontal displacements can be obtained from the expression for the radial force:

$$F_r = \frac{2\pi R}{c} \oint (B_z - B_{z,\text{ext}}). \qquad (4.104)$$

The stability condition reduces to

$$\partial F_r/\partial R < 0. \qquad (4.105)$$

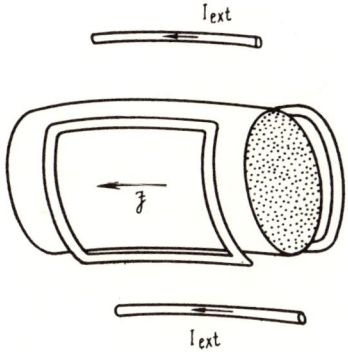

Fig. 20. Passive closed conductors in the form of a vertical frame for stabilization against vertical instability. Due to the shift of the plasma column a horizontal magnetic field is produced by the stabilizing conductors. The quadrupole maintaining field which serves to elongate the plasma cross section is not affected by the stabilizing conductors.

When the quantity $\left(\ln \frac{8R}{a} + \beta_{\mathscr{J}} + \frac{l_i}{2} - \frac{3}{2}\right)^{-1}$ can be neglected, this condition reads as follows:

$$\frac{\partial F}{\partial R} = \frac{2\pi}{c} \mathscr{J} |B_{z,\text{ext}}| \left(n - 1 + \frac{R}{\mathscr{J}} \frac{\delta \mathscr{J}}{\delta \xi_r} + \frac{R}{|B_{z,\text{ext}}|} \frac{\delta B_z}{\delta \xi_r} \right) < 0. \quad (4.106)$$

The variation $\delta \mathscr{J}/\delta \xi_r$ of the total current associated with the displacement of the plasma depends on the current drive scheme. The last term in Eq. (4.106) includes the variation of all of the transverse fields associated with the magnetic screens, the iron core, and the feedback system. As follows from Eq. (4.106) the plasma column can become unstable because the maintaining field has an undesirable geometry (the decay index is too big) and also because of the presence of an iron core transformer, which has been observed experimentally on the TFR device [67].

4.5. Equilibrium of a Plasma Column with a Nonplanar Axis

Suppose that the axis of a plasma column is determined by the equation $r = r_0(s)$ (s is the length of the arc). A curve is uniquely defined by its curvature k(s) and torsion $\varkappa(s)$. For physical problems a complex curvature K(s) is more convenient to use. The latter is defined as

$$K(s) = k(s) \exp\{-[\varkappa_0 s - \alpha(s)]\}, \quad \alpha(s) = \int_0^s \varkappa(s)\, ds. \quad (4.107)$$

The parameter $\varkappa_0$ in the definition for K(s) is determined by the periodicity condition

$$\varkappa_0 = [\alpha(L) + l]/L, \quad (4.108)$$

where L is the total length of the curve and l is an integer. Instead of two functions k(s) and $\varkappa(s)$ one can use the complex curvature K(s) and $\varkappa_0$ to describe any curve. When K(s) and $\varkappa_0$ are given, the usual curvature and the torsion are given by the following formulas:

$$k(s) = \sqrt{K_1^2 + K_2^2}, \quad \varkappa(s) = \varkappa_0 + (K_1' K_2 - K_2' K_1)/(K_1^2 + K_2^2), \quad (4.109)$$

where K_1 and K_2 are the real and imaginary parts of K(s):

$$K(s) = K_1(s) + iK_2(s). \quad (4.110)$$

Because of the periodicity of K(s) it is convenient to use its Fourier expansion

$$K(s) = \sum_{n=-\infty}^{\infty} K_n \exp\left[i\frac{2\pi n}{L} s\right], \quad (4.111)$$

Here we describe a curve by a new set of invariants, namely, the set of numbers $\{\varkappa_0, k_n\}$.

Note that the representation of curves as a set of Fourier coefficients $\{\varkappa_0, k_n\}$ of the complex curvature represents the simplest closed curves naturally as those characterized by the smallest number of parameters k_n. For example, a circle is characterized by a single nonvanishing parameter $k_0 = 1/R$, a helix which is the simplest unclosed curve has two parameters $\varkappa_0, k_0$. It is natural to consider curves with a complex curvature containing the smallest number of Fourier harmonics as elementary closed spatial curves. As shown in [73] the condition for closure can be met in the case where the complex curvature includes only two terms in the Fourier expansion:

$$K(s) = k_0 + k_N \exp\left[iN\frac{2\pi}{L}s\right], \quad N = 2, 3... \quad (4.112)$$

The condition for closure determines the form of the dependences $k_0 = k_0(\varkappa_0)$ and $k_N = k_N(\varkappa_0)$. For N = 2 the functions $k_0(\varkappa_0)$ and $k_2(\varkappa_0)$ are shown in Fig. 21. Curves with N = 2 represent a family of figure-eight shaped curves with $\varkappa_0 = 0-4\pi/L$. A detailed description of curves with K(s) of the form (4.112) referred to as spirals (an elementary spiral) can be found in [73].

Write τ, ν, and β for unit vectors along the tangent, normal, and binormal to the axis. They are related through the Serret-Frenet relations

$$d\tau/ds = k\nu, \quad d\nu/ds = -k\tau + \varkappa\beta, \quad d\beta/ds = -\varkappa\nu. \quad (4.113)$$

To describe an equilibrium it is convenient to use a local orthogonal coordinate system ρ, ω, s [74] which can be defined in the vicinity of a spatial curve. The coordinates ρ, ω are conventional polar coordinates in the plane perpendicular to the axis at the point s, with ρ being equal to the distance from the curve, and the angle $\omega - \alpha$ being the angle between the main normal and the direction to the point. The correction of the angle ω by the value α is made to orthogonalize the coordinate system. In this case one should remember that the conditions for the uniqueness of physical quantities in the coordinates ρ, ω, s are as follows:

$$f(\rho, \omega, s) = f(\rho, \omega + 2\pi m + n\alpha(L), s + nL), \quad (4.114)$$

where m and n are integers.

The unit vectors $\mathbf{e}_\rho$ and $\mathbf{e}_\omega$ are related to ν and β by

$$\left.\begin{array}{l}\mathbf{e}_\rho = \nu\cos(\omega-\alpha) + \beta\sin(\omega-\alpha); \\ \mathbf{e}_\omega = \partial\mathbf{e}_\rho/\partial\omega = -\nu\sin(\omega-\alpha) + \beta\cos(\omega-\alpha).\end{array}\right\} \quad (4.115)$$

The solution of the equilibrium equations for a plasma column with a nonplanar axis by means of an expansion in the curvature is given in [75, 76] and is described in review [2]. As in the case of

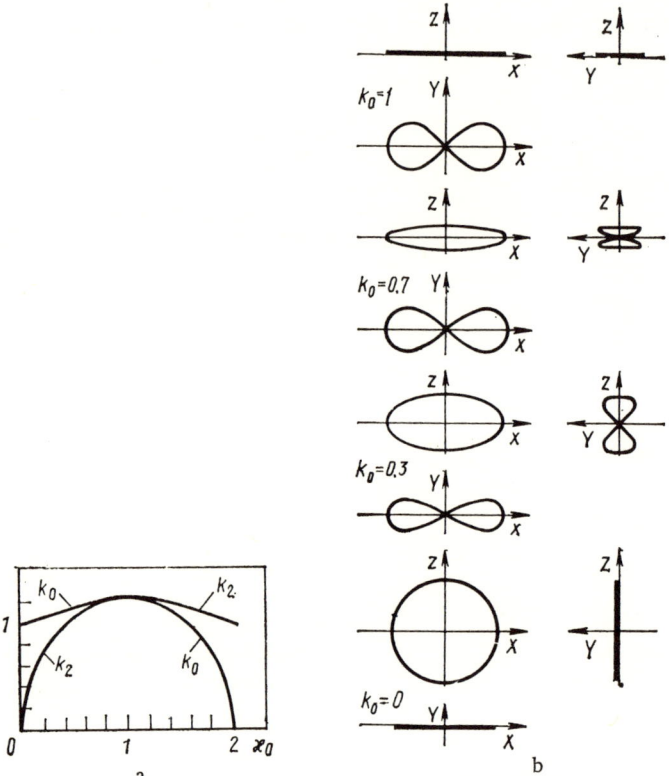

Fig. 21. a) The dependences of the coefficients $k_0(\varkappa_0)$ and $k_2(\varkappa_0)$ on the mean torsion for spirals with $N = 2$. The length of the spirals is normalized to 2π; b) the form of elementary closed curves (spirals) with $N = 2$.

a tokamak with a circular cross section, it is possible to obtain expressions for the shift of magnetic surfaces and for the total magnetic field for an arbitrary current and pressure distribution in the plasma column.

For smooth spatial curves without discontinuities in $k(s)$ and $\alpha(s)$ the magnetic field outside the plasma column is described by the following scalar potential ($\mathbf{B} = \nabla\varphi$):

$$\varphi = B_s s + \frac{2\mathscr{J}}{c}\omega + \frac{2\mathscr{J}}{c}\left[k\rho \ln\frac{a}{\rho} + \frac{3}{2}k\rho - b_1\frac{a}{\rho} + \frac{1}{2}k\frac{a^2}{\rho}\right) \times$$

$$\times \sin(\omega - \alpha) + \frac{2\mathcal{J}}{c} \tilde{b}_1 \left(\frac{\rho}{a} + \frac{a}{\rho}\right) \cos(\omega - \alpha)$$

$$-\frac{1}{8} B_s (\rho^3 - 3a^2 \rho) \frac{\partial}{\partial s} k \cos(\omega - \alpha). \tag{4.116}$$

Here $\mathcal{J}$ is the total current in the column, and a is the minor radius. The functions $b_1(s)$ and $\tilde{b}_1(s)$ depend on the current density and pressure distributions in the plasma column and are defined by

$$b_1(s) - i\tilde{b}_1(s) = -\sum_{n=-\infty}^{\infty} b_{\omega n} \exp[i(\alpha - \varkappa_n s)], \quad \varkappa_n = \varkappa_0 - \frac{2\pi}{L} n; \tag{4.117}$$

$$b_{\omega n} = \frac{8\pi \langle p \rangle + \langle B_\omega^2 \rangle/2 + \langle \varkappa_n \rho B_\omega B_s \rangle - \frac{3}{4} \varkappa_n B_s B_\omega(a)}{B_\omega(a)(B_\omega(a) - \varkappa_n a B_s)} k_n a. \tag{4.118}$$

Here $\langle \ \rangle$ denotes averaging over the cross section of the column. The number of terms in the series in Eq. (4.117) is the same as the number of Fourier coefficients in the complex curvature. For elementary closed curves this is two.

The poloidal magnetic field on the surface of the plasma column is of the form

$$B_\omega(s, \omega) = \frac{2\mathcal{J}}{ca} [1 + b_1(s) \cos(\omega - \alpha) + \tilde{b}_1(s) \sin(\omega - \alpha) + ka \cos(\omega - \alpha)]$$

$$-\frac{a^2}{4} B_s \frac{\partial}{\partial s} k(s) \sin(\omega - \alpha). \tag{4.119}$$

Using the local coordinate system ρ, ω, s only one cannot extract the maintaining field from the total field of the equilibrium configuration (4.116) since Eq. (4.116) is only valid in the vicinity of the coordinate axis. As in the case of an axisymmetric plasma it is necessary to determine the form of the functions which decreases as $|\mathbf{r}| \to \infty$ in terms of which the potential of the self-field of the plasma is expressed. The vector-potential of a current-carrying filament coinciding with the plasma axis is an analog of the nonplanar column of the function Ψ^0 (4.5). The vector-potential is defined by the Biot–Savart integral

$$\mathbf{A} = \int \frac{\boldsymbol{\tau}' \, ds'}{|\mathbf{r}_0 - \mathbf{r}'|}, \quad \boldsymbol{\tau}' = \boldsymbol{\tau}(s'), \tag{4.120}$$

where $\mathbf{r}_0$ is the radius-vector of the point of observation:

$$\mathbf{r}_0 = \mathbf{r}(s) + \rho \mathbf{e}_\rho; \qquad (4.121)$$

$\mathbf{r}' = \mathbf{r}(s')$ is the radius-vector of an elementary arc of the curve. Let $\mathbf{r}$ be the vector joining the point of observation s and the point of integration s' on the curve:

$$\mathbf{r} \equiv \mathbf{r}(s') - \mathbf{r}(s). \qquad (4.122)$$

The following relationships are useful for further analysis:

$$\frac{d\mathbf{r}}{ds} = -\tau, \quad \frac{d\mathbf{r}}{ds'} = \tau', \quad \frac{dr}{ds} = -\frac{(\mathbf{r}\tau)}{r}, \quad \frac{dr}{ds'} = \frac{(\mathbf{r}\tau')}{r}, \qquad (4.123)$$

Formula (4.120) can be rewritten in the form

$$\mathbf{A} = \frac{\mathcal{I}}{c} \oint \frac{\tau' \, ds'}{\sqrt{r^2 + \rho^2 - 2\rho(\mathbf{r}\mathbf{e}_\rho)}} \qquad (4.124)$$

and to a linear approximation in the curvature we have

$$\mathbf{A} = \frac{\mathcal{I}}{c} \oint \frac{\tau' \, ds'}{\sqrt{r^2 + \rho^2}} \left(1 + \rho \frac{(\mathbf{r}\mathbf{e}_\rho)}{r^2 + \rho^2} \right). \qquad (4.125)$$

Consider the main term in the longitudinal component of the vector-potential

$$A_s = \frac{\mathcal{I}}{c} \oint \frac{(\tau\tau') \, ds'}{\sqrt{r^2 + \rho^2}}. \qquad (4.126)$$

One can clearly see that as $\rho \to 0$ this expression has a logarithmic singularity. It can be expressed in explicit form if one takes into account that

$$\int_s^{s_1} \frac{(\mathbf{r}\tau')}{r\sqrt{r^2 + \rho^2}} \, ds = \ln \frac{r_1 + \sqrt{r_1^2 + \rho^2}}{\rho},$$

$$r_1 = |\mathbf{r}(s_1) - \mathbf{r}(s)|. \qquad (4.127)$$

We split the integral into two parts and add Eq. (4.127) with the appropriate sign. Then we have

$$\oint \frac{(\boldsymbol{\tau}\boldsymbol{\tau}')\, ds'}{\sqrt{r^2+\rho^2}} = \int_s^{s_1} \left[\frac{(\boldsymbol{\tau}\boldsymbol{\tau}')}{r} - \frac{(\mathbf{r}\boldsymbol{\tau}')}{r^2} \right] ds'$$

$$+ \int_{s_1}^{s} \left[\frac{\boldsymbol{\tau}\boldsymbol{\tau}'}{r} + \frac{(\mathbf{r}\boldsymbol{\tau}')}{r^2} \right] ds' + 2 \ln \frac{2r_1}{\rho}, \qquad (4.128)$$

In Eq. (4.128) we have dropped the term contributing $O(k^2\rho^2)$. The integrals in Eq. (4.128) depend solely on the arc length, and therefore to a first approximation A_s can be expressed as follows:

$$A_s = \frac{2\mathcal{J}}{c}\left[\ln\frac{8R}{\rho} - 2 + A_{s0}(s)\right], \quad R \equiv L/2\pi, \qquad (4.129)$$

where

$$A_{s0}(s) = \frac{1}{2}\int_s^{s_1}\left[\frac{(\boldsymbol{\tau}\boldsymbol{\tau}')}{r} - \frac{(\mathbf{r}\boldsymbol{\tau}')}{r^2}\right] ds'$$

$$+ \frac{1}{2}\int_{s_1}^{s}\left[\frac{(\boldsymbol{\tau}\boldsymbol{\tau}')}{r} + \frac{(\mathbf{r}\boldsymbol{\tau}')}{r^2}\right] ds' + \ln\frac{r_1}{4R} + 2. \qquad (4.130)$$

For an axisymmetric ring $A_{s0} = 0$.

Using Eq. (4.129) one can find the inductance of a thin plasma column with a nonplanar axis:

$$L = \frac{c}{\mathcal{J}^2}\int \mathbf{A}\mathbf{j}\, dV = 4\pi R\left[\ln\frac{8R}{a} - 2\right] + 2\oint A_{s0}(s)\, ds. \qquad (4.131)$$

An expression for the scalar potential of a current-carrying filament can be found from the following expression:

$$\frac{1}{\rho}\frac{\partial \varphi^0}{\partial \omega} = B_\omega = \frac{\partial A_\rho}{\partial s} - \frac{\partial A_s}{\partial \rho} + k\cos(\omega-\alpha)A_s$$

$$= -\frac{\partial}{\partial\rho}\frac{\mathcal{J}}{c}\oint\frac{\boldsymbol{\tau}\boldsymbol{\tau}'}{\sqrt{r^2+\rho^2}}\,ds' + \frac{\mathcal{J}}{c}\oint\frac{\mathbf{e}_\omega\,\mathbf{r}\,\boldsymbol{\tau}'}{(r^2+\rho^2)^{3/2}}\,ds'$$

$$+ 3\rho^2\frac{\mathcal{J}}{c}\oint\frac{(\mathbf{r}\boldsymbol{\tau}')(\mathbf{r}\,\mathbf{e}_\rho)}{(r^2+\rho^2)^{5/2}}\,ds'. \qquad (4.132)$$

Using Eq. (4.129) the first term on the right-hand side of Eq. (4.132) is equal to $1/\rho$. After integration with respect to ω we get

$$\varphi^0 = \frac{2\mathcal{J}}{c}\omega + \frac{\mathcal{J}}{c}\rho \oint \frac{e_\rho \, \mathbf{r}\tau'}{(r^2+\rho^2)^{3/2}} \, ds'$$

$$-3\frac{\mathcal{J}}{c}\rho \oint \frac{(\tau\tau')(\tau e_\omega)\rho^2}{(r^2+\rho^2)^{5/2}} \, ds' + \varphi_1(s). \qquad (4.133)$$

At low values of ρ the integrand in the last integral has a δ-function dependence on $s - s'$ and this term is equal to $\frac{\mathcal{J}}{c}k\rho \sin(\omega - \alpha)$. Extracting the singularity in the second integral which is at $s = s'$ one can obtain the following expression for the scalar potential of a current-carrying filament:

$$\varphi^0 = \frac{\mathcal{J}}{c}\left[2\omega + \left(k\rho \ln \frac{8R}{\rho} + \rho f_1\right)\sin(\omega - \alpha) \right.$$

$$\left. + \rho \tilde{f}_1 \cos(\omega - \alpha)\right] + \frac{\mathcal{J}}{c}\int^s \hat{f}_1(s') \, ds', \qquad (4.134)$$

where

$$\left.\begin{aligned}
f_1(s) &= \int_s^{s_1}\left(\frac{\beta \, \mathbf{r}\tau'}{r^3} - \frac{1}{2}k \frac{\mathbf{r}\tau'}{r^2}\right)ds' + \int_{s_1}^s \left(\frac{\beta \, \mathbf{r}\tau'}{r^3}\right.\\
&\left.+ \frac{1}{2}k\frac{\mathbf{r}\tau'}{r^2}\right)ds' + k \ln \frac{r_1}{4R}, \\
\tilde{f}_1(s) &= \oint \frac{v \, \mathbf{r}\tau'}{r^3} \, ds', \quad \hat{f}_1(s) = \oint \frac{\tau \mathbf{r}\tau'}{r^3} \, ds'.
\end{aligned}\right\} \qquad (4.135)$$

The functions $f_1(s)$, $\tilde{f}_1(s)$, and $\hat{f}_1(s)$ are zero in the case of an axisymmetric ring. The last term in Eq. (4.134) determines the longitudinal magnetic field produced by the current-carrying filament.

If the form of the scalar magnetic potential of the current-carrying filament is known, one can extract the maintaining field from the equilibrium field (4.116). To do this it is sufficient to subtract Eq. (4.134) from Eq. (4.116) and to drop the terms which are singular as $\rho \to 0$. The result we obtain for the potential φ_{ext} of the maintaining field is

$$\varphi_{\text{ext}} = B_s s - \frac{1}{8} B_s (\rho^3 - 3a^2 \rho) \frac{\partial}{\partial s} k \cos(\omega - \alpha)$$

$$+ \frac{\mathcal{J}}{c} \left(-k\rho \ln \frac{8R}{a} + \frac{3}{2} k\rho + b_1 \frac{\rho}{a} - \rho \tilde{f}_1 \right) \sin(\omega - \alpha)$$

$$- \frac{\mathcal{J}}{c} \left(\tilde{b}_1 \frac{\rho}{a} + \rho \tilde{f}_1 \right) \cos(\omega - \alpha). \tag{4.136}$$

The term proportional to the plasma current is the potential of two mutually perpendicular magnetic fields which are homogeneous in the plane s = const. One of these is directed along the binormal and is equal to

$$B_{\beta, \text{ext}} = -\frac{\mathcal{J}}{ca} \left(ka \ln \frac{8R}{a} - \frac{3}{2} ka - b_1 + a f_1 \right), \tag{4.137}$$

while the second is directed along the normal:

$$B_{\nu, \text{ext}} = -\frac{\mathcal{J}}{ca} (\tilde{b}_1 + a\tilde{f}_1). \tag{4.138}$$

When we know the value of the maintaining field required for equilibrium, we can find the position of the vacuum magnetic axis and, thus, the displacement of the plasma column with respect to this axis. Figure 22 shows the result of calculations of the displacement of a plasma column with an axis in the form of an elementary spatial curve N = 2. Two cases are considered with the equilibrium maintained firstly by external fields and secondly by an ideally conducting casing. In the first case the displacement of the current-carrying column with respect to the vacuum magnetic axis is an order of magnitude larger than in the second case.

4.6. Probe Diagnostics in a Tokamak

Consider the basic consequences of the equilibrium theory which can be used to obtain information on the plasma parameters from magnetic measurements.

In a strong longitudinal magnetic field, $B_{se} \gg B_{se} - B_s$, the parameter $\mu_{\mathcal{J}}$ (2.8) characterizing equilibrium in the minor radial direction (2.9), (3.102) can be expressed in terms of measurable values:

$$\mu_{\mathcal{J}} = \frac{c^2}{4\pi \mathcal{J}^2} \int [B_{se}^2 - B_s^2] dS = \frac{c^2}{2\pi \mathcal{J}^2} B_{se} \delta\Phi, \tag{4.139}$$

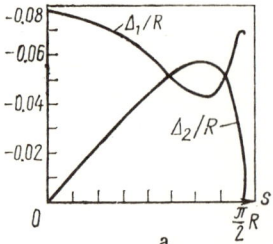 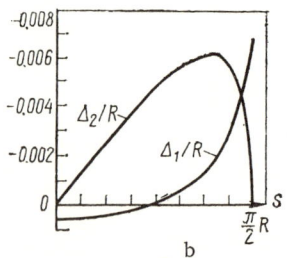

Fig. 22. Displacements of a nonplanar plasma column in the form of a spiral with $N = 2$, $\varkappa_0 = 0.3$. The current distribution is uniform, $\beta = 1\%$, $a/R = 0.1$, $aB_s/RB_\theta = 1$; Δ_1 is the displacement along the normal to the axis, Δ_2 is the displacement along the binormal: a) displacements in the vacuum magnetic field with respect to the vacuum magnetic axis; b) displacements of a plasma column situated in an ideally conducting casing ($b/a = 1.5$).

where $\delta\Phi$ denotes the flux forced out of the plasma:

$$\delta\Phi = \int (B_{se} - B_s) dS. \tag{4.140}$$

This is often simply called the diamagnetic signal.

Consider two concentric circular loops of radii b_1 and b_2 enclosing the plasma column (Fig. 23). With these loops one can measure the longitudinal magnetic flux linking cross sections $\rho < b_1$ and $\rho < b_2$:

$$\left. \begin{array}{l} \Phi(b_1) = B_{s0} \, 2\pi R \left(R - \sqrt{R^2 - b_1^2}\right) + \delta\Phi; \\ \Phi(b_2) = B_{s0} \, 2\pi R \left(R - \sqrt{R^2 - b_2^2}\right) + \delta\Phi, \end{array} \right\} \tag{4.141}$$

where R is the distance between the centers of the loops and the major axis of the torus, $B_{s0} = B_{se}(R)$. Hence we obtain the relation for the diamagnetic signal:

$$\delta\Phi = \Phi(b_1) - [\Phi(b_2) - \Phi(b_1)] \frac{b_1^2}{b_2^2 - b_1^2} \frac{\sqrt{R^2 - b_1^2} + \sqrt{R^2 - b_2^2}}{R + \sqrt{R^2 + b_1^2}}. \tag{4.142}$$

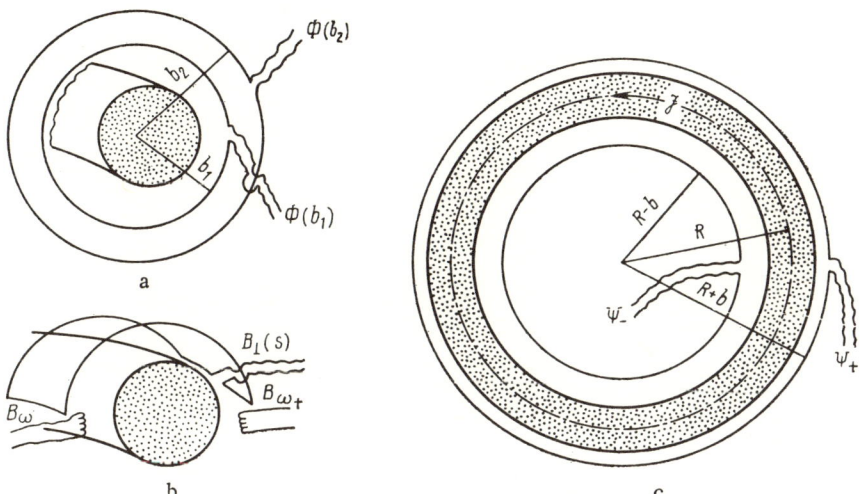

Fig. 23. Magnetic measurements for a plasma column with a circular cross section: a) loops enclosing the plasma column in the cross section ζ = const serving to measure the diamagnetic signal; b) longitudinal loops, serving to measure the flux functions Ψ_+ and Ψ_- and the mean magnetic field $\overline{B}_\perp$; c) local probes for measuring $B_{\omega+}$ and $B_{\omega-}$ and a local loop for measuring $B_\perp(s)$.

Making use of the integral relationships for the transverse equilibrium (2.9), (2.102) and having measured values for $\delta\Phi$ and the plasma current $\mathcal{J}$ one can determine the value of $\beta_{\mathcal{J}}$ or the value of $\beta_{\mathcal{J}\perp}$ for a plasma with anisotropic pressure.

Using Eqs. (4.36)-(4.38) for the flux function and the magnetic fields of the equilibrium plasma configuration, one can relate directly measured values to the displacement Δ of the plasma column and to the parameter $\beta_{\mathcal{J}} + l_i/2$. With a longitudinal loop $\rho = b$, ω = const located outside the plasma column (Fig. 23) one can measure the flux function $\Psi(b, \omega)$, and local coils measure the magnetic field B_ω. Let Ψ_+, $B_{\omega+}$, Ψ_-, and $B_{\omega-}$ be the values of the flux function and those of the tangential magnetic field components on the outer ($\omega = \pi$) and inner ($\omega = 0$) sides of the torus with radii R and b, respectively. Then we obtain from Eqs. (4.36)-(4.38)

$$\frac{2\mathcal{J}}{cR}\left(\ln\frac{b}{a} + \beta_{\mathcal{J}} + \frac{l_i}{2} - 1\right) = \frac{B_{\omega+} - B_{\omega-}}{2} + B_\perp; \qquad (4.143)$$

$$\frac{\Delta}{b} = \frac{b}{2R}\left[\frac{a^2}{b^2}\ln\frac{b}{a} + \frac{1}{2}\left(1 - \frac{a^2}{b^2}\right)\right] +$$

$$+ \frac{cb}{4\mathscr{J}}\left[\frac{B_{\omega+} - B_{\omega-}}{2}\left(1 - \frac{a^2}{b^2}\right) - B_\perp\left(1 + \frac{a^2}{b^2}\right)\right], \quad (4.144)$$

where

$$B_\perp \equiv (\Psi_- - \Psi_+)/4\pi Rb. \quad (4.145)$$

Equations (4.143)-(4.144) can be generalized for the case when the maintaining field is modulated along the s axis. The modulation can be due to the presence of an asymmetric iron core transformer, gaps in the casing, etc. In this case the plasma column suffers a deviation from the equatorial plane:

$$\delta z/R \approx \delta B_{z,\text{ext}}/B_s, \quad (4.146)$$

where $\delta B_{z,\text{ext}}$ is the deviation of the maintaining field from its averaged value. If we consider the perturbation of symmetry to be smooth $\rho\partial/\partial s \ll 1$, then Eqs. (4.143) and (4.144) require the substitution [77]

$$\frac{B_{\omega+} - B_{\omega-}}{2} \rightarrow \frac{B_{\omega+} - B_{\omega-}}{2} + \overline{B}_\perp - B_\perp(s). \quad (4.147)$$

Here $\overline{B}_\perp = (\Psi_- - \Psi_+)/4\pi Rb$ is the mean value of the transverse magnetic field in the region between two toroidal loops at $(\omega = 0)$ and $(\omega = \pi)$ and $B_\perp(s)$ is the local value of the transverse magnetic field which can be measured by a local loop located above the probes measuring the field B_ω.

In combination with the diamagnetic measurements Eq. (4.143) makes it possible to determine the internal inductance l_i characterizing the current density distribution and the energy of the poloidal magnetic field in the plasma column. The latter is required in order to calculate the ohmic heating power in the expression for the energy confinement time. When the pressure anisotropy is appreciable, $p_\parallel \neq p_\perp$ and $(\beta_{\mathscr{J}\parallel} + \beta_{\mathscr{J}\perp})/2$ enter Eqs. (4.143) and (4.144) instead of $\beta_\mathscr{J}$. Then there are insufficient equations to determine the value of l_i.

For an axisymmetric plasma column with a noncircular cross section one can obtain information on the internal plasma parameters by using the integral relations (Section 2.6). Thus, Eq. (2.90) permits us to determine both the relationship between the parameter β_χ and the diamagnetic signal [Eq. (2.102)] and the moments of the quantity $P = p + (B_s^2 - B_{se}^2)/8\pi$: $\int P x dV$, $\int P(x^2 - z^2)dV$, etc., using magnetic field measurements. The integral relations (2.115) allow the measurement of the moments of the current density: $\int j_s x dS$, $\int j_s(x^2 - z^2)dS$, etc. It should be noted that in addition to measurements of the tangential component of the magnetic field a knowledge of the normal component is required. Currently developed numerical methods make it possible to directly compare measurements of the magnetic fields and the flux function $\Psi(r, z)$ with model calculations [28]. For a plasma column with a noncircular cross section the poloidal field distribution on magnetic surfaces outside the plasma is explicitly dependent on the current density distribution. In particular, the poloidal magnetic field decreases rapidly at the vertices of an elongated cross section. This permits one to distinguish between skin, uniform, and peaked current distributions in the plasma even without diamagnetic measurements (Fig. 24).

In conclusion we note that measurements of the flux function $\Psi(l)$ and the tangential component of the poloidal magnetic field $B_\tau(l) = -(1/2\pi r)(\partial\Psi/\partial n)$ on a contour enclosing an axisymmetric plasma column are sufficient to determine the geometry of the vacuum magnetic surfaces, and, in particular, the shape of the current channel boundary. In its turn this makes it possible to control the position and the shape of the plasma column using a feedback system without information on the internal distribution of the plasma parameters [53].

CHAPTER 5

THE EVOLUTION OF A TOROIDAL EQUILIBRIUM

When the parameters of an equilibrium configuration, such as the plasma pressure and the current density, as well as the external fields, are changing in time, the position and the shape of the plasma column follow these changes. To estimate the consequences of such a process a numerical simulation of the evolution of an equilibrium is required.

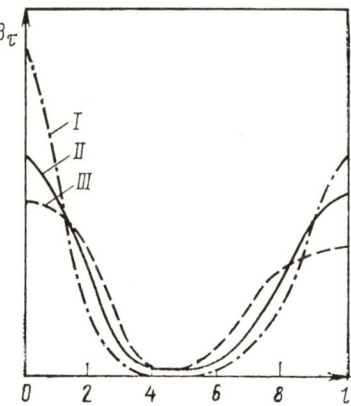

Fig. 24. The distribution of the poloidal field $B_\tau(l)$ on the inner surface of the casing in the finger-ring tokamak T-9 [28]. The point $l = 0$ corresponds to the inner side of the torus, $l = 10$ corresponds to the outer side. The curves are distinguished by the effective elongation of the current channel l_z/l_r: I) $l_z/l_r = 1$, highly peaked current distribution; II) $l_z/l_r = 2$, approximately uniform current distribution; III) $l_z/l_r = 3$, current distribution of the skin type.

For solving the problem of the evolution of an equilibrium it is necessary to know the time dependences of the gas-kinetic plasma pressure $p(a, t)$ and the poloidal current $F(a, t)$ entering the right-hand side of the equilibrium equation (2.21). Instead of these variables some other quantities, for example, the longitudinal current $\mathcal{J}(a, t)$, the poloidal flux $\Psi(a, t)$, the safety factor $q(a, t) = -d\Phi/d\Psi$, can be used in the problem of evolution.

This chapter gives a general formulation of the problem of evolution of equilibria.

5.1. One-Dimensional Transport Equations

The plasma pressure is determined by the distributions of the density and temperature of the plasma components (electrons and ions of different kinds):

$$p = \sum nT. \qquad (5.1)$$

Each of these distributions is determined by the appropriate equations of continuity and of thermal conductivity. It should be noted that equalization of both the temperature and the density along magnetic surfaces is much faster than their differentiation due to the averaged transport in the radial direction. Therefore, one-dimensional (averaged over magnetic surfaces) equations of evolution can be used, where T and n are taken to be functions of a and t.

To write these equations correctly we should take into account that the magnetic surfaces move in space with velocity

$$\mathbf{v}_a = d\mathbf{r}/dt, \qquad (5.2)$$

where $\mathbf{r}(t)$ is determined from the equation $a(\mathbf{r}, t) = \text{const}$. Since for a magnetic surface labeled by a we have by definition

$$\frac{d}{dt} = \frac{\partial a}{\partial t}\bigg|_\mathbf{r} + \frac{d\mathbf{r}}{dt}\nabla a = 0, \qquad (5.3)$$

then the velocity $\mathbf{v}_a$ of the magnetic surface is related to the variation of a at a fixed point by

$$\frac{\partial a}{\partial t}\bigg|_\mathbf{r} = -v_a^1, \qquad (5.4)$$

where $v_a^1 = \mathbf{v}_a \nabla a$ is a contravariant component of the vector $\mathbf{v}_a$.

Note that the velocity v_a^1 essentially depends on the choice of the coordinate $a(\mathbf{r}, t)$ which can mean any surface quantity, e.g., V, p, Ψ, Φ, etc. In a dissipative plasma the velocity $v_\Psi^1 = -\partial \Psi/\Psi'\partial t|_\mathbf{r}$ of the surfaces $\Psi = \text{const}$ of the poloidal flux differs from the velocity v_Φ^1 of the surfaces $\Phi = \text{const}$ of the longitudinal flux and from the velocity v_n^1 of the surfaces n = const. Sets of these surfaces seem to pass through each other corresponding to the different dependence of the functions $p(a, t)$, $\Phi(a, t)$, and $\Psi(a, t)$ on time.

Transport equations written in the variables a and t will contain both the velocity v of particles and the velocity v_a of magnetic surfaces. To obtain these equations we note that the time derivative $\partial/\partial t|_{\mathbf{r}}$ should be equal to

$$\frac{\partial}{\partial t}\bigg|_{\mathbf{r}} = \frac{\partial}{\partial t}\bigg|_a + \frac{\partial a}{\partial t}\bigg|_{\mathbf{r}} \frac{\partial}{\partial a} = \frac{\partial}{\partial t} - v_a^1 \frac{\partial}{\partial a} \qquad (5.5)$$

due to the motion of the surface $a(\mathbf{r}, t)$.

Let us denote by $\langle \ \rangle_V$ an averaging over the volume between the infinitesimally close magnetic surfaces $a = \text{const}$, $a + da = \text{const}$:

$$\langle f \rangle_V = \frac{d}{dV} \int f \sqrt{g}\, d\theta\, d\zeta = \langle f\sqrt{g}\rangle / \langle\sqrt{g}\rangle. \qquad (5.6)$$

(Recall that $\langle \ \rangle$ denotes an averaging with respect to θ and ζ). The derivative $\partial f/\partial t|_{\mathbf{r}}$ of surface variables averaged over infinitesimal volumes is determined by the operator

$$\frac{D}{Dt} = \frac{\partial}{\partial t}\bigg|_a + \left\langle 4\pi^2 \sqrt{g}\, \frac{\partial a}{\partial t} \right\rangle \frac{\partial}{\partial V}, \qquad (5.7)$$

where

$$\frac{\partial}{\partial V} = \frac{1}{4\pi^2 \langle\sqrt{g}\rangle}\frac{\partial}{\partial a}. \qquad (5.8)$$

The averaged expression for div $\mathbf{A}$ in accordance with Eq. (A.32) is of the form

$$\langle \text{div }\mathbf{A}\rangle_V = \frac{\partial}{\partial V}\langle 4\pi^2 \sqrt{g}\, A^1\rangle. \qquad (5.9)$$

Let us denote by v and q the contravariant components of the velocity and heat flux vectors averaged over θ and ζ with weight $\sqrt{g}$:

$$v \equiv \langle \sqrt{g}\, \mathbf{v}\nabla a\rangle, \quad q = \langle \sqrt{g}\, \mathbf{q}\nabla a\rangle, \qquad (5.10)$$

Then using Eqs. (5.7), (5.8), and (5.10) the transport equations can be written as follows:

$$\frac{Dn}{Dt} + \frac{\partial}{\partial V}(nv) = \Gamma; \qquad (5.11)$$

$$\frac{3}{2} n \left(\frac{DT}{Dt} + v \frac{\partial T}{\partial V} \right) + nT \frac{\partial v}{\partial V} + \frac{\partial q}{\partial V} = Q. \qquad (5.12)$$

Here Γ and Q are the particle sources and heat sources per unit volume averaged over a magnetic layer for each species.

Equations (5.11) and (5.12) represent the first two even moments (multiplication by $v^0 = 1$ and $|v|^2$ followed by integration with respect to velocities) of the kinetic equation, which express the laws of conservation of particles and energy. The first two odd moments (v and $|v|^2 v$) which correspond to the conservation of momentum and energy flux can be used to find the hydrodynamic velocity and heat flux in Eqs. (5.11) and (5.12). The determination of these expressions for a particular geometry is the subject of neoclassical transport theory (for microscopically stable plasmas) and of turbulence theory (for microscopically unstable plasma). As is shown by the experimental thermonuclear studies of many years standing, heat and particle transport in high-temperature plasma confined in a magnetic field is usually anomalous. Thus, we have in fact no accurate and reliable transport equations for the evaluation of the particle density, temperature, and pressure. When numerical simulations are carried out, semiempirical expressions for heat and particle fluxes are usually used. The one-dimensional form of the equations simplifies the use of these semiempirical dependences even for pronounced two-dimensional cases (elliptical, D-shaped cross sections of the plasma column).

The question of the particle flux also arises when the diffusion of magnetic fluxes is considered.

5.2. Equation for Evolution of Magnetic Fluxes

The function $F(\Psi)$, which with $p'(\Psi)$ enters the right-hand side of the equilibrium equation (2.21), can be expressed in terms of derivatives of the longitudinal flux. In fact, the longitudinal flux between two magnetic surfaces enclosing a volume dV is equal to

$$d\Phi = \frac{4\pi}{c} \int \frac{F}{2\pi r} dS = \frac{4\pi}{c} dV \left\langle \frac{1}{4\pi^2 r^2} \right\rangle_V, \qquad (5.13)$$

so [see also Eq. (2.63)]

$$\frac{4\pi}{c} F = \left\langle \frac{1}{4\pi^2 r^2} \right\rangle_V^{-1} \frac{d\Phi}{dV} = \frac{1}{\langle \sqrt{g}/g_{33} \rangle} \frac{d\Phi}{da} , \qquad (5.14)$$

Using the definition of the safety factor

$$q = -d\Phi/d\Psi = -\Phi'/\Psi', \qquad (5.15)$$

the function F can be expressed in terms of q:

$$\frac{4\pi}{c} F = -\frac{q}{\langle \sqrt{g}/g_{33} \rangle} \frac{d\Psi}{da} . \qquad (5.16)$$

It is convenient to express F in terms of q because for fast processes, when one can neglect dissipation, $q(\Psi)$ is conserved in time, $\partial q(\Psi, t)/\partial t = 0$, and is determined by the initial data, $q(a, t) = q_0(a)$. This fact is a consequence of flux conservation in a nondissipative plasma.

For a given $q(a)$ the evolution problem reduces to solving the equilibrium equation for a given $p(a,t)$ and external conditions as functions of time.

For dissipative plasmas an equation for the evolution of magnetic fluxes should be added to the set of equilibrium and transport equations. It may be represented by an evolution equation for $\mu = 1/q$. This equation together with those for the radial velocities of different components of the plasma is obtained from the momentum balance equations for different species (the first moment of the kinetic equation). It was shown in Chapter 1 how one can obtain this equation in the drift approximation from the law of particle motion in a strong magnetic field. Neglecting inertia this equation can be written in the form of a vector equation for each species of the plasma

$$-\nabla p_\| + \frac{en}{c}[\mathbf{vB}] - \frac{\lambda}{c}[\mathbf{jB}] + \frac{B^2}{4\pi}\nabla\lambda - \frac{\mathbf{B}(\mathbf{B}\nabla\lambda)}{4\pi}$$
$$+ \lambda\nabla\frac{B^2}{8\pi} + n(e\mathbf{E} - \mathbf{R}) = 0. \qquad (5.17)$$

Here

$$\lambda = \frac{4\pi(p_\| - p_\perp)}{B^2} ,$$

$n, e, p_{\parallel}, p_{\perp},$ and R are the characteristics of a given species. Let us multiply Eq. (5.17) by $\mathbf{B}$/ne and average over the magnetic layer:

$$\langle \mathbf{EB} \rangle_V = \frac{1}{e} \langle \mathbf{RB} \rangle_V + \frac{1}{e} \left\langle \frac{\mathbf{B} \nabla p_{\parallel}}{n} \right\rangle_V - \left\langle \frac{\lambda}{ne} \mathbf{B} \nabla \frac{|B|^2}{8\pi} \right\rangle_V. \quad (5.18)$$

The covariant components of the vector $\mathbf{E} = -\partial \mathbf{A}/c\partial t - \nabla \varphi^E$ in accordance with Eq. (2.53) are determined by

$$E_2 = -\frac{\partial \Phi}{c \partial t} - \frac{\partial \varphi^E}{\partial \theta}, \quad E_3 = -\frac{\partial \Psi}{c \partial t} - \frac{\partial \varphi^E}{\partial \zeta} \quad (5.19)$$

in terms of derivatives of magnetic fluxes. The contravariant components of the vector $\mathbf{B}$ are given by Eq. (2.51). As a result of the averaging we get

$$\langle \mathbf{EB} \rangle_V = \frac{1}{c} \left(\frac{\partial \Phi}{\partial t} \frac{\partial \Psi}{\partial a} - \frac{\partial \Psi}{\partial t} \frac{\partial \Phi}{\partial a} \right). \quad (5.20)$$

In the regime of high collision frequency (where the mean free path of charged particles is less than the length of the torus) the right-hand side of Eq. (5.18) reduces to the following:

$$\frac{1}{c} \langle \mathbf{RB} \rangle = \frac{1}{\sigma_{\parallel}} \langle \mathbf{jB} \rangle = \frac{4\pi}{c\sigma_{\parallel}} (F \mathcal{J}' - \mathcal{J} F'), \quad (5.21)$$

where $\sigma_{\parallel}(a, t)$ is the effective conductivity of the plasma, and the prime denotes a derivative with respect to a.

In the collisionless regime additional terms on the right-hand side of Eq. (5.18) are required. The general equation can be written in the form

$$\frac{\partial \Phi}{\partial t} \Psi' - \frac{\partial \Psi}{\partial t} \Phi' = \frac{4\pi}{\sigma_{\text{eff}}} F[F (\mathcal{J}/F)' - \mathcal{J}_B'], \quad (5.22)$$

where $\mathcal{J}_B$ is the so-called "bootstrap" current induced by diffusion of "banana" orbits of trapped particles in the poloidal magnetic field. It is convenient to rewrite Eq. (5.22) in terms of the rotational transform $\mu = 1/q = -\Psi'/\Phi'$. Dividing Eq. (5.22) by Φ' and differentiating with respect to a, we obtain

$$\frac{\partial \mu}{\partial t} - \frac{\dot{\Phi}}{\Phi'} \frac{\partial \mu}{\partial a} = \frac{4\pi}{\Phi'} \frac{\partial}{\partial a} \left\{ \frac{F}{\sigma_{\text{eff}} \Phi'} [F (\mathcal{J}/F)' - \mathcal{J}_B'] \right\}, \quad (5.23)$$

where $\dot{\Phi} = \partial \Phi(a, t)/\partial t$.

In the case of axial symmetry in accordance with Eqs. (2.57) and (2.58)

$$\mathcal{J} = \frac{c}{4\pi} \left\langle \frac{g_{22}}{\sqrt{g}} \right\rangle \Phi' \mu, \quad F = \frac{c}{4\pi} \Phi' / \langle \sqrt{g}/g_{33} \rangle. \quad (5.24)$$

Equation (5.22) and its equivalent (5.23) represent the equations for diffusion of magnetic fluxes. Let us recall that a is a surface variable and the derivatives with respect to time are taken at constant a. If, for example, a is taken to be the toroidal flux, then $\dot{\Phi} = 0$, $\Phi' = 1$. The left-hand sides of Eqs. (5.22) and (5.23) represent the variation of Ψ or μ relative to the longitudinal magnetic flux. It may seem strange that the equation for diffusion of the poloidal magnetic flux does not contain a term corresponding to the convection of the magnetic flux with the diffusing plasma. But this term is contained on the right-hand side of these equations. To confirm this let us consider a covariant ζ component of Eq. (5.17). In the case of axial symmetry, $\partial/\partial\zeta = 0$, it has the form

$$n(eE_3 - R_3) + \frac{en}{c}\sqrt{g}\, v^1 B^2 - \frac{B_3}{4\pi}(B_\nabla \lambda) = 0. \quad (5.25)$$

According to Eq. (2.54) $B_3 = B_3(a)$, and Eq. (5.25) gives

$$\langle n(eE_3 - R_3)\rangle_V - \frac{e}{c}\langle nv^1\rangle_V \Psi'/2\pi = 0 \quad (5.26)$$

after averaging over the magnetic layer. With Eq. (5.19) for E_3 it is equivalent to

$$\frac{\partial \Psi}{\partial t} + v \frac{\partial \Psi}{\partial a} = \frac{1}{e}\langle nR_3\rangle_V. \quad (5.27)$$

The mean velocity of the plasma convection, $v = \langle nv_V^1 \rangle / \langle n \rangle_V$, which enters Eq. (5.27) and which is also required to solve the transport equations for plasma particles and energy (5.11), (5.12), can be obtained in a formal way from the θ component of Eq. (5.17). It also includes the rate of compression of the toroidal flux $-\dot{\Phi}/\Phi'$ and the diffusion. It is clear that Eq. (5.22) can be considered as the result of substituting such an expression for the convective velocity into Eq. (5.27).

Let us note that since the diffusion arises not only due to a drag force between different components of the plasma but also due to viscosity [the terms with λ in Eq. (5.17)], then the local velocities of electrons and ions may, in principle, be unequal. However, it is seen from Eq. (5.26) that in the case of axial symmetry the diffusion is ambipolar, $\Sigma e \langle nv^1 \rangle = 0$, because $\Sigma n(eE_3 - R_3) = 0$. This resolves the question as to which plasma component the velocity in Eq. (5.27) refers.

As seen from Eq. (5.17) the neoclassical effects of additional transport of particles across the magnetic surfaces, due to the terms with finite λ, are associated with top−bottom asymmetry of the equilibrium plasma. Producing an artificial anisotropy of the stress tensor which is asymmetric with respect to the equatorial plane of the torus can, in principle, affect the direction of the effective diffusion.

Expressions for plasma convection and hence for the current $\mathcal{J}$ which apply to real plasmas may differ from those given by the neoclassical transport theory [4]. As was mentioned above, at the present time there is no concensus on the processes determining the observed transport. For this reason we do not list concrete expressions for the particle and heat fluxes and have considered only the general structure of the equations describing the evolution of equilibria. This general consideration illustrates the physics of the equations which should be used for two-dimensional continually varying equilibrium configurations.

APPENDIX

For a description of three-dimensional MHD equilibrium configurations it is important to use an appropriate coordinate system a, θ, ζ, where the coordinate surfaces a = const coincide with magnetic surfaces while the cyclic variables θ and ζ, corresponding to the poloidal and toroidal directions respectively, are chosen arbitrarily. The coordinates a, θ, ζ are not orthogonal. Such coordinates are also useful for two-dimensional configurations. Here we shall describe the main points in the theory of curvilinear coordinates related to the theory of equilibrium magnetic configurations.

a) <u>Basic Vectors.</u> A system of curvilinear coordinates $a \equiv x^1$, $\theta \equiv x^2$, $\zeta \equiv x^3$ may be introduced both by the transformation

$$\mathbf{r} = \mathbf{r}(x^1, x^2, x^3), \qquad (A.1)$$

and by the equations of coordinate surfaces

$$x^i(\mathbf{r}) = \text{const.} \tag{A.2}$$

Correspondingly, there are two sets of basic and reciprocal vectors. The basic vectors (with a subscript)

$$\mathbf{e}_i = \partial \mathbf{r}/\partial x^i \tag{A.3}$$

define the direction of variation of a coordinate x^i in accordance with the expression for a directed line element

$$d\mathbf{l} = \frac{\partial \mathbf{r}}{\partial x^i} dx^i \tag{A.4}$$

(repeated indices indicate a sum). Thus, the vectors $\mathbf{e}_2$ and $\mathbf{e}_3$ are tangential to the coordinate surface $x^1 = \text{const}$ and so on.

The reciprocal vectors (with a superscript)

$$\mathbf{e}^k = \nabla x^k, \tag{A.5}$$

in accordance with the definition are directed along the normal to the corresponding coordinate surfaces $x^k = \text{const}$. Thus, the vector $\mathbf{e}^1$ is orthogonal to the vectors $\mathbf{e}_2$, $\mathbf{e}_3$, etc.

Similarly, the vector $\mathbf{e}_1$, being tangential to the line of intersection of the coordinate surfaces $x^2 = \text{const}$, $x^3 = \text{const}$, is orthogonal to the vectors $\mathbf{e}_3$ and $\mathbf{e}_2$. Therefore, we can write

$$\begin{aligned}\mathbf{e}_1 &= C_1\,[\mathbf{e}^2\,\mathbf{e}^3], \quad \mathbf{e}_2 = C_2\,[\mathbf{e}^3\,\mathbf{e}^1], \quad \mathbf{e}_3 = C_3\,[\mathbf{e}^1\,\mathbf{e}^2], \\ \mathbf{e}^1 &= C^1\,[\mathbf{e}_2\,\mathbf{e}_3], \quad \mathbf{e}^2 = C^2\,[\mathbf{e}_3\,\mathbf{e}_1], \quad \mathbf{e}^3 = C^3\,[\mathbf{e}_1\,\mathbf{e}_2]. \end{aligned} \tag{A.6}$$

Note that the expression for the differential of a scalar function

$$d\varphi = \nabla \varphi \, d\mathbf{l} = \left(\frac{\partial \varphi}{\partial x^k}\nabla x^k\right)\left(\frac{\partial \mathbf{r}}{\partial x^i} dx^2\right) = \frac{\partial \varphi}{\partial x^k} \mathbf{e}_i \mathbf{e}^k\, dx^i \tag{A.7}$$

results in

$$\mathbf{e}_i \mathbf{e}^k = \delta_{ik}, \tag{A.8}$$

and consequently

$$C_1 = C_2 = C_3 = \frac{1}{C^1} = \frac{1}{C^2} = \frac{1}{C^3} = \sqrt{g}, \tag{A.9}$$

where
$$\sqrt{g} = \mathbf{e}_1 \mathbf{e}_2 \mathbf{e}_3 = \frac{1}{\mathbf{e}^1 \mathbf{e}^2 \mathbf{e}^3} \qquad (A.10)$$

is the Jacobian of the transformation $\mathbf{r}(x^1, x^2, x^3)$ in accordance with the definition of vectors $\mathbf{e}_i$:

$$\sqrt{g} = \frac{\partial \mathbf{r}}{\partial x^1} \frac{\partial \mathbf{r}}{\partial x^2} \frac{\partial \mathbf{r}}{\partial x^3} = \frac{1}{\nabla x^1 \nabla x^2 \nabla x^3}. \qquad (A.11)$$

b) Metric Tensor. All the intrinsic features of the geometry of a coordinate system x^1, x^2, x^3 are determined by the expression for the square of the line element

$$dl^2 = g_{ik} dx^i dx^k. \qquad (A.12)$$

The coefficients g_{ik} are the components of the fundamental metric tensor. From the expression for dl it follows that

$$g_{ik} = \mathbf{e}_i \mathbf{e}_k. \qquad (A.13)$$

The square of a gradient is expressed in terms of another tensor g^{ik}:

$$(\nabla \varphi)^2 = g^{ik} \frac{\partial \varphi}{\partial x^i} \frac{\partial \varphi}{\partial x^k}, \qquad (A.14)$$

where

$$g^{ik} = \mathbf{e}^i \mathbf{e}^k. \qquad (A.15)$$

Let us denote the algebraic adjuncts of the elements g_{ik} and g^{ik} by G_{ik} and G^{ik}, respectively. Using Eq. (A.3) for $\mathbf{e}_i$ and Eqs. (A.6) and (A.8), we can easily obtain

$$g_{ik} = g G^{ik}. \qquad (A.16)$$

Correspondingly, using Eq. (A.5) for $\mathbf{e}^i$ we have

$$g^{ik} = g G_{ik}. \qquad (A.17)$$

Producing a square of the expression $\sqrt{g} = \mathbf{e}_1 \mathbf{e}_2 \mathbf{e}_3$ and using the rule of the multiplication of determinants, we get

$$g = \begin{vmatrix} \mathbf{e}_1 \mathbf{e}_1 & \mathbf{e}_1 \mathbf{e}_2 & \mathbf{e}_1 \mathbf{e}_3 \\ \mathbf{e}_2 \mathbf{e}_1 & \mathbf{e}_2 \mathbf{e}_2 & \mathbf{e}_2 \mathbf{e}_3 \\ \mathbf{e}_3 \mathbf{e}_1 & \mathbf{e}_3 \mathbf{e}_2 & \mathbf{e}_3 \mathbf{e}_3 \end{vmatrix} = \begin{vmatrix} g_{11} & g_{12} & g_{13} \\ g_{21} & g_{22} & g_{23} \\ g_{31} & g_{32} & g_{33} \end{vmatrix}, \qquad (A.18)$$

i.e., $g = \text{Det } g_{ik}$. The square of the expression $(\sqrt{g})^{-1} = \mathbf{e}^1 \mathbf{e}^2 \mathbf{e}^3$ gives $\text{Det } g^{ik} = 1/g$.

c) Components of Vectors. Any physical vector **A** (a magnetic field **B**, a velocity **v**, etc.) can be written in terms of the components of both the basic and the reciprocal vectors

$$\mathbf{A} = A^i \mathbf{e}_i = A_i \mathbf{e}^i, \tag{A.19}$$

where A_i is a covariant and A^i is a contravariant projection of the vector.

Using the condition (A.8), we find

$$A^i = \mathbf{A}\mathbf{e}^i = \mathbf{A}\nabla x^i, \quad A_i = \mathbf{A}\mathbf{e}_i = \mathbf{A}\partial \mathbf{r}/\partial x^i. \tag{A.20}$$

For the scalar product of two vectors we have

$$\mathbf{A}\mathbf{B} = A^i \, \mathbf{e}_i \mathbf{e}^k B_k = A^i B_i = A_i B^i. \tag{A.21}$$

The square of the length of a vector is

$$|\mathbf{A}|^2 = A_i A^i. \tag{A.22}$$

The vector product of two vectors $\mathbf{A} = A_1 \mathbf{e}^1 + A_2 \mathbf{e}^2 + A_3 \mathbf{e}^3$ and $\mathbf{B} = B_1 \mathbf{e}^1 + B_2 \mathbf{e}^2 + B_3 \mathbf{e}^3$ is

$$[\mathbf{A}\mathbf{B}]^1 = [\mathbf{A}\mathbf{B}]\mathbf{e}^1 = (A_2 B_3 - A_3 B_2)[\mathbf{e}^2 \mathbf{e}^3]\mathbf{e}^1 = \frac{1}{\sqrt{g}}(A_2 B_3 - A_3 B_2). \tag{A.23}$$

Correspondingly

$$[\mathbf{A}\mathbf{B}]_1 = [\mathbf{A}\mathbf{B}]\mathbf{e}_1 = (A^2 B^3 - A^3 B^2)[\mathbf{e}_2 \mathbf{e}_3]\mathbf{e}_1 = \sqrt{g}\,(A^2 B^3 - A^3 B^2). \tag{A.24}$$

The expressions for other components of the vector product may be obtained by cyclic permutation of the indices.

If one represents the vector **A** in the expressions $A^i = \mathbf{A}\mathbf{e}^i$ and $A_i = \mathbf{A}\mathbf{e}_i$ in the form $\mathbf{A} = A_k \mathbf{e}^k$ and $\mathbf{A} = A^k \mathbf{e}_k$, then the relationship between the co- and contravariant components of a vector can be obtained as follows:

$$A^i = g^{ik} A_k, \quad A_i = g_{ik} A^k. \tag{A.25}$$

Thus, for example

$$|\mathbf{A}|^2 = g_{ik} A^i A^k = g^{ik} A_i A_k. \tag{A.26}$$

In an orthogonal coordinate system we have

$$|A|^2 = g_{11} A^1 A^1 + g_{22} A^2 A^2 + g_{33} A^3 A^3 = g^{11} A_1 A_1 + g^{22} A_2 A_2 + g^{33} A_3 A_3. \quad (A.27)$$

The expressions for the "physical" components of a vector in terms of the contra- and covariant components are as follows:

$$A_i^{phys} = \sqrt{g_{ii}} A^i = \sqrt{g^{ii}} A_i \text{ etc.} \quad (A.28)$$

d) <u>Differential Operators</u>. For solving physical problems using a curvilinear coordinate system it is necessary to know the expressions for commonly used operators such as ∇, div, curl, and ∇^2. Due to the distinction between co- and contravariant components of the vectors these expressions may be represented in different forms. It is expedient to use the simplest one. In some expressions it is convenient to use the covariant components, while for other operators — the contravariant ones.

From the expression $\nabla \varphi = (\partial \varphi / \partial x^i) \nabla x^i = (\partial \varphi / \partial x^i) e^i$ it is seen that the covariant components of a gradient simply coincide with the derivatives with respect to the corresponding coordinates:

$$(\nabla \varphi)_i = \partial \varphi / \partial x^i. \quad (A.29)$$

The contraviant components are expressed in terms of $\partial \varphi / \partial x^i$ by the general rules and are quite complicated. Thus, in an equation containing gradients it is expedient to use the covariant components. For example, the radial component of the equilibrium equation has the form

$$c \frac{\partial p}{\partial x^1} = \sqrt{g} (j^2 B^3 - j^3 B^2). \quad (A.30)$$

Let us apply the operator div to a vector $\mathbf{A} = A^i \mathbf{e}_i$:

$$\text{div } \mathbf{A} = \mathbf{e}_i \nabla A^i + A^i \text{ div } \mathbf{e}_i. \quad (A.31)$$

Taking into account that $\mathbf{e}_i \nabla A^i = \partial A^i / \partial x^i$ and $\text{div } \mathbf{e}_1 = \text{div } \sqrt{g} [\nabla x^2 \nabla x^3] = [\nabla x^2 \nabla x^3] \nabla \sqrt{g} = [\nabla x^2 \nabla x^3] \nabla x^2 \partial \sqrt{g}/\partial x^1 = \partial \ln(\sqrt{g})/\partial x^1$, we get

$$\text{div } \mathbf{A} = \frac{1}{\sqrt{g}} \frac{\partial}{\partial x^i} (\sqrt{g} A^i). \quad (A.32)$$

Applying the operator curl to a vector $\mathbf{A} = A_i \mathbf{e}^i = A_i \nabla x^i$ we find

$$\text{curl } \mathbf{A} = [\nabla A_i \nabla x^i]. \quad (A.33)$$

In accordance with the expression for a vector product we obtain

$$(\operatorname{curl} \mathbf{A})^1 = \frac{1}{\sqrt{g}} \left(\frac{\partial A_3}{\partial x^2} - \frac{\partial A_2}{\partial x^3} \right). \tag{A.34}$$

e) <u>Integral Operators.</u> Let us give the expressions for integrals in terms of curvilinear coordinates.

In accordance with the definition (A.12) for the square of the line element, a contour integral of the first kind along the coordinate line $x^2 = \text{const}$, $x^3 = \text{const}$ is written in the form

$$\int f dl_{(1)} = \int f \sqrt{g_{11}} \, dx^1. \tag{A.35}$$

A contour integral of the second kind in accordance with the expression for the scalar product of two vectors with components

$$\mathbf{A} = \{A_1, A_2, A_3\}, \quad d\mathbf{l} = \{dx^1, dx^2, dx^3\}$$

has the form

$$\int \mathbf{A} d\mathbf{l} = \int A_i dx^i. \tag{A.36}$$

The directed area element of the coordinate surface $x^1 = \text{const}$ is

$$d\mathbf{S}_{(1)} = [d\mathbf{l}_{(2)} \, d\mathbf{l}_{(3)}] = [\mathbf{e}_2 \, \mathbf{e}_3] \, dx_2 \, dx_3 = \sqrt{g} \, \nabla x^1 \, dx^2 \, dx^3. \tag{A.37}$$

Thus, the volume element in terms of curvilinear coordinates is

$$dV = d\mathbf{l}_{(1)} \, d\mathbf{S}_{(1)} = \sqrt{g} \, dx^1 \, dx^2 \, dx^3; \tag{A.38}$$

$$\int f dV = \int f \sqrt{g} \, dx^1 \, dx^2 \, dx^3. \tag{A.39}$$

Since we have $|\nabla x^1|^2 = g^{11}$, a surface integral of the first kind over the coordinate surface $x^1 = \text{const}$ is

$$\int f dS_{(1)} = \int f \sqrt{g g^{11}} \, dx^2 \, dx^3. \tag{A.40}$$

Expressing the arbitrary directed area element in terms of its covariant components with the help of Eq. (A.37), we have

$$dS_i = \sqrt{g} \, \{dx^2 \, dx^3, \ dx^3 \, dx^1, \ dx^1 \, dx^2\}. \tag{A.41}$$

From this we obtain the expression for the flux of a vector $\mathbf{A}$ through an arbitrary surface:

$$\int \mathbf{A} d\mathbf{S} = \int \sqrt{g} \, (A^1 \, dx^2 \, dx^3 + A^2 \, dx^1 \, dx^3 + A^3 dx^1 \, dx^2). \tag{A.42}$$

REFERENCES

1. V. D. Shafranov, in: Reviews of Plasma Physics, Vol. 2, M. A. Leontovich (ed.), Consultants Bureau, New York (1966).
2. L. S. Solov'ev and V. D. Shafranov, Reviews of Plasma Physics, Vol. 5, M. A. Leontovich (ed.), Consultants Bureau, New York (1970).
3. V. S. Mukhovatov, in: Advances in Science and Technology, Plasma Physics [in Russian], Vol. 1, Part 1, VINITI, Moscow (1980), pp. 3-118.
4. A. A. Galeev and R. Z. Sagdeev, in: Reviews of Plasma Physics, Vol. 7, M. A. Leontovich (ed.), Consultants Bureau, New York (1979).
5. F. L. Hinton and R. D. Hazeltine, Rev. Mod. Phys., $\underline{48}$, 239 (1976).
6. D. V. Sivukhin, in: Reviews of Plasma Physics, Vol. 1, M. A. Leontovich (ed.), Consultants Bureau, New York (1965).
7. G. Chew, M. Goldberger, and M. Low, Proc. Roy. Soc., $\underline{A236}$, 112 (1956).
8. A. P. Popryadukhin, Zh. Tekh. Fiz., $\underline{35}$, 927 (1965).
9. T. F. Volkov, in: Reviews of Plasma Physics, Vol. 4, M. A. Leontovich (ed.), Consultants Bureau, New York (1966).
10. J. B. Taylor, Plasma Physics. Lectures at a Seminar, Triest, 5-31 October, 1964, IAEA, Vienna (1965), p. 449.
11. S. Hirshman, Phys. Fluids, $\underline{21}$, 224 (1978).
12. S. I. Braginskii and V. D. Shafranov, in: Plasma Physics and the Problems of Controlled Thermonuclear Reactions, Vol. 2, Pergamon (1959).
13. V. D. Shafranov, Zh. Eksp. Teor. Fiz., $\underline{33}$, 710 (1957).
14. H. Grad and H. Rubin, IAEA Salzburg Congress, Vol. 31 (1958), p. 190.
15. R. Lüst and A. Schlüter, Z. Naturforsch., $\underline{12a}$, 850 (1957).
16. H. Grad, P. N. Hu, and D. A. Stevens, Proc. Nat. Acad. Sci. USA, $\underline{72}$, 3789 (1975).
17. G. Bateman and Y.-K. M. Peng, Phys. Rev. Lett., $\underline{38}$, 829 (1977).
18. J. L. Johnson et al., Phys. Fluids, $\underline{1}$, 281 (1958).
19. B. B. Kadomtsev, Zh. Eksp. Teor. Fiz., $\underline{37}$, 1352 (1959).
20. B. B. Kadomtsev and O. P. Pogutse, Zh. Eksp. Teor. Fiz., $\underline{65}$, 575 (1973).
21. M. N. Rosenbluth, R. Z. Sagdeev, J. B. Taylor, et al., Nucl. Fusion, $\underline{6}$, 297 (1966).

22. J. D. Callen, B. V. Waddell, et al., in: Plasma Physics and Controlled Nuclear Fusion Research, Vol. 1, Vienna, IAEA (1979), p. 415.
23. S. Hamada, Nucl. Fusion, 1-2, 23 (1963).
24. J. Green and J. L. Johnson, Phys. Fluids, 5, 510 (1962).
25. V. D. Shafranov and L. E. Zakharov, Nucl. Fusion, 12, 599 (1972).
26. V. D. Shafranov, Plasma Phys., 13, 757 (1971).
27. L. E. Zakharov and V. D. Shafranov, Zh. Tekh. Fiz., 43, 225 (1973).
28. A. V. Bortnikov, Yu. T. Baiborodov, N. N. Brevnov, et al., Sixth European Conference on Controlled Fusion and Plasma Physics, Vol. 1, Moscow (1973), p. 165.
29. F. L. Helton and T. S. Wang, Nucl. Fusion, 18, 1523 (1978).
30. A. J. Wootton, Nucl. Fusion, 19, 987 (1979).
31. R. Gajewskii, Phys. Fluids, 15, 70 (1972).
32. L. E. Zakharov and D. V. Orlinskii, Preprint of the I. V. Kurchatov Institute of Atomic Energy, IAE-2536, Moscow (1977).
33. H. R. Strauss, Phys. Fluids, 17, 1040 (1974).
34. G. Bateman, MHD Instabilities, MIT Press, Cambridge (1978).
35. B. A. Abramov, V. V. Vikhrev, and O. P. Pogutse, Fiz. Plazmy, 4, 633 (1978).
36. J. P. Goedbloed and L. E. Zakharov, Nucl. Fusion, 20, 1100 (1980).
37. L. E. Zakharov, Fiz. Plazmy, 7, 18 (1981).
38. V. D. Shafranov, Zh. Eksp. Teor. Fiz., 37, 1088 (1959).
39. A. E. Bajanova and V. D. Shafranov, Zh. Tekh. Fiz., 41, 1357 (1971).
40. L. V. Leites, Elektrichestvo, 11, 76 (1960).
41. V. D. Shafranov, Zh. Tekh. Fiz., 42, 1785 (1972).
42. L. E. Zakharov, Nucl. Fusion, 13, 595 (1973).
43. L. E. Zakharov, Zh. Tekh. Fiz., 45, 1049 (1975).
44. A. N. Tikhonov, Dokl. Akad. Nauk SSSR, 151, 501 (1963).
45. E. M. Kiuru and A. S. Mechenov, Tables for Solving Linear Fredholm Equations of the First Kind by the Regularization Method [in Russian], Computer Center of Moscow State University (1970), Issue 45.
46. W. Kerner, D. Pfirsch, and H. Tasso, Nucl. Fusion, 12, 433 (1972).
47. M. A. Leontovich, in: Plasma Physics and the Problems of Controlled Thermonuclear Reactions, Vol. 1, Pergamon (1960).

48. S. V. Putvinskii, Zh. Tekh. Fiz., 49, 1834 (1979).
49. V. E. Bykov et al., Preprint of the Kharkov Physicotechnical Institute, KhFTI-76-39, Kharkov (1976).
50. Methods in Computational Physics, Vol. 16, Controlled Fusion, Academic Press, New York–San Francisco–London (1976).
51. L. E. Zakharov, Zh. Tekh. Fiz., 44, 1606 (1974).
52. Y. Suzuki, Nucl. Fusion, 14, 346 (1974).
53. L. E. Zakharov, in: Plasma Physics and Controlled Nuclear Fusion Research, Vol. 1, Vienna, IAEA, p. 421.
54. W. Feneberg and K. Lackner, Nucl. Fusion, 13, 549 (1973).
55. J. D. Callen and R. A. Dory, Phys. Fluids, 15, 1523 (1972).
56. M. Chu, et al., Preprint of GALF, GA-A 12726.
57. K. Lackner, Computer Phys. Commun., 12, 33 (1973).
58. M. S. Chance et al., in: Plasma Physics and Controlled Nuclear Fusion Research, Vol. 1, Vienna, IAEA (1975), p. 463.
59. P. N. Vabischevich, L. M. Degtyarev, and A. P. Favorskii, Fiz. Plasmy, 4, 995 (1978).
60. J. F. Clark and D. J. Sigmar, Phys. Rev. Lett., 38, 70 (1977).
61. V. D. Khait, Fiz. Plasmy, 6, 871 (1980).
62. H. Grad, Phys. Fluids, 10, 137 (1967).
63. G. O. Spies and D. B. Nelson, Phys. Fluids, 17, 1879 (1974).
64. A. Sestero and A. Taroni, Nucl. Fusion, 16, 164 (1976).
65. W. A. Cooper, G. Bateman, and D. B. Nelson, Nucl. Fusion, 20, 985 (1980).
66. V. S. Mukhovatov and V. D. Shafranov, Nucl. Fusion, 11, 605 (1971).
67. J. Blum, G. Cissoko, R. Cas Dei, et al., in: Plasma Physics and Controlled Nuclear Fusion Research, Vol. 1, Vienna, IAEA (1979), p. 521.
68. V. D. Shafranov, At. Energ., 18, 255 (1965).
69. V. S. Mukhovatov, in: Plasma Physics and Controlled Nuclear Fusion Research, Vol. 2, Vienna, IAEA (1966), p. 577.
70. V. A. Abramov, O. P. Pogutse, and E. I. Yurchenko, Fiz. Plasmy, 1, 536 (1975).
71. J. A. Wesson, Nucl. Fusion, 18, 87 (1978).
72. S. M. Osovets, in: Plasma Physics and the Problems of Controlled Thermonuclear Reactions, Vol. 2, Pergamon (1959).
73. V. D. Zakharov and V. D. Shafranov, Preprint of the I. V. Kurchatov Institute of Atomic Energy, IAE-2789, Moscow (1977).
74. C. Mercier, Nucl. Fusion, 3, 9 (1963).

75. V. D. Shafranov, Nucl. Fusion, 4, 114 (1964).
76. V. D. Shafranov, Nucl. Fusion, 4, 232 (1964).
77. V. D. Shafranov and L. E. Zakharov, in: Plasma Physics and Controlled Nuclear Fusion Research, Vol. 2, Vienna, IAEA (1977), p. 155.
78. L. E. Zakharov and V. D. Shafranov, Zh. Tekh. Fiz., 48, 1156 (1978).

INDEX

Ambipolar diffusion, 293
Active resistance, 42, 44
Adiabatic invariant, 13
Alfvén
  current, 14
  oscillations, 67, 81–82, 137
  time, 141
  waves, 86
Ampere force, 154
Anisotropic pressure, 116, 156, 159, 245–248, 269, 283, 284
Anomalous
  Doppler effect, 21, 24, 30
  electron thermal conductivity, 52, 147, 148
Approximation,
  drift, 290
  large aspect ratio, 183, 204
  longwave, 196, 200
  low toroidicity, 183, 185, 249
Associated Legendre function, 203, 207
Axis, nonplanar, 248, 273
Axisymmetric mode, 271

Ballooning
  criterion, 100
  mode, 65, 88
  transformation, 90, 91
Banana
  current, 158
  gas, 158
  orbits, 291
  trajectories, 14

Bateman, 70, 72
Bernstein mode, 20
Beta,
  critical, 66, 252
  current, 163
  plasma, 65
Biot–Savart, 229, 275
Bootstrap current, 161, 291
Boundary conditions, 167

Casing,
  finite conductivity, 265
  principle of virtual, 179, 234, 258
Cauchy, 179
Cerenkov
  effect, 35
  electron resonance, 20
  ion resonance, 20–22, 24, 30
Circle, Larmor, 156, 157
Clark–Sigmar model, 239
Collision, integral, 4
  Landau form of, 10
Complex curvature, 274
Condition, boundary, 167
Conductivity, thermal, 287
Configuration, force-free, 226
Confinement time, 2, 147
Conformal mapping, 221
Connor, 69
Conservation,
  energy, 289
  flux, 290
  heat flux, 289
  momentum, 161, 289

Conservation (continued),
    particles, 289
Continuity equation, 287
Coordinates, natural, 177,
      178, 237, 240, 243,
      253, 269, 270
Coppi, 66
Core, iron, 260, 263, 273
Coulomb
  collisions, 1, 9
  interaction, 16
  logarithm, 10
Criterion, stability, 272
Critical
  beta, 66, 252
  pressure, 66
  velocity, 4
Cross-section, elliptical, 190
Current
  banana, 158
  beta, 163
  bootstrap, 161, 291
  Foucault, 260
  function, 165
  profile, longitudinal, 163
Curvature, complex, 274
Curvilinear coordinates, 73

Debye radius, 5
Decay index, 272
Diamagnetic
  plasma, 2, 17, 39
  signal, 283
  temperature, 1, 17
Diamagnetism, 163, 186
Diffusion, 147, 148
  ambipolar, 293
Displacement, rigid vertical
    and horizontal, 271
Dissipative instability, 71–72,
    136 ff., 147
Distribution function,
  electron, 3
  Maxwellian, 2, 155

Doppler,
  anomalous, effect, 21, 24,
    30
  resonance, 20, 21, 22, 30
Dreicer field, 1, 3
Drift
  approximation, 13, 290
  motion, 161

Eigenfunction, 227, 228, 232,
    271
Electric field, toroidal, 1, 12
Electron
  distribution function, 4
  energy, 2
  energy confinement time,
    52
  epithermal, 44
  superthermal, 4
  temperature, 2
  thermal conductivity, 52,
    147, 148
  trapped, 15, 40
  runaway, 1, 54
Electronic spectrum of solids,
    70, 122
Elliptical cross-section, 190
Elongation, 239
Energy, conservation of, 289
Epithermal
  electrons, 44
  ions, 42
Equation,
  continuity, 287
  Euler, 156
  Euler–Lagrange, 244
  Euler–Ostrogradskii, 243
  Fredholm, 217
  Laplace, 179, 221
  Maxwell, 156
  momentum balance, 290
  transport, 288, 289
Equilibrium
  evolution, 237
  stability of, 271
  stellarator, 172, 173

INDEX

Euler
  equation, 156
  mesh, 228
  method, 228, 236
Euler–Ostrogradskii equation, 243
Evolution of equilibrium, 237
Expansion, Fourier, 274
Extraordinary wave, 46

Factor, safety, 156, 168, 255
Fan instability, 2, 17, 25
Feedback, 272, 273, 285
Feneberg, 234
Field, maintaining, 178, 180, 191
Fielding, 70
Finite
  conductivity casing, 265
  difference method, 227, 234
Flute instability, 65
Flux
  conservation, 290
  function, 165, 166
  helical, 196
Fock function, 202
Force, ampere, 154
Force-free configuration, 226
Foucault currents, 260
Fourier expansion, 88, 91, 144, 274
Fredholm equations, 217
Freidberg, 70
Function,
  associated Legendre, 203
  current, 165
  eigen, 227, 228, 232, 271
  flux, 165, 166
  Fock, 202
  Green's, 227, 228
  helical flux, 196
  Legendre, 207
Functional, 125, 144
Furth, 66, 71

Galeev, 154, 161
Gas, banana, 158
Gap, 222, 226
Glasser, 69
Grad, 245
Green, 71
Green's function, 227, 228, 229
Grossman, 70
Gurevich, 5, 7–8

Haas, 67, 70
Hastie, 69
Hazeltine, 154
Heat flux conservation, 289
Heating,
  anomalous, 41, 55
  ohmic, 284
Helical
  flux function, 196
  instability, 172
  symmetry, 169
Hill vortex, 190, 208
Hinton, 154
Hypergeometric series, 234

Index, decay, 272
Induced scattering, 56
Inductance, 229
  internal, 284
Inductive effects, 43
Instability,
  dissipative, 71–72, 136 ff., 147
  electrostatic, 19
  explosive, 18
  fan, 2, 17, 25
  flute, 65, 88, 89
  gravitational dissipative, 143, 145
  helical, 172
  ion-acoustic, 20, 67, 137, 144
  kink, 65, 70, 81, 88, 120, 199

Instability (continued),
   magnetosonic, 67
   MHD, 236, 237
Integral, Biot–Savart, 277
Inverse variables method, 236
Ion
   -acoustic mode, 20, 55, 66
   anomalous heating, 41, 55
   Cerenkov, resonance, 20–22, 24
   epithermal, 39, 42
   trapped, 15
Iron core, 260, 263, 273
Island, magnetic, 173

Johnson, 70
Joule heating, 2, 54

Kadomtsev, 67
Khait, 243
Killeen, 66, 71
Kinematic moment, 157
Kinetic theory, 1
Kink instability, 65, 70, 81, 88, 120, 199
Kruskol–Shafranov criterion, 132

Lackner, 234
Lagrangian, 82, 85, 93
Landau,
   damping, 27, 35, 49, 56
   form of collision integral, 10
Langmuir, magnetized oscillations, 20
Laplace transform, 179, 221
Large aspect ratio approximation, 183, 204
Larmor
   circles, 156, 157
   radius, 155
Lebedev, 8
Legendre functions, 207
Longitudinal current, 163

Longwave approximation, 196, 200
Low-toroidicity approximation, 183, 185, 249

Magnetic
   axis, nonplanar, 248
   islands, 147, 173
   surfaces, 165
   well, 67, 70
Magnetized, Langmuir oscillations, 20
Magnetosonic oscillations, 67, 82, 86, 137
Maintaining field, 178, 180, 191
Mapping, conformal, 221
Maxwell equations, 156
Maxwellian distribution, 2, 155
Mercier, 70
   criterion, 66, 68, 95, 143
Method,
   Euler, 236
   finite difference, 227, 234
   inverse variables, 236
   moments, 239
   numerical, 227
   successive relaxation, 239
MHD
   equations, 67
   instability, 236, 237
Mikhailovsky, 71
Minimum action principle, 82
Model, Clark–Sigmar, 239
Moment, kinematic, 157
Moments, method of, 239
Momentum balance, 290
   conservation, 161, 289

Natural coordinates, 177, 178, 237, 240, 243, 253, 269, 270
Nelson, 72
Neoclassical transport theory, 154, 289, 293

INDEX

Nonplanar axis, 248, 273
Normal mode analysis, 78, 83, 92
Numerical methods, 227

Oak Ridge, 70
Ohkawa diffusion formula, 147
Ohmic heating, 284
Ohm's law, 160
Orbits, banana, 291
Oscillation, 161

Paramagnetism, 163
Particles,
  conservation of, 289
  passing, 155, 158
  trapped, 155, 158
Peng, 68
PEST code, 111
Photoneutrons, 2, 12
Pinch,
  stabilized, 163
  z, Θ, 163
Plasma
  diamagnetism, 163, 186
  diffusion, 154, 160, 161, 164
Plasmon, 25, 55
Pogutse, 67, 69, 72
Pressure anisotropy, 156, 159, 245-248, 269, 284
Princeton group, 68, 70, 71, 110
Principle, virtual casing, 179, 234, 258
Profile, pressure, 163
Pseudoclassical losses, 147

Quasi-modes, 93
Quasi-neutrality, 159

Relativistic corrections, 9
Resistive modes (see Dissipative Instability), 71
Rigid vertical and horizontal displacement, 271

Ripple trapping, 40, 54
Rosenbluth, 66, 71
Rotational transform, 68, 168
Runaway electrons, 1

Safety factor, 66, 156, 168, 255
Sagdeev, 154, 160
Scott–Trubnikov formula, 46
Second stability, 70, 134
Separatrix, 191
Series, hypergometric, 234
Serret–Frenet relations, 275
Shafranov, 67, 70
Shear, stabilizing effect, 66, 69
Sigmar, 239
Skin time, 138, 144
Solov'ev, 67
Stabilized pinch, 163
Stability of equilibrium, 271
Stellarator, 200
  equilibrium, 172, 173
Stochasticity, 173
  of magnetic field, 147
Successive relaxation method, 239
Surface, magnetic, 165
Suydam criterion, 65, 67, 95, 102
Symmetry, helical, 169, 200
Synchrotron radiation, 2, 17, 40, 44, 46

Taylor, 69
Tensor, stress, 293
Theorem, virial, 181
Thermal conductivity, 287
Thomson scattering, 1
Tokamak, 163
  T-6, 2
  T-10, 2, 14
  TFR, 2
  TM-3, 18, 29, 55
Transform, rotational, 68, 168

Transport equations, 288, 289
  neoclassical, 154, 289, 293
Trapped particles, 155, 158
Triangularity, 239
Trivelpiece–Gould mode, 20, 21, 31, 33, 40, 53, 55
Turbulence theory, 289

Upper hybrid oscillations, 20

Variational method, 82, 96, 98, 100, 105, 120, 144
Van Dam, 69
Van der Pol averaging method, 94, 141
Virial theorem, 181
Virtual casing principle, 179, 234, 258
Vortex, Hill's, 190, 208

Ware, 67
Wronskian, 203

X-ray radiation, 2, 17, 47

Yurchenko, 67, 68, 69, 72

Zakharov, 70
Z-pinch, 163